Beyond Multiple Linear Regression

CHAPMAN & HALL/CRC
Texts in Statistical Science Series

Joseph K. Blitzstein, *Harvard University, USA*
Julian J. Faraway, *University of Bath, UK*
Martin Tanner, *Northwestern University, USA*
Jim Zidek, *University of British Columbia, Canada*

Recently Published Titles

Time Series
A Data Analysis Approach Using R
Robert H. Shumway and David S. Stoffer

Practical Multivariate Analysis, Sixth Edition
Abdelmonem Afifi, Susanne May, Robin A. Donatello, and Virginia A. Clark

Time Series: A First Course with Bootstrap Starter
Tucker S. McElroy and Dimitris N. Politis

Probability and Bayesian Modeling
Jim Albert and Jingchen Hu

Surrogates
Gaussian Process Modeling, Design, and Optimization for the Applied Sciences
Robert B. Gramacy

Statistical Analysis of Financial Data
With Examples in R
James Gentle

Statistical Rethinking
A Bayesian Course with Examples in R and STAN, Second Edition
Richard McElreath

Statistical Machine Learning
A Model-Based Approach
Richard Golden

Randomization, Bootstrap and Monte Carlo Methods in Biology, Fourth Edition
Bryan F. J. Manly, Jorje A. Navarro Alberto

Principles of Uncertainty, Second Edition
Joseph B. Kadane

Beyond Multiple Linear Regression
Applied Generalized Linear Models and Multilevel Models in R
Paul Roback, Julie Legler

For more information about this series, please visit:
https://www.crcpress.com/Chapman--HallCRC-Texts-in-Statistical-Science/book-series/CHTEXSTASCI

Beyond Multiple Linear Regression
Applied Generalized Linear Models and Multilevel Models in R

Paul Roback
Julie Legler

CRC Press
Taylor & Francis Group
Boca Raton London New York

CRC Press is an imprint of the
Taylor & Francis Group, an **informa** business

A CHAPMAN & HALL BOOK

First edition published 2021
by CRC Press
6000 Broken Sound Parkway NW, Suite 300, Boca Raton, FL 33487-2742

and by CRC Press
2 Park Square, Milton Park, Abingdon, Oxon, OX14 4RN

© 2021 Taylor & Francis Group, LLC

CRC Press is an imprint of Taylor & Francis Group, LLC

Reasonable efforts have been made to publish reliable data and information, but the author and publisher cannot assume responsibility for the validity of all materials or the consequences of their use. The authors and publishers have attempted to trace the copyright holders of all material reproduced in this publication and apologize to copyright holders if permission to publish in this form has not been obtained. If any copyright material has not been acknowledged please write and let us know so we may rectify in any future reprint.

Except as permitted under U.S. Copyright Law, no part of this book may be reprinted, reproduced, transmitted, or utilized in any form by any electronic, mechanical, or other means, now known or hereafter invented, including photocopying, microfilming, and recording, or in any information storage or retrieval system, without written permission from the publishers.

For permission to photocopy or use material electronically from this work, access www.copyright.com or contact the Copyright Clearance Center, Inc. (CCC), 222 Rosewood Drive, Danvers, MA 01923, 978-750-8400. For works that are not available on CCC please contact mpkbookspermissions@tandf.co.uk

Trademark notice: Product or corporate names may be trademarks or registered trademarks and are used only for identification and explanation without intent to infringe.

Library of Congress Control Number: 2020950457

ISBN: 978-1-4398-8538-3 (hbk)
ISBN: 978-0-4290-6666-5 (ebk)

To our families:
Karen, Samantha, Timothy, and Sophie
Paul, Ali, Mark, Sean, Lila, and Eva

Contents

Preface xv

1 Review of Multiple Linear Regression 1
- 1.1 Learning Objectives 1
- 1.2 Introduction to Beyond Multiple Linear Regression 1
- 1.3 Assumptions for Linear Least Squares Regression 3
  - 1.3.1 Cases Without Assumption Violations 4
  - 1.3.2 Cases With Assumption Violations 6
- 1.4 Review of Multiple Linear Regression 8
  - 1.4.1 Case Study: Kentucky Derby 8
- 1.5 Initial Exploratory Analyses 8
  - 1.5.1 Data Organization 8
  - 1.5.2 Univariate Summaries 9
  - 1.5.3 Bivariate Summaries 9
- 1.6 Multiple Linear Regression Modeling 12
  - 1.6.1 Simple Linear Regression with a Continuous Predictor 12
  - 1.6.2 Linear Regression with a Binary Predictor 17
  - 1.6.3 Multiple Linear Regression with Two Predictors .. 18
  - 1.6.4 Inference in Multiple Linear Regression: Normal Theory 19
  - 1.6.5 Inference in Multiple Linear Regression: Bootstrapping 20
  - 1.6.6 Multiple Linear Regression with an Interaction Term 22
  - 1.6.7 Building a Multiple Linear Regression Model 24
- 1.7 Preview of Remaining Chapters 26
  - 1.7.1 Soccer 26
  - 1.7.2 Elephant Mating 27
  - 1.7.3 Parenting and Gang Activity 28
  - 1.7.4 Crime 28
- 1.8 Exercises 29
  - 1.8.1 Conceptual Exercises 29
  - 1.8.2 Guided Exercises 32
  - 1.8.3 Open-Ended Exercises 36

2 Beyond Least Squares: Using Likelihoods — 39
- 2.1 Learning Objectives — 39
- 2.2 Case Study: Does Sex Run in Families? — 40
  - 2.2.1 Research Questions — 41
- 2.3 Model 0: Sex Unconditional, Equal Probabilities — 42
- 2.4 Model 1: Sex Unconditional, Unequal Probabilities — 43
  - 2.4.1 What Is a Likelihood? — 43
  - 2.4.2 Finding MLEs — 46
  - 2.4.3 Summary — 49
  - 2.4.4 Is a Likelihood a Probability Function? (optional) — 50
- 2.5 Model 2: Sex Conditional — 50
  - 2.5.1 Model Specification — 50
  - 2.5.2 Application to Hypothetical Data — 51
- 2.6 Case Study: Analysis of the NLSY Data — 53
  - 2.6.1 Model Building Plan — 53
  - 2.6.2 Exploratory Data Analysis — 54
  - 2.6.3 Likelihood for the Sex Unconditional Model — 55
  - 2.6.4 Likelihood for the Sex Conditional Model — 56
  - 2.6.5 Model Comparisons — 58
- 2.7 Model 3: Stopping Rule Model (waiting for a boy) — 61
  - 2.7.1 Non-nested Models — 63
- 2.8 Summary of Model Building — 64
- 2.9 Likelihood-Based Methods — 65
- 2.10 Likelihoods and This Course — 66
- 2.11 Exercises — 67
  - 2.11.1 Conceptual Exercises — 67
  - 2.11.2 Guided Exercises — 67
  - 2.11.3 Open-Ended Exercises — 68

3 Distribution Theory — 71
- 3.1 Learning Objectives — 71
- 3.2 Introduction — 71
- 3.3 Discrete Random Variables — 72
  - 3.3.1 Binary Random Variable — 72
  - 3.3.2 Binomial Random Variable — 73
  - 3.3.3 Geometric Random Variable — 74
  - 3.3.4 Negative Binomial Random Variable — 75
  - 3.3.5 Hypergeometric Random Variable — 77
  - 3.3.6 Poisson Random Variable — 79
- 3.4 Continuous Random Variables — 80
  - 3.4.1 Exponential Random Variable — 80
  - 3.4.2 Gamma Random Variable — 81
  - 3.4.3 Normal (Gaussian) Random Variable — 83
  - 3.4.4 Beta Random Variable — 84

3.5	Distributions Used in Testing		85
	3.5.1	χ^2 Distribution	86
	3.5.2	Student's t-Distribution	87
	3.5.3	F-Distribution	87
3.6	Additional Resources		88
3.7	Exercises		88
	3.7.1	Conceptual Exercises	88
	3.7.2	Guided Exercises	90

4 Poisson Regression — 93

4.1	Learning Objectives		93
4.2	Introduction to Poisson Regression		94
	4.2.1	Poisson Regression Assumptions	95
	4.2.2	A Graphical Look at Poisson Regression	95
4.3	Case Studies Overview		96
4.4	Case Study: Household Size in the Philippines		97
	4.4.1	Data Organization	98
	4.4.2	Exploratory Data Analyses	98
	4.4.3	Estimation and Inference	102
	4.4.4	Using Deviances to Compare Models	104
	4.4.5	Using Likelihoods to Fit Models (optional)	106
	4.4.6	Second Order Model	107
	4.4.7	Adding a Covariate	109
	4.4.8	Residuals for Poisson Models (optional)	110
	4.4.9	Goodness-of-Fit	112
4.5	Linear Least Squares vs. Poisson Regression		113
4.6	Case Study: Campus Crime		114
	4.6.1	Data Organization	114
	4.6.2	Exploratory Data Analysis	115
	4.6.3	Accounting for Enrollment	117
4.7	Modeling Assumptions		118
4.8	Initial Models		118
	4.8.1	Tukey's Honestly Significant Differences	119
4.9	Overdispersion		121
	4.9.1	Dispersion Parameter Adjustment	121
	4.9.2	No Dispersion vs. Overdispersion	123
	4.9.3	Negative Binomial Modeling	123
4.10	Case Study: Weekend Drinking		125
	4.10.1	Research Question	125
	4.10.2	Data Organization	126
	4.10.3	Exploratory Data Analysis	126
	4.10.4	Modeling	127
	4.10.5	Fitting a ZIP Model	129
	4.10.6	The Vuong Test (optional)	131

		4.10.7	Residual Plot	132

- 4.10.7 Residual Plot 132
- 4.10.8 Limitations 132
- 4.11 Exercises 133
  - 4.11.1 Conceptual Exercises 133
  - 4.11.2 Guided Exercises 136
  - 4.11.3 Open-Ended Exercises 142

5 Generalized Linear Models: A Unifying Theory   145
- 5.1 Learning Objectives 145
- 5.2 One-Parameter Exponential Families 145
  - 5.2.1 One-Parameter Exponential Family: Poisson 146
  - 5.2.2 One-Parameter Exponential Family: Normal 147
- 5.3 Generalized Linear Modeling 148
- 5.4 Exercises 149

6 Logistic Regression   151
- 6.1 Learning Objectives 151
- 6.2 Introduction to Logistic Regression 151
  - 6.2.1 Logistic Regression Assumptions 152
  - 6.2.2 A Graphical Look at Logistic Regression 153
- 6.3 Case Studies Overview 153
- 6.4 Case Study: Soccer Goalkeepers 154
  - 6.4.1 Modeling Odds 155
  - 6.4.2 Logistic Regression Models for Binomial Responses   155
  - 6.4.3 Theoretical Rationale (optional) 158
- 6.5 Case Study: Reconstructing Alabama 159
  - 6.5.1 Data Organization 159
  - 6.5.2 Exploratory Analyses 160
  - 6.5.3 Initial Models 161
  - 6.5.4 Tests for Significance of Model Coefficients 162
  - 6.5.5 Confidence Intervals for Model Coefficients 163
  - 6.5.6 Testing for Goodness-of-Fit 164
  - 6.5.7 Residuals for Binomial Regression 166
  - 6.5.8 Overdispersion 167
  - 6.5.9 Summary 170
- 6.6 Linear Least Squares vs. Binomial Regression 170
- 6.7 Case Study: Trying to Lose Weight 171
  - 6.7.1 Data Organization 172
  - 6.7.2 Exploratory Data Analysis 173
  - 6.7.3 Initial Models 176
  - 6.7.4 Drop-in-Deviance Tests 179
  - 6.7.5 Model Discussion and Summary 180
- 6.8 Exercises 181
  - 6.8.1 Conceptual Exercises 181

		6.8.2	Guided Exercises	182
		6.8.3	Open-Ended Exercises	189

7 Correlated Data — 193
7.1 Learning Objectives — 193
7.2 Introduction — 193
7.3 Recognizing Correlation — 194
7.4 Case Study: Dams and Pups — 195
7.5 Sources of Variability — 195
7.6 Scenario 1: No Covariates — 196
7.7 Scenario 2: Dose Effect — 199
7.8 Case Study: Tree Growth — 203
7.8.1 Format of the Data Set — 204
7.8.2 Sources of Variability — 205
7.8.3 Analysis Preview: Accounting for Correlation — 206
7.9 Summary — 207
7.10 Exercises — 207
7.10.1 Conceptual Exercises — 207
7.10.2 Guided Exercises — 209
7.10.3 Note on Correlated Binary Outcomes — 210

8 Introduction to Multilevel Models — 211
8.1 Learning Objectives — 211
8.2 Case Study: Music Performance Anxiety — 212
8.3 Initial Exploratory Analyses — 213
8.3.1 Data Organization — 213
8.3.2 Exploratory Analyses: Univariate Summaries — 214
8.3.3 Exploratory Analyses: Bivariate Summaries — 216
8.4 Two-Level Modeling: Preliminary Considerations — 220
8.4.1 Ignoring the Two-Level Structure (not recommended) — 220
8.4.2 A Two-Stage Modeling Approach (better but imperfect) — 221
8.5 Two-Level Modeling: A Unified Approach — 225
8.5.1 Our Framework — 225
8.5.2 Random vs. Fixed Effects — 227
8.5.3 Distribution of Errors: Multivariate Normal — 227
8.5.4 Technical Issues when Testing Parameters (optional) — 229
8.5.5 An Initial Model with Parameter Interpretations — 231
8.6 Building a Multilevel Model — 234
8.6.1 Model Building Strategy — 234
8.6.2 An Initial Model: Random Intercepts — 234
8.7 Binary Covariates at Level One and Level Two — 236
8.7.1 Random Slopes and Intercepts Model — 236
8.7.2 Pseudo R-squared Values — 239

		8.7.3 Adding a Covariate at Level Two	240

    8.8 Adding Further Covariates 242
        8.8.1 Interpretation of Parameter Estimates 243
        8.8.2 Model Comparisons 245
    8.9 Centering Covariates 246
    8.10 A Final Model for Music Performance Anxiety 248
    8.11 Modeling Multilevel Structure: Is It Necessary? 251
    8.12 Notes on Using R (optional) 254
    8.13 Exercises 255
        8.13.1 Conceptual Exercises 255
        8.13.2 Guided Exercises 256
        8.13.3 Open-Ended Exercises 258

9 Two-Level Longitudinal Data   263

    9.1 Learning Objectives 263
    9.2 Case Study: Charter Schools 264
    9.3 Initial Exploratory Analyses 265
        9.3.1 Data Organization 265
        9.3.2 Missing Data 266
        9.3.3 Exploratory Analyses for General Multilevel Models   268
        9.3.4 Exploratory Analyses for Longitudinal Data 269
    9.4 Preliminary Two-Stage Modeling 273
        9.4.1 Linear Trends Within Schools 273
        9.4.2 Effects of Level Two Covariates on Linear Time Trends 274
        9.4.3 Error Structure Within Schools 279
    9.5 Initial Models 279
        9.5.1 Unconditional Means Model 280
        9.5.2 Unconditional Growth Model 281
        9.5.3 Modeling Other Trends over Time 284
    9.6 Building to a Final Model 286
        9.6.1 Uncontrolled Effects of School Type 286
        9.6.2 Add Percent Free and Reduced Lunch as a Covariate 289
        9.6.3 A Final Model with Three Level Two Covariates . . . 291
        9.6.4 Parametric Bootstrap Testing 294
    9.7 Covariance Structure among Observations 301
        9.7.1 Standard Covariance Structure 302
        9.7.2 Alternative Covariance Structures 305
        9.7.3 Non-longitudinal Multilevel Models 306
        9.7.4 Final Thoughts Regarding Covariance Structures . 306
        9.7.5 Details of Covariance Structures (optional) 307
    9.8 Notes on Using R (optional) 308
    9.9 Exercises 309
        9.9.1 Conceptual Exercises 309
        9.9.2 Guided Exercises 312

		9.9.3	Open-Ended Exercises	316

10 Multilevel Data With More Than Two Levels — 321
- 10.1 Learning Objectives — 321
- 10.2 Case Studies: Seed Germination — 322
- 10.3 Initial Exploratory Analyses — 323
  - 10.3.1 Data Organization — 323
  - 10.3.2 Exploratory Analyses — 325
- 10.4 Initial Models — 332
  - 10.4.1 Unconditional Means — 333
  - 10.4.2 Unconditional Growth — 335
- 10.5 Encountering Boundary Constraints — 337
- 10.6 Parametric Bootstrap Testing — 343
- 10.7 Exploding Variance Components — 349
- 10.8 Building to a Final Model — 352
- 10.9 Covariance Structure (optional) — 358
  - 10.9.1 Details of Covariance Structures — 361
- 10.10 Notes on Using R (optional) — 363
- 10.11 Exercises — 364
  - 10.11.1 Conceptual Exercises — 364
  - 10.11.2 Guided Exercises — 368
  - 10.11.3 Open-Ended Exercises — 370

11 Multilevel Generalized Linear Models — 373
- 11.1 Learning Objectives — 373
- 11.2 Case Study: College Basketball Referees — 374
- 11.3 Initial Exploratory Analyses — 374
  - 11.3.1 Data Organization — 374
  - 11.3.2 Exploratory Analyses — 376
- 11.4 Two-Level Modeling with a Generalized Response — 380
  - 11.4.1 A GLM Approach — 380
  - 11.4.2 A Two-Stage Modeling Approach — 381
  - 11.4.3 A Unified Multilevel Approach — 384
- 11.5 Crossed Random Effects — 386
- 11.6 Parametric Bootstrap for Model Comparisons — 390
- 11.7 A Final Model for Examining Referee Bias — 394
- 11.8 Estimated Random Effects — 398
- 11.9 Notes on Using R (optional) — 399
- 11.10 Exercises — 401
  - 11.10.1 Conceptual Exercises — 401
  - 11.10.2 Open-Ended Exercises — 405

Bibliography — 409

Index — 417

Preface

Beyond Multiple Linear Regression: Applied Generalized Linear Models and Multilevel Models in R [R Core Team, 2020] is intended to be accessible to undergraduate students who have successfully completed a regression course through, for example, a textbook like *Stat2* [Cannon et al., 2019]. We started teaching this course at St. Olaf College in 2003 so students would be able to deal with the non-normal, correlated world we live in. It has been offered at St. Olaf every year since. Even though there is no mathematical prerequisite, we still introduce fairly sophisticated topics such as likelihood theory, zero-inflated Poisson, and parametric bootstrapping in an intuitive and applied manner. We believe strongly in case studies featuring real data and real research questions; thus, most of the data in the textbook (and available at our GitHub repo[1]) arises from collaborative research conducted by the authors and their students, or from student projects. Our goal is that, after working through this material, students will develop an expanded toolkit and a greater appreciation for the wider world of data and statistical modeling.

When we teach this course at St. Olaf, we are able to cover Chapters 1-11 during a single semester, although in order to make time for a large, open-ended group project we sometimes cover some chapters in less depth (e.g., Chapters 3, 7, 10, or 11). How much you cover will certainly depend on the background of your students (ours have seen both multiple linear and logistic regression), their sophistication level (we have statistical but no mathematical prerequisites), and time available (we have a 14-week semester). It will also depend on your choice of topics; in our experience, we have found that generalized linear models (GLMs) and multilevel models nicely build on students' previous regression knowledge and allow them to better model data from many real contexts, but we also acknowledge that there are other good choices of topics for an applied "Stat3" course. The short chapter guide below can help you thread together the material in this book to create the perfect course for you:

- Chapter 1: Review of Multiple Linear Regression. We've found that our students really benefit from a review in the first week or so, plus in this initial chapter we introduce our approach to exploratory data analysis (EDA) and model building while reminding students about concepts like indicators, interactions, and bootstrapping.

---

[1] https://github.com/proback/BeyondMLR

- Chapter 2: Beyond Least Squares: Using Likelihoods. This chapter builds intuition for likelihoods and their usefulness in testing and estimation; any section involving calculus is optional. Chapter 2 could be skipped at the risk that later references to likelihoods become more blurry and understanding more shallow.
- Chapter 3: Distribution Theory. A quick summary of key discrete and continuous probability distributions, this chapter can be used as a reference as needed.
- Chapter 4: Poisson Regression. This is the most important chapter for generalized linear models, where each of the three case studies introduces new ideas such as coefficient interpretation, Wald-type and drop-in-deviance tests, Wald-type and profile likelihood confidence intervals, offsets, overdispersion, quasilikelihood, zero-inflation, and alternatives like negative binomial.
- Chapter 5: Generalized Linear Models: A Unifying Theory. Chapter 5 is short, but it importantly shows how linear, logistic, binomial, Poisson, and other regression methods are connected. We believe it's important that students appreciate that GLMs aren't just a random collection of modeling approaches.
- Chapter 6: Logistic Regression. We begin with two case studies involving binomial regression, drawing connections with Chapters 4 and 5, before a third case study involving binary logistic regression.
- Chapter 7: Correlated Data. This is the transition chapter, building intuition about correlated data through an extended simulation and a real case study, although you can jump right to Chapter 8 if you wish. Chapters 8-11 contain the multilevel model material and, for the most part, they do not depend on earlier chapters (except for generalized responses in Chapter 11 and references to ideas such as likelihoods, inferential approaches, etc.). In fact, during one semester we taught the multilevel material before the GLM material to facilitate academic civic engagement projects that needed multilevel models (during that semester our order of chapters was: 1, 2, 7, 8, 9, 10, 3, 4, 5, 6, 11).
- Chapter 8: Introduction to Multilevel Models. As we go through a comprehensive case study, several important ideas are motivated, including EDA for multilevel data, the two-stage approach, multivariate normal distributions, coefficient interpretations, fixed and random effects, random slopes and intercepts, and more. Another simulation illustrates the effect of inappropriately using regression methods that assume independence for correlated data.
- Chapter 9: Two-Level Longitudinal Data. This chapter covers the special case of Chapter 8 models where there are multiple measurements over time for each subject. New topics include longitudinal-specific EDA, missing data methods, parametric bootstrap inference, and covariance structure.
- Chapter 10: Multilevel Data with More Than Two Levels. The ideas from

Chapters 8 and 9 are extended to a three-level case study. New ideas include boundary constraints and exploding numbers of variance components and fixed effects.
- Chapter 11: Multilevel Generalized Linear Models. This chapter brings everything together, combining multilevel data with non-normal responses. Crossed random effects and random effects estimates are both introduced here.

Three types of exercises are available for each chapter. **Conceptual Exercises** ask about key ideas in the contexts of case studies from the chapter and additional research articles where those ideas appear. **Guided Exercises** provide real data sets with background descriptions and lead students step-by-step through a set of questions to explore the data, build and interpret models, and address key research questions. Finally, **Open-Ended Exercises** provide real data sets with contextual descriptions and ask students to explore key questions without prescribing specific steps. A solutions manual with solutions to all exercises will be available to qualified instructors at our book's website[2].

This work is licensed under a Creative Commons Attribution-NonCommercial-ShareAlike 4.0 International License.

Acknowledgments. We would like to thank students of Stat 316 at St. Olaf College since 2010 for their patience as this book has taken shape with their feedback. We would especially like to thank these St. Olaf students for their summer research efforts which significantly improved aspects of this book: Cecilia Noecker, Anna Johanson, Nicole Bettes, Kiegan Rice, Anna Wall, Jack Wolf, Josh Pelayo, Spencer Eanes, and Emily Patterson. Early editions of this book also benefitted greatly from feedback from instructors who used these materials in their classes, including Matt Beckman, Laura Boehm Vock, Beth Chance, Laura Chihara, Mine Dogucu, and Katie Ziegler-Graham. Finally, we have appreciated the support of two NSF grants (#DMS-1045015 and #DMS-0354308) and of our colleagues in the Department of Mathematics, Statistics, and Computer Science at St. Olaf. We are also thankful to Samantha Roback for developing the cover image.

---

[2] www.routledge.com

1
Review of Multiple Linear Regression

1.1 Learning Objectives

After finishing this chapter, you should be able to:
- Identify cases where linear least squares regression (LLSR) assumptions are violated.
- Generate exploratory data analysis (EDA) plots and summary statistics.
- Use residual diagnostics to examine LLSR assumptions.
- Interpret parameters and associated tests and intervals from multiple regression models.
- Understand the basic ideas behind bootstrapped confidence intervals.

```
# Packages required for Chapter 1
library(knitr)
library(gridExtra)
library(GGally)
library(kableExtra)
library(jtools)
library(rsample)
library(broom)
library(tidyverse)
```

1.2 Introduction to Beyond Multiple Linear Regression

Ecologists count species, criminologists count arrests, and cancer specialists count cases. Political scientists seek to explain who is a Democrat, pre-med students are curious about who gets into medical school, and sociologists study which people get tattoos. In the first case, ecologists, criminologists and cancer specialists are concerned about outcomes which are counts. The political scientists', pre-med students' and sociologists' interest centers on binary responses: Democrat or not, accepted or not, and tattooed or not. We can model these non-Gaussian (non-normal) responses in a more natural way by fitting **generalized linear models (GLMs)** as opposed to using **linear least squares regression (LLSR)** models.

When models are fit to data using linear least squares regression (LLSR), inferences are possible using traditional statistical theory under certain conditions: if we can assume that there is a linear relationship between the response (Y) and an explanatory variable (X), the observations are independent of one another, the responses are approximately normal for each level of the X, and the variation in the responses is the same for each level of X. If we intend to make inferences using GLMs, necessary assumptions are different. First, we will not be constrained by the normality assumption. When conditions are met, GLMs can accommodate non-normal responses such as the counts and binary data in our preceding examples. While the observations must still be independent of one another, the variance in Y at each level of X need not be equal nor does the assumption of linearity between Y and X need to be plausible.

However, GLMs cannot be used for models in the following circumstances: medical researchers collect data on patients in clinical trials weekly for 6 months; rat dams are injected with teratogenic substances and their offspring are monitored for defects; and, musicians' performance anxiety is recorded for several performances. Each of these examples involves correlated data: the same patient's outcomes are more likely to be similar from week-to-week than outcomes from different patients; litter mates are more likely to suffer defects at similar rates in contrast to unrelated rat pups; and, a musician's anxiety is more similar from performance to performance than it is with other musicians. Each of these examples violate the independence assumption of simpler linear models for LLSR or GLM inference.

The **Generalized Linear Models** in the book's title extends least squares methods you may have seen in linear regression to handle responses that are non-normal. The **Multilevel Models** in the book's title will allow us to create models for situations where the observations are not independent of one another. Overall, these approaches will permit us to get much more out of data and may be more faithful to the actual data structure than models based on ordinary least squares. These models will allow you to expand *beyond multiple linear regression*.

In order to understand the motivation for handling violations of assumptions, it is helpful to be able to recognize the model assumptions for inference with LLSR in the context of different studies. While linearity is sufficient for fitting an LLSR model, in order to make inferences and predictions the observations must also be independent, the responses should be approximately normal at each level of the predictors, and the standard deviation of the responses at each level of the predictors should be approximately equal. After examining circumstances where inference with LLSR is appropriate, we will look for violations of these assumptions in other sets of circumstances. These are settings where we may be able to use the methods of this text. We've kept the examples in the exposition simple to fix ideas. There are exercises which describe more realistic and complex studies.

1.3 Assumptions for Linear Least Squares Regression

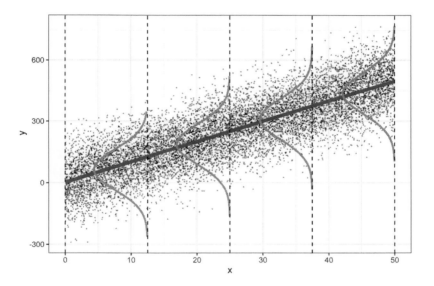

FIGURE 1.1: Assumptions for linear least squares regression (LLSR).

Recall that making inferences or predictions with models fit using linear least squares regression requires that the following assumptions be tenable. The acronym LINE can be used to recall the assumptions required for making inferences and predictions with models based on LLSR. If we consider a simple linear regression with just a single predictor X, then:

- **L:** There is a linear relationship between the mean response (Y) and the explanatory variable (X),
- **I:** The errors are independent—there's no connection between how far any two points lie from the regression line,

- **N:** The responses are normally distributed at each level of X, and
- **E:** The variance or, equivalently, the standard deviation of the responses is equal for all levels of X.

These assumptions are depicted in Figure 1.1.

- **L:** The mean value for Y at each level of X falls on the regression line.
- **I:** We'll need to check the design of the study to determine if the errors (vertical distances from the line) are independent of one another.
- **N:** At each level of X, the values for Y are normally distributed.
- **E:** The spread in the Y's for each level of X is the same.

1.3.1 Cases Without Assumption Violations

It can be argued that the following studies do not violate assumptions for inference in linear least squares regression. We begin by identifying the response and the explanatory variables followed by describing each of the LINE assumptions in the context of the study, commenting on possible problems with the assumptions.

1) **Reaction times and car radios.** A researcher suspects that loud music can affect how quickly drivers react. She randomly selects drivers to drive the same stretch of road with varying levels of music volume. Stopping distances for each driver are measured along with the decibel level of the music on their car radio.

    - *Response variable:* Reaction time
    - *Explanatory variable:* Decibel level of music

    The assumptions for inference in LLSR would apply if:

    - **L:** The mean reaction time is linearly related to decibel level of the music.
    - **I:** Stopping distances are independent. The random selection of drivers should assure independence.
    - **N:** The stopping distances for a given decibel level of music vary and are normally distributed.
    - **E:** The variation in stopping distances should be approximately the same for each decibel level of music.

    There are potential problems with the linearity and equal standard deviation assumptions. For example, if there is a threshold for the volume of music where the effect on reaction times remains the same, mean reaction times would not be a linear function of music.

1.3 Assumptions for Linear Least Squares Regression

Another problem may occur if a few subjects at each decibel level took a really long time to react. In this case, reaction times would be right skewed and the normality assumption would be violated. Often we can think of circumstances where the LLSR assumptions may be suspect. Later in this chapter we will describe plots which can help diagnose issues with LLSR assumptions.

2) **Crop yield and rainfall.** The yield of wheat per acre for the month of July is thought to be related to the rainfall. A researcher randomly selects acres of wheat and records the rainfall and bushels of wheat per acre.

- *Response variable:* Yield of wheat measured in bushels per acre for July
- *Explanatory variable:* Rainfall measured in inches for July
- **L:** The mean yield per acre is linearly related to rainfall.
- **I:** Field yields are independent; knowing one (X, Y) pair does not provide information about another.
- **N:** The yields for a given amount of rainfall are normally distributed.
- **E:** The standard deviation of yields is approximately the same for each rainfall level.

Again we may encounter problems with the linearity assumption if mean yields increase initially as the amount of rainfall increases after which excess rainfall begins to ruin crop yield. The random selection of fields should assure independence if fields are not close to one another.

3) **Heights of sons and fathers.** Sir Francis Galton suspected that a son's height could be predicted using the father's height. He collected observations on heights of fathers and their firstborn sons [Stigler, 2002].

- *Response variable:* Height of the firstborn son
- *Explanatory variable:* Height of the father
- **L:** The mean height of firstborn sons is linearly related to heights of fathers.
- **I:** The height of one firstborn son is independent of the heights of other firstborn sons in the study. This would be the case if firstborn sons were randomly selected.
- **N:** The heights of firstborn sons for a given father's height are normally distributed.

- **E:** The standard deviation of firstborn sons' heights at a given father's height is the same.

Heights and other similar measurements are often normally distributed. There would be a problem with the independence assumption if multiple sons from the same family were selected. Or, there would be a problem with equal variance if sons of tall fathers had much more variety in their heights than sons of shorter fathers.

1.3.2 Cases With Assumption Violations

1) **Grades and studying.** Is the time spent studying predictive of success on an exam? The time spent studying for an exam, in hours, and success, measured as Pass or Fail, are recorded for randomly selected students.

    - *Response variable:* Exam outcome (Pass or Fail)
    - *Explanatory variable:* Time spent studying (in hours)

    Here the response is a binary outcome which violates the assumption of a normally distributed response at each level of X. In Chapter 6, we will see logistic regression, which is more suitable for models with binary responses.

2) **Income and family size.** Do wealthy families tend to have fewer children compared to lower income families? Annual income and family size are recorded for a random sample of families.

    - *Response variable:* Family size, number of children
    - *Explanatory variable:* Annual income, in dollars

    Family size is a count taking on integer values from 0 to (technically) no upper bound. The normality assumption may be problematic again because the distribution of family size is likely to be skewed, with more families having one or two children and only a few with a much larger number of children. Both of these concerns, along with the discrete nature of the response, lead us to question the validity of the normality assumption. In fact, we might consider Poisson models discussed in Chapter 4. Study design should also specify that families are done adding children to their family.

3) **Exercise, weight, and sex.** Investigators collected the weight, sex, and amount of exercise for a random sample of college students.

    - *Response variable:* Weight
    - *Explanatory variables:* Sex and hours spent exercising in a typical week

1.4 Assumptions for Linear Least Squares Regression

With two predictors, the assumptions now apply to the combination of sex and exercise. For example, the linearity assumption implies that there is a linear relationship in mean weight and amount of exercise for males and, similarly, a linear relationship in mean weight and amount of exercise for females. This data may not be appropriate for LLSR modeling because the standard deviation in weight for students who do not exercise for each sex is likely to be considerably greater than the standard deviation in weight for students who follow an exercise regime. We can assess this potential problem by plotting weight by amount of exercise for males and females separately. There may also be a problem with the independence assumption because there is no indication that the subjects were randomly selected. There may be subgroups of subjects likely to be more similar, e.g., selecting students at a gym and others in a TV lounge.

4) **Surgery outcome and patient age.** Medical researchers investigated the outcome of a particular surgery for patients with comparable stages of disease but different ages. The ten hospitals in the study had at least two surgeons performing the surgery of interest. Patients were randomly selected for each surgeon at each hospital. The surgery outcome was recorded on a scale of 1-10.

- *Response variable:* Surgery outcome, scale 1-10
- *Explanatory variable:* Patient age, in years

Outcomes for patients operated on by the same surgeon are more likely to be similar and have similar results. For example, if surgeons' skills differ or if their criteria for selecting patients for surgery vary, individual surgeons may tend to have better or worse outcomes, and patient outcomes will be dependent on surgeon. Furthermore, outcomes at one hospital may be more similar, possibly due to factors associated with different patient populations. The very structure of this data suggests that the independence assumption will be violated. Multilevel models, which we begin discussing in Chapter 8, will explicitly take this structure into account for a proper analysis of this study's results.

While we identified possible violations of assumptions for inference in LLSR for each of the examples in this section, there may be violations of the other assumptions that we have not pointed out. Prior to reading this book, you have presumably learned some ways to handle these violations such as applying variance stabilizing transformations or logging responses, but you will discover other models in this text that may be more appropriate for the violations we have presented.

1.4 Review of Multiple Linear Regression

1.4.1 Case Study: Kentucky Derby

Before diving into generalized linear models and multilevel modeling, we review key ideas from multiple linear regression using an example from horse racing. The Kentucky Derby is a 1.25-mile horse race held annually at the Churchill Downs race track in Louisville, Kentucky. Our data set `derbyplus.csv` contains the `year` of the race, the winning horse (`winner`), the `condition` of the track, the average `speed` (in feet per second) of the winner, and the number of `starters` (field size, or horses who raced) for the years 1896-2017 [Wikipedia contributors, 2018]. The track `condition` has been grouped into three categories: fast, good (which includes the official designations "good" and "dusty"), and slow (which includes the designations "slow", "heavy", "muddy", and "sloppy"). We would like to use least squares linear regression techniques to model the speed of the winning horse as a function of track condition, field size, and trends over time.

1.5 Initial Exploratory Analyses

1.5.1 Data Organization

The first five and last five rows from our data set are illustrated in Table 1.1. Note that, in certain cases, we created new variables from existing ones:

- `fast` is an **indicator variable**, taking the value 1 for races run on fast tracks, and 0 for races run under other conditions,
- `good` is another indicator variable, taking the value 1 for races run under good conditions, and 0 for races run under other conditions,
- `yearnew` is a **centered variable**, where we measure the number of years since 1896, and
- `fastfactor` replaces `fast = 0` with the description "not fast", and `fast = 1` with the description "fast". Changing a numeric categorical variable to descriptive phrases can make plot legends more meaningful.

1.5 Initial Exploratory Analyses

TABLE 1.1: The first five and the last five observations from the Kentucky Derby case study.

year	winner	condition	speed	starters	fast	good	yearnew	fastfactor
1896	Ben Brush	good	51.66	8	0	1	0	not fast
1897	Typhoon II	slow	49.81	6	0	0	1	not fast
1898	Plaudit	good	51.16	4	0	1	2	not fast
1899	Manuel	fast	50.00	5	1	0	3	fast
1900	Lieut. Gibson	fast	52.28	7	1	0	4	fast
2013	Orb	slow	53.71	19	0	0	117	not fast
2014	California Chrome	fast	53.37	19	1	0	118	fast
2015	American Pharoah	fast	53.65	18	1	0	119	fast
2016	Nyquist	fast	54.41	20	1	0	120	fast
2017	Always Dreaming	fast	53.40	20	1	0	121	fast

1.5.2 Univariate Summaries

With any statistical analysis, our first task is to explore the data, examining distributions of individual responses and predictors using graphical and numerical summaries, and beginning to discover relationships between variables. This should *always* be done *before* any model fitting! We must understand our data thoroughly before doing anything else.

First, we will examine the response variable and each potential covariate individually. Continuous variables can be summarized using histograms and statistics indicating center and spread; categorical variables can be summarized with tables and possibly bar charts.

In Figure 1.2(a), we see that the primary response, winning speed, follows a distribution with a slight left skew, with a large number of horses winning with speeds between 53-55 feet per second. Plot (b) shows that the number of starters is mainly distributed between 5 and 20, with the largest number of races having between 15 and 20 starters.

The primary categorical explanatory variable is track condition, where 88 (72%) of the 122 races were run under fast conditions, 10 (8%) under good conditions, and 24 (20%) under slow conditions.

1.5.3 Bivariate Summaries

The next step in an initial exploratory analysis is the examination of numerical and graphical summaries of relationships between model covariates and responses. Figure 1.3 is densely packed with illustrations of bivariate relationships. The relationship between two continuous variables is depicted with scatterplots below the diagonal and correlation coefficients above the diagonal.

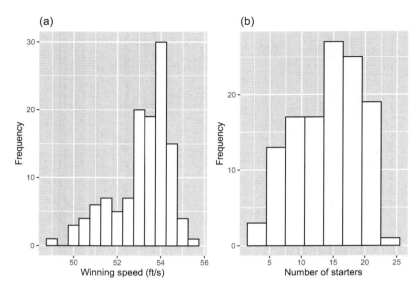

FIGURE 1.2: Histograms of key continuous variables. Plot (a) shows winning speeds, while plot (b) shows the number of starters.

Here, we see that higher winning speeds are associated with more recent years, while the relationship between winning speed and number of starters is less clear cut. We also see a somewhat strong correlation between year and number of starters—we should be aware of highly correlated explanatory variables whose contributions might overlap too much.

Relationships between categorical variables like track condition and continuous variables can be illustrated with side-by-side boxplots as in the top row, or with stacked histograms as in the first column. As expected, we see evidence of higher speeds on fast tracks and also a tendency for recent years to have more fast conditions. These observed trends can be supported with summary statistics generated by subgroup. For instance, the mean speed under fast conditions is 53.6 feet per second, compared to 52.7 ft/s under good conditions and 51.7 ft/s under slow conditions. Variability in winning speeds, however, is greatest under slow conditions (SD = 1.36 ft/s) and least under fast conditions (0.94 ft/s).

Finally, notice that the diagonal illustrates the distribution of individual variables, using density curves for continuous variables and a bar chart for categorical variables. Trends observed in the last two diagonal entries match trends observed in Figure 1.2.

By using shape or color or other attributes, we can incorporate the effect of a third or even fourth variable into the scatterplots of Figure 1.3. For example,

1.6 Initial Exploratory Analyses

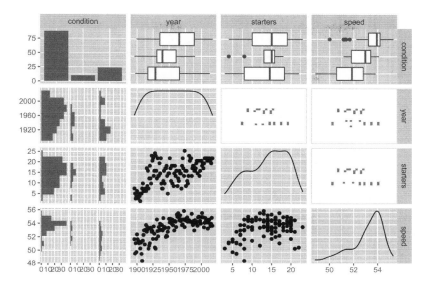

FIGURE 1.3: Relationships between pairs of variables in the Kentucky Derby data set.

in the **coded scatterplot** of Figure 1.4 we see that speeds are generally faster under fast conditions, but the rate of increasing speed over time is greater under good or slow conditions.

Of course, any graphical analysis is exploratory, and any notable trends at this stage should be checked through formal modeling. At this point, a statistician begins to ask familiar questions such as:

- are winning speeds increasing in a linear fashion?
- does the rate of increase in winning speed depend on track condition or number of starters?
- after accounting for other explanatory variables, is greater field size (number of starters) associated with faster winning speeds (because more horses in the field means a greater chance one horse will run a very fast time) or slower winning speeds (because horses are more likely to bump into each other or crowd each others' attempts to run at full gait)?
- are any of these associations statistically significant?
- how well can we predict the winning speed in the Kentucky Derby?

As you might expect, answers to these questions will arise from proper consideration of variability and properly identified statistical models.

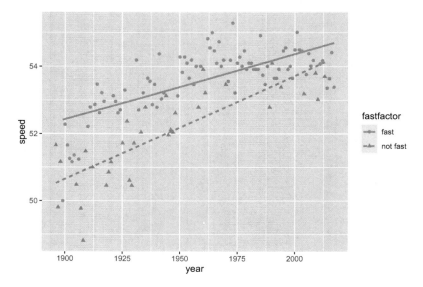

FIGURE 1.4: Linear trends in winning speeds over time, presented separately for fast conditions vs. good or slow conditions.

1.6 Multiple Linear Regression Modeling

1.6.1 Simple Linear Regression with a Continuous Predictor

We will begin by modeling the winning speed as a function of time; for example, have winning speeds increased at a constant rate since 1896? For this initial model, let Y_i be the speed of the winning horse in year i. Then, we might consider Model 1:

$$Y_i = \beta_0 + \beta_1 \text{Year}_i + \epsilon_i \quad \text{where} \quad \epsilon_i \sim N(0, \sigma^2). \tag{1.1}$$

In this case, β_0 represents the true intercept—the expected winning speed during Year 0. β_1 represents the true slope—the expected increase in winning speed from one year to the next, assuming the rate of increase is linear (i.e., constant with each successive year since 1896). Finally, the **error** (ϵ_i) terms represent the deviations of the actual winning speed in Year i (Y_i) from the expected speeds under this model ($\beta_0 + \beta_1 \text{Year}_i$)—the part of a horse's winning speed that is not explained by a linear trend over time. The variability in these deviations from the regression model is denoted by σ^2.

The parameters in this model (β_0, β_1, and σ^2) can be estimated through

1.6 Multiple Linear Regression Modeling

ordinary least squares methods; we will use hats to denote estimates of population parameters based on empirical data. Values for $\hat{\beta}_0$ and $\hat{\beta}_1$ are selected to minimize the sum of squared residuals, where a **residual** is simply the observed prediction error—the actual winning speed for a given year minus the winning speed predicted by the model. In the notation of this section,

- Predicted speed: $\hat{Y}_i = \hat{\beta}_0 + \hat{\beta}_1 \text{Year}_i$
- Residual (estimated error): $\hat{\epsilon}_i = Y_i - \hat{Y}_i$
- Estimated variance of points around the line: $\hat{\sigma}^2 = \sum \hat{\epsilon}_i^2 / (n-2)$

Using Kentucky Derby data, we estimate $\hat{\beta}_0 = 2.05$, $\hat{\beta}_1 = 0.026$, and $\hat{\sigma} = 0.90$. Thus, according to our simple linear regression model, winning horses of the Kentucky Derby have an estimated winning speed of 2.05 ft/s in Year 0 (more than 2000 years ago!), and the winning speed improves by an estimated 0.026 ft/s every year. With an R^2 of 0.513, the regression model explains a moderate amount (51.3%) of the year-to-year variability in winning speeds, and the trend toward a linear rate of improvement each year is statistically significant at the 0.05 level (t(120) = 11.251, p < .001).

```
model1 <- lm(speed ~ year, data = derby.df)
```

```
##             Estimate Std. Error  t value Pr(>|t|)
## (Intercept) 2.05347  4.543754    0.4519   6.521e-01
## year        0.02613  0.002322    11.2515  1.717e-20

## R squared =        0.5134
## Residual standard error =  0.9032
```

You may have noticed in Model 1 that the intercept has little meaning in context, since it estimates a winning speed in Year 0, when the first Kentucky Derby run at the current distance (1.25 miles) was in 1896. One way to create more meaningful parameters is through **centering**. In this case, we could create a centered year variable by subtracting 1896 from each year for Model 2:

$$Y_i = \beta_0 + \beta_1 \text{Yearnew}_i + \epsilon_i \quad \text{where} \quad \epsilon_i \sim N(0, \sigma^2)$$
$$\text{and} \quad \text{Yearnew} = \text{Year} - 1896.$$

Note that the only thing that changes from Model 1 to Model 2 is the estimated intercept; $\hat{\beta}_1$, R^2, and $\hat{\sigma}$ all remain exactly the same. Now $\hat{\beta}_0$ tells us that the estimated winning speed in 1896 is 51.59 ft/s, but estimates of the linear rate of improvement or the variability explained by the model remain the same. As Figure 1.5 shows, centering year has the effect of shifting the y-axis from year 0 to year 1896, but nothing else changes.

```
model2 <- lm(speed ~ yearnew, data = derby.df)
```

```
##               Estimate Std. Error  t value    Pr(>|t|)
## (Intercept)   51.58839   0.162549   317.37   2.475e-177
## yearnew        0.02613   0.002322    11.25    1.717e-20

##   R squared =    0.5134
##   Residual standard error =   0.9032
```

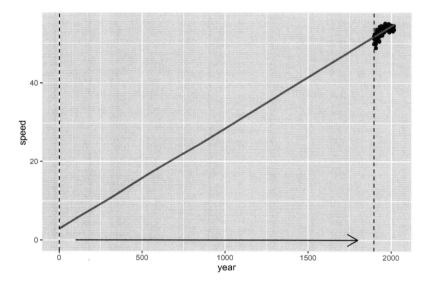

FIGURE 1.5: Compare Model 1 (with intercept at 0) to Model 2 (with intercept at 1896).

We should also attempt to verify that our LINE linear regression model assumptions fit for Model 2 if we want to make inferential statements (hypothesis tests or confidence intervals) about parameters or predictions. Most of these assumptions can be checked graphically using a set of residual plots as in Figure 1.6:

- The upper left plot, Residuals vs. Fitted, can be used to check the Linearity assumption. Residuals should be patternless around Y = 0; if not, there is a pattern in the data that is currently unaccounted for.
- The upper right plot, Normal Q-Q, can be used to check the Normality assumption. Deviations from a straight line indicate that the distribution of residuals does not conform to a theoretical normal curve.

1.6 Multiple Linear Regression Modeling

- The lower left plot, Scale-Location, can be used to check the Equal Variance assumption. Positive or negative trends across the fitted values indicate variability that is not constant.
- The lower right plot, Residuals vs. Leverage, can be used to check for influential points. Points with high leverage (having unusual values of the predictors) and/or high absolute residuals can have an undue influence on estimates of model parameters.

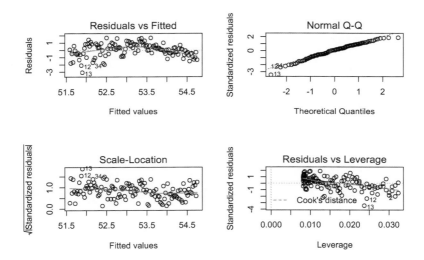

FIGURE 1.6: Residual plots for Model 2.

In this case, the Residuals vs. Fitted plot indicates that a quadratic fit might be better than the linear fit of Model 2; other assumptions look reasonable. Influential points would be denoted by high values of Cook's Distance; they would fall outside cutoff lines in the northeast or southeast section of the Residuals vs. Leverage plot. Since no cutoff lines are even noticeable, there are no potential influential points of concern.

We recommend relying on graphical evidence for identifying regression model assumption violations, looking for highly obvious violations of assumptions before trying corrective actions. While some numerical tests have been devised for issues such as normality and influence, most of these tests are not very reliable, highly influenced by sample size and other factors. There is typically no residual plot, however, to evaluate the Independence assumption; evidence for lack of independence comes from knowing about the study design and methods of data collection. In this case, with a new field of horses each year, the assumption of independence is pretty reasonable.

Based on residual diagnostics, we should test Model 2Q, in which a quadratic term is added to the linear term in Model 2.

$$Y_i = \beta_0 + \beta_1 \text{Yearnew}_i + \beta_2 \text{Yearnew}_i^2 + \epsilon_i \quad \text{where} \quad \epsilon_i \sim N(0, \sigma^2).$$

This model could suggest, for example, that the rate of increase in winning speeds is slowing down over time. In fact, there is evidence that the quadratic model improves upon the linear model (see Figure 1.7). R^2, the proportion of year-to-year variability in winning speeds explained by the model, has increased from 51.3% to 64.1%, and the pattern in the Residuals vs. Fitted plot of Figure 1.6 has disappeared in Figure 1.8, although normality is a little sketchier in the left tail, and the larger mass of points with fitted values near 54 appears to have slightly lower variability. The significantly negative coefficient for β_2 suggests that the rate of increase is indeed slowing in more recent years.

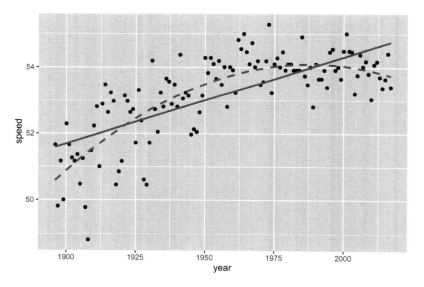

FIGURE 1.7: Linear (solid) vs. quadratic (dashed) fit.

```
derby.df <- mutate(derby.df, yearnew2 = yearnew^2)
model2q <- lm(speed ~ yearnew + yearnew2, data = derby.df)
```

```
##                  Estimate  Std. Error  t value    Pr(>|t|)
## (Intercept)   50.5874566   2.082e-01  243.010   2.615e-162
## yearnew        0.0761728   7.950e-03    9.581    1.839e-16
## yearnew2      -0.0004136   6.359e-05   -6.505    1.921e-09
##   R squared =    0.641
```

1.6 Multiple Linear Regression Modeling

```
## Residual standard error =   0.779
```

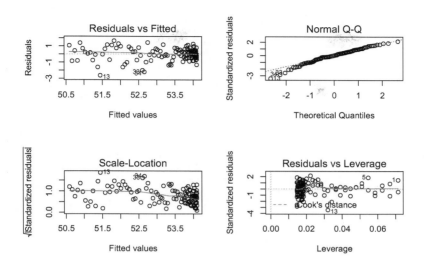

FIGURE 1.8: Residual plots for Model 2Q.

1.6.2 Linear Regression with a Binary Predictor

We also may want to include track condition as an explanatory variable. We could start by using `fast` as the lone predictor: Do winning speeds differ for fast and non-fast conditions? `fast` is considered an **indicator variable**—it takes on only the values 0 and 1, where 1 indicates presence of a certain attribute (like fast racing conditions). Since `fast` is numeric, we can use simple linear regression techniques to fit Model 3:

$$Y_i = \beta_0 + \beta_1 \text{Fast}_i + \epsilon_i \quad \text{where} \quad \epsilon_i \sim N(0, \sigma^2). \tag{1.2}$$

Here, it's easy to see the meaning of our slope and intercept by writing out separate equations for the two conditions:

- Good or slow conditions (`fast = 0`)

$$Y_i = \beta_0 + \epsilon_i$$

- Fast conditions (`fast = 1`)

$$Y_i = (\beta_0 + \beta_1) + \epsilon_i$$

β_0 is the expected winning speed under good or slow conditions, while β_1 is the difference between expected winning speeds under fast conditions vs. non-fast conditions. According to our fitted Model 3, the estimated winning speed under non-fast conditions is 52.0 ft/s, while mean winning speeds under fast conditions are estimated to be 1.6 ft/s higher.

```
model3 <- lm(speed ~ fast, data = derby.df)
```

```
##              Estimate Std. Error t value   Pr(>|t|)
## (Intercept)   51.994    0.1826   284.698  1.117e-171
## fast           1.629    0.2150     7.577   8.166e-12
```

```
##  R squared =  0.3236
##  Residual standard error =  1.065
```

You might be asking at this point: If we simply wanted to compare mean winning speeds under fast and non-fast conditions, why didn't we just run a two-sample t-test? The answer is: we did! The t-test corresponding to β_1 is equivalent to an independent-samples t-test under equal variances. Convince yourself that this is true, and that the equal variance assumption is needed.

1.6.3 Multiple Linear Regression with Two Predictors

The beauty of the linear regression framework is that we can add explanatory variables in order to explain more variability in our response, obtain better and more precise predictions, and control for certain covariates while evaluating the effect of others. For example, we could consider adding `yearnew` to Model 3, which has the indicator variable `fast` as its only predictor. In this way, we would estimate the difference between winning speeds under fast and non-fast conditions *after accounting for the effect of time*. As we observed in Figure 1.3, recent years have tended to have more races under fast conditions, so Model 3 might overstate the effect of fast conditions because winning speeds have also increased over time. A model with terms for both year and track condition will estimate the difference between winning speeds under fast and non-fast conditions *for a fixed year*; for example, if it had rained in 2016 and turned the track muddy, how much would we have expected the winning speed to decrease?

1.6 Multiple Linear Regression Modeling

Our new model (Model 4) can be written:

$$Y_i = \beta_0 + \beta_1 \text{Yearnew}_i + \beta_2 \text{Fast}_i + \epsilon_i \quad \text{where} \quad \epsilon_i \sim N(0, \sigma^2). \quad (1.3)$$

and linear least squares regression (LLSR) provides the following parameter estimates:

```
model4 <- lm(speed ~ yearnew + fast, data = derby.df)
```

```
##              Estimate Std. Error t value   Pr(>|t|)
## (Intercept) 50.91782   0.154602  329.35  5.360e-178
## yearnew      0.02258   0.001919   11.77   1.117e-21
## fast         1.22685   0.150721    8.14   4.393e-13

##   R squared =  0.6874
##   Residual standard error =   0.7269
```

Our new model estimates that winning speeds are, on average, 1.23 ft/s faster under fast conditions after accounting for time trends, which is down from an estimated 1.63 ft/s without accounting for time. It appears our original model (Model 3) may have overestimated the effect of fast conditions by conflating it with improvements over time. Through our new model, we also estimate that winning speeds increase by 0.023 ft/s per year, after accounting for track condition. This yearly effect is also smaller than the 0.026 ft/s per year we estimated in Model 1, without adjusting for track condition. Based on the R^2 value, Model 4 explains 68.7% of the year-to-year variability in winning speeds, a noticeable increase over using either explanatory variable alone.

1.6.4 Inference in Multiple Linear Regression: Normal Theory

So far we have been using linear regression for descriptive purposes, which is an important task. We are often interested in issues of statistical inference as well—determining if effects are statistically significant, quantifying uncertainty in effect size estimates with confidence intervals, and quantifying uncertainty in model predictions with prediction intervals. Under LINE assumptions, all of these inferential tasks can be completed with the help of the t-distribution and estimated standard errors.

Here are examples of inferential statements based on Model 4:

- We can be 95% confident that average winning speeds under fast conditions are between 0.93 and 1.53 ft/s higher than under non-fast conditions, after accounting for the effect of year.
- Fast conditions lead to significantly faster winning speeds than non-fast conditions (t = 8.14 on 119 df, p < .001), holding year constant.

- Based on our model, we can be 95% confident that the winning speed in 2017 under fast conditions will be between 53.4 and 56.3 ft/s. Note that Always Dreaming's actual winning speed barely fit within this interval—the 2017 winning speed was a borderline outlier on the slow side.

```
confint(model4)
```

```
                2.5 %    97.5 %
(Intercept) 50.61169 51.22395
yearnew      0.01878  0.02638
fast         0.92840  1.52529
```

```
new.data <- data.frame(yearnew = 2017 - 1896, fast = 1)
predict(model4, new = new.data, interval = "prediction")
```

```
    fit   lwr   upr
1 54.88 53.41 56.34
```

1.6.5 Inference in Multiple Linear Regression: Bootstrapping

Remember that you must check LINE assumptions using the same residual plots as in Figure 1.6 to ensure that the inferential statements in the previous section are valid. In cases when model assumptions are shaky, one alternative approach to statistical inference is **bootstrapping**; in fact, bootstrapping is a robust approach to statistical inference that we will use frequently throughout this book because of its power and flexibility. In bootstrapping, we use only the data we've collected and computing power to estimate the uncertainty surrounding our parameter estimates. Our primary assumption is that our original sample represents the larger population, and then we can learn about uncertainty in our parameter estimates through repeated samples (with replacement) from our original sample.

If we wish to use bootstrapping to obtain confidence intervals for our coefficients in Model 4, we could follow these steps:

- take a (bootstrap) sample of 122 years of Derby data with replacement, so that some years will get sampled several times and others not at all. This is **case resampling**, so that all information from a given year (winning speed, track condition, number of starters) remains together.

1.6 Multiple Linear Regression Modeling

- fit Model 4 to the bootstrap sample, saving $\hat{\beta}_0$, $\hat{\beta}_1$, and $\hat{\beta}_2$.
- repeat the two steps above a large number of times (say 1000).
- the 1000 bootstrap estimates for each parameter can be plotted to show the **bootstrap distribution** (see Figure 1.9).
- a 95% confidence interval for each parameter can be found by taking the middle 95% of each bootstrap distribution—i.e., by picking off the 2.5 and 97.5 percentiles. This is called the **percentile method**.

```
# updated code from tobiasgerstenberg on github
set.seed(413)
bootreg <- derby.df %>%
  bootstraps(1000) %>%
  pull(splits) %>%
  map_dfr(~lm(speed ~ yearnew + fast, data = .) %>%
          tidy())
bootreg %>%
  group_by(term) %>%
  dplyr::summarize(low=quantile(estimate, .025),
          high=quantile(estimate, .975))
```

```
# A tibble: 3 x 3
  term           low     high
  <chr>         <dbl>    <dbl>
1 (Intercept)   50.6     51.3
2 fast          0.909    1.57
3 yearnew       0.0182   0.0265
```

In this case, we see that 95% bootstrap confidence intervals for β_0, β_1, and β_2 are very similar to the normal-theory confidence intervals we found earlier. For example, the normal-theory confidence interval for the effect of fast tracks is 0.93 to 1.53 ft/s, while the analogous bootstrap confidence interval is 0.91 to 1.57 ft/s.

There are many variations on this bootstrap procedure. For example, you could sample residuals rather than cases, or you could conduct a parametric bootstrap in which error terms are randomly chosen from a normal distribution. In addition, researchers have devised other ways of calculating confidence intervals besides the percentile method, including normality, studentized, and bias-corrected and accelerated methods (Hesterberg [2015]; Efron and Tibshirani [1993]; Davison and Hinkley [1997]). We will focus on case resampling and percentile confidence intervals for now for their understandability and wide applicability.

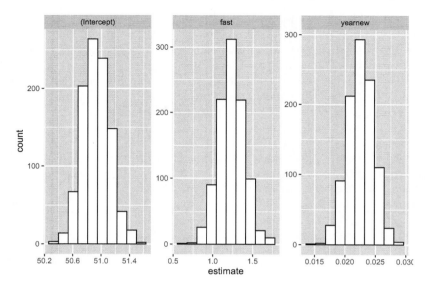

FIGURE 1.9: Bootstrapped distributions for Model 4 coefficients.

1.6.6 Multiple Linear Regression with an Interaction Term

Adding terms to form a multiple linear regression model as we did in Model 4 is a very powerful modeling tool, allowing us to account for multiple sources of uncertainty and to obtain more precise estimates of effect sizes after accounting for the effect of important covariates. One limitation of Model 4, however, is that we must assume that the effect of track condition has been the same for 122 years, or conversely that the yearly improvements in winning speeds are identical for all track conditions. To expand our modeling capabilities to allow the effect of one predictor to change depending on levels of a second predictor, we need to consider **interaction terms**. Amazingly, if we create a new variable by taking the product of yearnew and fast (i.e., the **interaction** between yearnew and fast), adding that variable into our model will have the desired effect.

Thus, consider Model 5:

$$Y_i = \beta_0 + \beta_1 \text{Yearnew}_i + \beta_2 \text{Fast}_i$$
$$+ \beta_3 \text{Yearnew}_i \times \text{Fast}_i + \epsilon_i \quad \text{where} \quad \epsilon_i \sim N(0, \sigma^2)$$

where LLSR provides the following parameter estimates:

1.6 Multiple Linear Regression Modeling

```
model5 <- lm(speed ~ yearnew + fast + yearnew:fast,
             data=derby.df)
```

```
##                  Estimate Std. Error t value   Pr(>|t|)
## (Intercept)      50.52863  0.205072  246.394  6.989e-162
## yearnew           0.03075  0.003471    8.859  9.839e-15
## fast              1.83352  0.262175    6.994  1.730e-10
## yearnew:fast     -0.01149  0.004117   -2.791  6.128e-03

##   R squared =  0.7068
##   Residual standard error =   0.7071
```

According to our model, estimated winning speeds can be found by:

$$\hat{Y}_i = 50.53 + 0.031 \text{Yearnew}_i + 1.83 \text{Fast}_i - 0.011 \text{Yearnew}_i \times \text{Fast}_i. \quad (1.4)$$

Interpretations of model coefficients are most easily seen by writing out separate equations for fast and non-fast track conditions:

Fast = 0 :
$$\hat{Y}_i = 50.53 + 0.031 \text{Yearnew}_i$$
Fast = 1 :
$$\hat{Y}_i = (50.53 + 1.83) + (0.031 - 0.011) \text{Yearnew}_i$$

leading to the following interpretations for estimated model coefficients:

- $\hat{\beta}_0 = 50.53$. The expected winning speed in 1896 under non-fast conditions was 50.53 ft/s.
- $\hat{\beta}_1 = 0.031$. The expected yearly increase in winning speeds under non-fast conditions is 0.031 ft/s.
- $\hat{\beta}_2 = 1.83$. The winning speed in 1896 was expected to be 1.83 ft/s faster under fast conditions compared to non-fast conditions.
- $\hat{\beta}_3 = -0.011$. The expected yearly increase in winning speeds under fast conditions is 0.020 ft/s, compared to 0.031 ft/s under non-fast conditions, a difference of 0.011 ft/s.

In fact, using interaction allows us to model the relationships we noticed in Figure 1.4, where both the intercept and slope describing the relationships between speed and year differ depending on whether track conditions were fast or not. Note that we interpret the coefficient for the interaction term by comparing slopes under fast and non-fast conditions; this produces a much more understandable interpretation for a reader than attempting to interpret the -0.011 directly.

1.6.7 Building a Multiple Linear Regression Model

We now begin iterating toward a "final model" for these data, on which we will base conclusions. Typical features of a "final multiple linear regression model" include:

- explanatory variables allow one to address primary research questions
- explanatory variables control for important covariates
- potential interactions have been investigated
- variables are centered where interpretations can be enhanced
- unnecessary terms have been removed
- LINE assumptions and the presence of influential points have both been checked using residual plots
- the model tells a "persuasive story parsimoniously"

Although the process of reporting and writing up research results often demands the selection of a sensible final model, it's important to realize that (a) statisticians typically will examine and consider an entire taxonomy of models when formulating conclusions, and (b) different statisticians sometimes select different models as their "final model" for the same set of data. Choice of a "final model" depends on many factors, such as primary research questions, purpose of modeling, tradeoff between parsimony and quality of fitted model, underlying assumptions, etc. Modeling decisions should never be automated or made completely on the basis of statistical tests; subject area knowledge should always play a role in the modeling process. You should be able to defend any final model you select, but you should not feel pressured to find the one and only "correct model", although most good models will lead to similar conclusions.

Several tests and measures of model performance can be used when comparing different models for model building:

- R^2. Measures the variability in the response variable explained by the model. One problem is that R^2 always increases with extra predictors, even if the predictors add very little information.
- adjusted R^2. Adds a penalty for model complexity to R^2 so that any increase in performance must outweigh the cost of additional complexity. We should ideally favor any model with higher adjusted R^2, regardless of size, but the penalty for model complexity (additional terms) is fairly ad-hoc.
- AIC (Akaike Information Criterion). Again attempts to balance model performance with model complexity, with smaller AIC levels being preferable, regardless of model size. The BIC (Bayesian Information Criterion) is similar to the AIC, but with a greater penalty for additional model terms.

1.6 Multiple Linear Regression Modeling

- extra sum of squares F test. This is a generalization of the t-test for individual model coefficients which can be used to perform significance tests on **nested models**, where one model is a reduced version of the other. For example, we could test whether our final model (below) really needs to adjust for track condition, which is comprised of indicators for both fast condition and good condition (leaving slow condition as the reference level). Our null hypothesis is then $\beta_3 = \beta_4 = 0$. We have statistically significant evidence (F = 57.2 on 2 and 116 df, p < .001) that track condition is associated with winning speeds, after accounting for quadratic time trends and number of starters.

One potential final model for predicting winning speeds of Kentucky Derby races is:

$$Y_i = \beta_0 + \beta_1 \text{Yearnew}_i + \beta_2 \text{Yearnew}_i^2 + \beta_3 \text{Fast}_i \\ + \beta_4 \text{Good}_i + \beta_5 \text{Starters}_i + \epsilon_i \quad \text{where} \quad \epsilon_i \sim N(0, \sigma^2) \tag{1.5}$$

and LLSR provides the following parameter estimates:

```
model0 <- lm(speed ~ yearnew + yearnew2 + fast + good +
                 starters, data = derby.df)
```

```
##                 Estimate  Std. Error   t value   Pr(>|t|)
## (Intercept)  50.0203151   1.946e-01   256.980  1.035e-161
## yearnew       0.0700341   6.130e-03    11.424   1.038e-20
## yearnew2     -0.0003697   4.598e-05    -8.041   8.435e-13
## fast          1.3926656   1.305e-01    10.670   6.199e-19
## good          0.9156982   2.077e-01     4.409   2.331e-05
## starters     -0.0252836   1.360e-02    -1.859   6.559e-02

##   R squared =  0.8267
##   Residual standard error =  0.5483
```

```
# Compare models with and without terms for track condition
model0_reduced <- lm(speed ~ yearnew + yearnew2 +
                       starters, data = derby.df)
drop_in_dev <- anova(model0_reduced, model0, test = "F")
```

```
  ResidDF   RSS    SS Df        pval
1     118 69.26    NA NA          NA
2     116 34.87  57.2  2   5.194e-18
```

This model accounts for the slowing annual increases in winning speed with a negative quadratic term, adjusts for baseline differences stemming from track conditions, and suggests that, for a fixed year and track condition, a larger field is associated with slower winning times (unlike the positive relationship we saw between speed and number of starters in our exploratory analyses). The model explains 82.7% of the year-to-year variability in winning speeds, and residual plots show no serious issues with LINE assumptions. We tested interaction terms for different effects of time or number of starters based on track condition, but we found no significant evidence of interactions.

1.7 Preview of Remaining Chapters

Having reviewed key ideas from multiple linear regression, you are now ready to extend those ideas, especially to handle non-normal responses and lack of independence. This section provides a preview of the type of problems you will encounter in the book. For each journal article cited, we provide an abstract in the authors' words, a description of the type of response and, when applicable, the structure of the data. Each of these examples appears later as an exercise, where you can play with the actual data or evaluate the analyses detailed in the articles.

1.7.1 Soccer

Roskes et al. [2011] The right side? Under time pressure, approach motivation leads to right-oriented bias. *Psychological Science* [Online] **22(11)**:1403-7. DOI: 10.1177/0956797611418677, October 2011.

> **Abstract:** Approach motivation, a focus on achieving positive outcomes, is related to relative left-hemispheric brain activation, which translates to a variety of right-oriented behavioral biases. [...] In our analysis of all Federation Internationale de Football Association (FIFA) World Cup penalty shoot-outs, we found that goalkeepers were two times more likely to dive to the right than to the left when their team was behind, a situation that we conjecture induces approach motivation. Because penalty takers shot toward the two sides of the goal equally often, the goalkeepers' right-oriented bias was dysfunctional, allowing more goals to be scored.

1.7 Preview of Remaining Chapters

The response for this analysis is the direction of the goalkeeper dive, a binary variable. For example, you could let Y=1 if the dive is to the right and Y=0 if the dive is to the left. This response is clearly not normally distributed. One approach to the analysis is logistic regression as described in Chapter 6. A binomial random variable could also be created for this application by summing the binary variables for each game so that Y= the number of dives right out of the number of dives the goalkeeper makes during a game. [Thought question: Do you believe the last line of the abstract?]

1.7.2 Elephant Mating

Poole [1989] Mate guarding, reproductive success and female choice in African elephants. *Animal Behavior* **37**:842-49.

Abstract: Male guarding of females, male mating success and female choice were studied for 8 years among a population of African elephants, *Loxodonta africana*. Males were not able to compete successfully for access to oestrous females until approximately 25 years of age. Males between 25 and 35 years of age obtained matings during early and late oestrus, but rarely in mid-oestrus. Large musth males over 35 years old guarded females in mid-oestrus. Larger, older males ranked above younger, smaller males and the number of females guarded by males increased rapidly late in life. Body size and longevity are considered important factors in determining the lifetime reproductive success of male elephants...

Poole and her colleagues recorded, for each male elephant, his age (in years) and the number of matings for a given year. The researchers were interested in how age affects the males' mating patterns. Specifically, questions concern whether there is a steady increase in mating success as an elephant ages or if there is an optimal age after which the number of matings decline. Because the responses of interest are counts (number of matings for each elephant for a given year), we will consider a Poisson regression (see Chapter 4). The general form for Poisson responses is the number of events for a specified time, volume, or space.

1.7.3 Parenting and Gang Activity

Walker-Barnes and Mason [2001] Ethnic differences in the effect of parenting on gang involvement and gang delinquency: a longitudinal, hierarchical linear modeling perspective. *Child Development* **72(6)**:1814-31.

Abstract: This study examined the relative influence of peer and parenting behavior on changes in adolescent gang involvement and gang-related delinquency. An ethnically diverse sample of 300 ninth-grade students was recruited and assessed on eight occasions during the school year. Analyses were conducted using hierarchical linear modeling. Results indicated that, in general, adolescents decreased their level of gang involvement over the course of the school year, whereas the average level of gang delinquency remained constant over time. As predicted, adolescent gang involvement and gang-related delinquency were most strongly predicted by peer gang involvement and peer gang delinquency, respectively. Nevertheless, parenting behavior continued to significantly predict change in both gang involvement and gang delinquency, even after controlling for peer behavior. A significant interaction between parenting and ethnic and cultural heritage found the effect of parenting to be particularly salient for Black students, for whom higher levels of behavioral control and lower levels of lax parental control were related to better behavioral outcomes over time, whereas higher levels of psychological control predicted worse behavioral outcomes.

The response for this study is a gang activity measure which ranges from 1 to 100. While it may be reasonable to assume this measure is approximately normal, the structure of this data implies that it is not a simple regression problem. Individual students have measurements made at 8 different points in time. We cannot assume that we have 2400 independent observations because the same measurements on one individual are more likely to be similar than a measurement of another student. Multilevel modeling as discussed in Chapter 9 can often be used in these situations.

1.7.4 Crime

Gelman et al. [2007] An analysis of the NYPD's stop-and-frisk policy in the context of claims of racial bias. *Journal of the American Statistical Association* **102(479)**:813-823.

Abstract: Recent studies by police departments and researchers confirm that police stop racial and ethnic minority citizens more often than whites, relative to their proportions in the population. However, it has been argued stop rates more accurately reflect rates of crimes committed by each ethnic group, or that stop rates reflect elevated rates in specific social areas such as neighborhoods or precincts. Most of the research on stop rates and police-citizen interactions has focused on traffic stops, and analyses of pedestrian stops are rare. In this paper, we analyze data from 175,000 pedestrian stops by the New York Police Department over a fifteen-month period. We disaggregate stops by police precinct, and compare stop rates by racial and ethnic group controlling for previous race-specific arrest rates. We use hierarchical multilevel models to adjust for precinct-level variability, thus directly addressing the question of geographic heterogeneity that arises in the analysis of pedestrian stops. We find that persons of African and Hispanic descent were stopped more frequently than whites, even after controlling for precinct variability and race-specific estimates of crime participation.

This application involves both non-normal data (number of stops by ethnic group can be modeled as a Poisson response) and multilevel data (number of stops within precincts will likely be correlated due to characteristics of the precinct population). This type of analysis will be the last type you encounter, multilevel generalized linear modeling, as addressed in Chapter 11.

1.8 Exercises

1.8.1 Conceptual Exercises

1. **Applications that do not violate assumptions for inference in LLSR.** Identify the response and explanatory variable(s) for each

problem. Write the LLSR assumptions for inference in the context of each study.

    a. **Cricket Chirps.** Researchers record the number of cricket chirps per minute and temperature during that time to investigate whether the number of chirps varies with the temperature.

    b. **Women's Heights.** A random selection of women aged 20-24 years are selected and their shoe size is used to predict their height.

2. **Applications that do violate assumptions for inference in LLSR.** All of the examples in this section have at least one violation of the LLSR assumptions for inference. Begin by identifying the response and explanatory variables. Then, identify which model assumption(s) are violated.

    a. **Low Birthweights.** Researchers are attempting to see if socioeconomic status and parental stability are predictive of low birthweight. They classify a child as having a low birthweight if their birthweight is less than 2,500 grams.

    b. **Clinical Trial I.** A Phase II clinical trial is designed to compare the number of patients getting relief at different dose levels. 100 patients get dose A, 100 get dose B, and 100 get dose C.

    c. **Canoes and Zip Codes.** For each of over 27,000 overnight permits for the Boundary Waters Canoe area, the zip code for the group leader has been translated to the distance traveled and socioeconomic data. Thus, for each zip code we can model the number of trips made as a function of distance traveled and various socioeconomic measures.

    d. **Clinical Trial II.** A randomized clinical trial investigated postnatal depression and the use of an estrogen patch. Patients were randomly assigned to either use the patch or not. Depression scores were recorded on 6 different visits.

3. **Kentucky Derby.** The next set of questions is related to the Kentucky Derby case study from this chapter.

    a. Discuss the pros and cons of using side-by-side boxplots vs. stacked histograms to illustrate the relationship between year and track condition in Figure 1.3.

    b. Why is a scatterplot more informative than a correlation coefficient to describe the relationship between speed of the winning horse and year in Figure 1.3.

    c. How might you incorporate a fourth variable, say number of starters, into Figure 1.4?

1.8 Exercises

  d. Explain why ϵ_i in Equation (1.1) measures the vertical distance from a data point to the regression line.
  e. In the first t-test in Section 1.6.1 (t = 11.251 for $H_0 : \beta_1 = 0$), notice that $t = \frac{\hat{\beta}_1}{SE(\hat{\beta}_1)} = \frac{.026}{.0023} = 11.251$. Why is the t-test based on the ratio of the estimated slope to its standard error?
  f. In Equation (1.2), explain why the t-test corresponding to β_1 is equivalent to an independent-samples t-test under equal variances. Why is the equal variance assumption needed?
  g. When interpreting β_2 in Equation (1.3), why do we have to be careful to say *for a fixed year* or *after adjusting for year*? Is it wrong to leave a qualifier like that off?
  h. Interpret in context a 95% confidence interval for β_0 in Model 4.
  i. State (in context) the result of a t-test for β_1 in Model 4.
  j. Why is there no ϵ_i term in Equation (1.4)?
  k. If you considered the interaction between two continuous variables (like `yearnew` and `starters`), how would you provide an interpretation for that coefficient in context?
  l. Interpret (in context) the LLSR estimates for β_3 and β_5 in Equation (1.5).

4. **Moneyball.** In a 2011 article in *The Sport Journal*, Farrar and Bruggink [2011] attempt to show that Major League Baseball general managers did not immediately embrace the findings of Michael Lewis's 2003 *Moneyball* book [Lewis, 2003]. They contend that players' on-base percentage remained relatively undercompensated compared to slugging percentage three years after the book came out. Two regression models are described: Team Run Production Model and Player Salary Model.

  a. Discuss potential concerns (if any) with the LINE assumptions for linear regression in each model.
  b. In Table 3, the authors contend that Model 1 is better than Model 3. Could you argue that Model 3 is actually better? How could you run a formal hypothesis test comparing Model 1 to Model 3?
  c. If authors had chosen Model 3 in Table 3 with the two interaction terms, how would that affect their final analysis, in which they compare coefficients of slugging and on-base percentage? (Hint: write out interpretations for the two interaction coefficients—the first one should be NL:OBP and the second one should be NL:SLG)
  d. The authors write that, "It should also be noted that the runs scored equation fit is better than the one Hakes and Sauer have

for their winning equation." What do you think they mean by this statement? Why might this comparison not be relevant?
e. In Table 4, Model 1 has a higher adjusted R^2 than Model 2, yet the extra term in Model 1 (an indicator value for the National League) is not significant at the 5% level. Explain how this is possible.
f. What limits does this paper have on providing guidance to baseball decision makers?

1.8.2 Guided Exercises

1. **Gender discrimination in bank salaries**. In the 1970's, Harris Trust was sued for gender discrimination in the salaries it paid its employees. One approach to addressing this issue was to examine the starting salaries of all skilled, entry-level clerical workers between 1965 and 1975. The following variables, which can be found in `banksalary.csv`, were collected for each worker [Ramsey and Schafer, 2002]:
   - `bsal` = beginning salary (annual salary at time of hire)
   - `sal77` = annual salary in 1977
   - `sex` = MALE or FEMALE
   - `senior` = months since hired
   - `age` = age in months
   - `educ` = years of education
   - `exper` = months of prior work experience

   Creating an indicator variable based on `sex` could be helpful.

   a. Identify observational units, the response variable, and explanatory variables.
   b. The mean starting salary of male workers ($5957) was 16% higher than the mean starting salary of female workers ($5139). Confirm these mean salaries. Is this enough evidence to conclude gender discrimination exists? If not, what further evidence would you need?
   c. How would you expect age, experience, and education to be related to starting salary? Generate appropriate exploratory plots; are the relationships as you expected? What implications does this have for modeling?
   d. Why might it be important to control for seniority (number of years with the bank) if we are only concerned with the salary when the worker started?
   e. By referring to exploratory plots and summary statistics, are

any explanatory variables (including sex) closely related to each other? What implications does this have for modeling?

f. Fit a simple linear regression model with starting salary as the response and experience as the sole explanatory variable (Model 1). Interpret the intercept and slope of this model; also interpret the R-squared value. Is there a significant relationship between experience and starting salary?

g. Does Model 1 meet all linear least squares regression assumptions? List each assumption and how you decided if it was met or not.

h. Is a model with all 4 confounding variables (Model 2, with senior, educ, exper, and age) better than a model with just experience (Model 1)? Justify with an appropriate significance test in addition to summary statistics of model performance.

i. You should have noticed that the term for age was not significant in Model 2. What does this imply about age and about future modeling steps?

j. Generate an appropriate coded scatterplot to examine a potential age-by-experience interaction. How would you describe the nature of this interaction?

k. A potential final model (Model 3) would contain terms for seniority, education, and experience in addition to sex. Does this model meet all regression assumptions? State a 95% confidence interval for sex and interpret this interval carefully in the context of the problem.

l. Based on Model 3, what conclusions can be drawn about gender discrimination at Harris Trust? Do these conclusions have to be qualified at all, or are they pretty clear cut?

m. Often salary data is logged before analysis. Would you recommend logging starting salary in this study? Support your decision analytically.

n. Regardless of your answer to the previous question, provide an interpretation for the coefficient for the male coefficient in a modified Model 3 after logging starting salary.

o. Build your own final model for this study and justify the selection of your final model. You might consider interactions with gender, since those terms could show that discrimination is stronger among certain workers. Based on your final model, do you find evidence of gender discrimination at Harris Trust?

2. **Sitting and MTL thickness.** Siddarth et al. [2018] researched relations between time spent sitting (sedentary behavior) and the thickness of a participant's medial temporal lobe (MTL) in a 2018 paper entitled, "Sedentary behavior associated with reduced medial

temporal lobe thickness in middle-aged and older adults". MTL volume is negatively associated with Alzheimer's disease and memory impairment. Their data on 35 adults can be found in `sitting.csv`. Key variables include:
- `MTL` = Medial temporal lobe thickness in mm
- `sitting` = Reported hours/day spent sitting
- `MET` = Reported metabolic equivalent unit minutes per week
- `age` = Age in years
- `sex` = Sex (`M` = Male, `F` = Female)
- `education` = Years of education completed

a. In their article's introduction, Siddarth et al. differentiate their analysis on sedentary behavior from an analysis on active behavior by citing evidence supporting the claim that, "one can be highly active yet still be sedentary for most of the day." Fit your own linear model with `MET` and `sitting` as your explanatory and response variables, respectively. Using R^2, how much of the subject to subject variability in hours/day spent sitting can be explained by MET minutes per week? Does this support the claim that sedentary behaviors may be independent from physical activity?

b. In the paper's section, "Statistical analysis", the authors report that, "Due to the skewed distribution of physical activity levels, we used log-transformed values in all analyses using continuous physical activity measures." Generate both a histogram of `MET` values and log–transformed `MET` values. Do you agree with the paper's decision to use a log-transformation here?

c. Fit a preliminary model with `MTL` as the response and `sitting` as the sole explanatory variable. Are LLSR conditions satisfied?

d. Expand on your previous model by including a centered version of `age` as a covariate. Interpret all three coefficients in this model.

e. One model fit in Siddarth et al. [2018] includes `sitting`, log–transformed `MET`, and `age` as explanatory variables. They report an estimate $\hat{\beta}_1 = -0.02$ with confidence interval $(-0.04, -0.002)$ for the coefficient corresponding to `sitting`, and $\hat{\beta}_2 = 0.007$ with confidence interval $(-0.07, 0.08)$ for the coefficient corresponding to `MET`. Verify these intervals and estimates on your own.

f. Based on your confidence intervals from the previous part, do you support the paper's claim that, "it is possible that sedentary behavior is a more significant predictor of brain structure, specifically MTL thickness [than physical activity]"? Why or why not?

g. A *New York Times* article was published discussing Siddarth et al. [2018] with the title "Standing Up at Your Desk Could Make You Smarter" [Friedman, 2018]. Do you agree with this headline choice? Why or why not?

3. **Housing prices and log transformations.** The dataset kingCountyHouses.csv contains data on over 20,000 houses sold in King County, Washington [Kaggle, 2018a]. The dataset includes the following variables:
   - price = selling price of the house
   - date = date house was sold, measured in days since January 1, 2014
   - bedrooms = number of bedrooms
   - bathrooms = number of bathrooms
   - sqft = interior square footage
   - floors = number of floors
   - waterfront = 1 if the house has a view of the waterfront, 0 otherwise
   - yr_built = year the house was built
   - yr_renovated = 0 if the house was never renovated, the year the house was renovated if else

   We wish to create a linear model to predict a house's selling price.

   a. Generate appropriate graphs and summary statistics detailing both price and sqft individually and then together. What do you notice?
   b. Fit a simple linear regression model with price as the response variable and sqft as the explanatory variable (Model 1). Interpret the slope coefficient β_1. Are all conditions met for linear regression?
   c. Create a new variable, logprice, the natural log of price. Fit Model 2, where logprice is now the response variable and sqft is still the explanatory variable. Write out the regression line equation.
   d. How does logprice change when sqft increases by 1?
   e. Recall that $\log(a) - \log(b) = \log\left(\frac{a}{b}\right)$, and use this to derive how price changes as sqft increases by 1.
   f. Are LLSR assumptions satisfied in Model 2? Why or why not?
   g. Create a new variable, logsqft, the natural log of sqft. Fit Model 3 where price and logsqft are the response and explanatory variables, respectively. Write out the regression line equation.
   h. How does predicted price change as logsqft increases by 1 in Model 3?
   i. How does predicted price change as sqft increases by 1%? As a hint, this is the same as multiplying sqft by 1.01.

j. Are LLSR assumptions satisfied in Model 3? Why or why not?
k. Fit Model 4, with `logsqft` and `logprice` as the response and explanatory variables, respectively. Write out the regression line equation.
l. In Model 4, what is the effect on `price` corresponding to a 1% increase in `sqft`?
m. Are LLSR assumptions satisfied in Model 4? Why or why not?
n. Find another explanatory variable which can be added to Model 4 to create a model with a higher adjusted R^2 value. Interpret the coefficient of this added variable.

1.8.3 Open-Ended Exercises

1. **The Bechdel Test.** In April, 2014, website FiveThirtyEight published the article, "The Dollar-And-Cents Case Against Hollywood's Exclusion of Women" [Hickey, 2014]. There, they analyze returns on investment for 1,615 films released between 1990 and 2013 based on the Bechdel test. The test, developed by cartoonist Alison Bechdel, measures gender bias in films by checking if a film meets three criteria:

    - there are at least two named women in the picture
    - they have a conversation with each other at some point
    - that conversation isn't about a male character

    While the test is not a perfect metric of gender bias, data from it does allow for statistical analysis. In the FiveThirtyEight article, they find that, "passing the Bechdel test had no effect on the film's return on investment." Their data can be found in `bechdel.csv`. Key variables include:

    - `year` = the year the film premiered
    - `pass` = 1 if the film passes the Bechdel test, 0 otherwise
    - `budget` = budget in 2013 U.S. dollars
    - `totalGross` = total gross earnings in 2013 U.S. dollars
    - `domGross` = domestic gross earnings in 2013 U.S. dollars
    - `intGross` = international gross earnings in 2013 U.S. dollars
    - `totalROI` = total return on investment (total gross divided by budget)
    - `domROI` = domestic return on investment
    - `intROI` = international return on investment

    With this in mind, carry out your own analysis. Does passing the Bechdel test have any effect on a film's return on investment?

2. **Waitress tips.** A student collected data from a restaurant where

1.8 Exercises

she was a waitress [Dahlquist and Dong, 2011]. The student was interested in learning under what conditions a waitress can expect the largest tips—for example: At dinner time or late at night? From younger or older patrons? From patrons receiving free meals? From patrons drinking alcohol? From patrons tipping with cash or credit? And should tip amount be measured as total dollar amount or as a percentage? Data can be found in `TipData.csv`. Here is a quick description of the variables collected:

- Day = day of the week
- Meal = time of day (Lunch, Dinner, Late Night)
- Payment = how bill was paid (Credit, Cash, Credit with Cash tip)
- Party = number of people in the party
- Age = age category of person paying the bill (Yadult, Middle, SenCit)
- GiftCard = was gift card used?
- Comps = was part of the meal complimentary?
- Alcohol = was alcohol purchased?
- Bday = was a free birthday meal or treat given?
- Bill = total size of the bill
- W.tip = total amount paid (bill plus tip)
- Tip = amount of the tip
- Tip.Percentage = proportion of the bill represented by the tip

2

Beyond Least Squares: Using Likelihoods

2.1 Learning Objectives

After finishing this chapter, you should be able to:
- Describe the concept of a likelihood, in words.
- Know and apply the Principle of Maximum Likelihood for a simple example.
- Identify three ways in which you can obtain or approximate an MLE.
- Use likelihoods to compare models.
- Construct a likelihood for a simple model.

This text encourages you to broaden your statistical horizons by moving beyond independent, identically distributed, normal responses (iidN). This chapter on likelihood focuses on ways to fit models, determine estimates, and compare models for a wide range of types of responses, not just iidN data. In your earlier study of statistics, you fit simple linear models using ordinary least squares (OLS). Fitting those models assumes that the mean value of a response, Y, is linearly related to some variable, X. However, often responses are not normally distributed. For example, a study in education may involve scoring responses on a test as correct or incorrect. This binary response may be explained by the number of hours students spend studying. However, we do not expect a variable which takes on only 0 or 1 to be a linear function of time spent studying (see Figure 2.1).

```
# Packages required for Chapter 2
library(gridExtra)
library(knitr)
library(mosaic)
library(xtable)
library(kableExtra)
library(tidyverse)
```

In this instance we'll use logistic regression instead of linear least squares regression. Fitting a logistic regression requires the use of likelihood methods. Another setting where likelihood methods come into play is when data

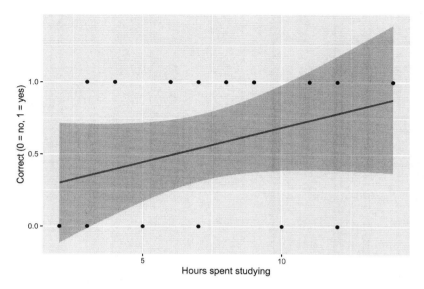

FIGURE 2.1: An attempt to fit a linear regression model to a binary response variable.

is produced from a complex structure which may imply correlation among outcomes. For example, test scores for students who have been taught by the same teacher may be correlated. We'll see that likelihood methods are useful when modeling correlated data. Likelihood methods not only provide a great deal of flexibility in the types of models we can fit, but they also provide ways in which to compare models as well. You might find likelihood methods a bit more complicated, but conceptually the approach is straightforward. As you go through the material here, worry less about calculus and computational details and focus on the concepts. You will have software to help you with computation, but model specification and interpretation will be up to you.

2.2 Case Study: Does Sex Run in Families?

Doesn't it seem that some families tend to have lots of boys, while others have more than their fair share of girls? Is it really the case that each child human couples produce is equally likely to be a male or female? Or does sex run in families? It can be argued that these kinds of questions have implications for population demographics and sibling harmony. For example, a 2009 study at

2.2 Case Study: Does Sex Run in Families?

the University of Ulster in Northern Ireland found that growing up with sisters, as compared to brothers, can enhance the quality of life of an adult [BBC News, 2009].

Sibling harmony aside, why do people care about gender imbalance? Comparisons of sex ratios between countries illustrate some compelling reasons. Some think that genetic or biological influences within families, such as "sex running in families," can affect sex ratios. Mating behavior such as waiting until the family includes a boy or both sexes affects sex ratios. Some believe that sex ratios point to the practice of sex selection in a country accomplished through abortion or infanticide. Furthermore, there is speculation that an excess of men could lead to unrest among young males unable to find marriage partners or start families.

In 1930, statistician R.A. Fisher posited a 50:50 equilibrium theory regarding sex ratios in terms of parental expenditure. Most often, in practice, sex ratios differ from what Fisher predicted. From 1970 to 2002, the sex ratio at birth in the US among white non-Hispanics was 105 boys to 100 girls, but only 103 boys to 100 girls among African Americans and Native Americans [Mathews and Hamilton, 2005]. A 1997 study in *Nature* reports evidence which suggests that the human sex ratio may be currently shifting in the United States toward more female babies, closer to Fisher's prediction! [Komdeur et al., 1997] Sex ratio comparisons between countries are also intriguing. For example, Switzerland has a sex ratio of 106 boys to 100 girls, whereas there are 112 boys to every 100 girls in China according to The World Factbook [Central Intelligence Agency, 2013]. In the next section, we bring the notion of gender imbalance closer to home by focusing on families instead of countries or sub-populations.

To investigate this question and others, we look at the gender composition of 5,626 families collected by the National Longitudinal Survey of Youth [Bureau of Labor Statistics, 1997]. We fit models to explore whether there is evidence sex runs in families, a model we refer to as a Sex Conditional Model. We also consider a separate but related question about whether couples are "waiting for a boy." [Rodgers and Doughty, 2001].

2.2.1 Research Questions

We specify several models related to gender balance in families. Our models liken having babies to flipping a coin (heads=boy, tails=girl), of course, recognizing that in truth there is a little more to having babies. The baseline model (Model 0) assumes that the probability of a boy is the same as the probability of a girl. The first model (Model 1) considers the situation that the coin is loaded and the probability of heads (a boy) is different than the probability of tails (a girl). Next, we consider a model (Model 2) that conditions on the previous number of boys or girls in a family to get at the question of whether sex runs in families. This data is also used for a different set of models that relate to couples' behavior. Specifically, we look to see if there is evidence that couples are waiting for a boy. Searching for evidence of waiting for a girl, or waiting for both a boy and a girl, are left as exercises.

Models 0 and 1 assume that having children is like flipping a coin. The gender of each child is independent of the gender of other children and the probability of a boy is the same for each new child. Let p_B be the probability a child is a boy.

1. **Model 0: Sex Unconditional Model (Equal probabilities).** Is a child just as likely to be a boy as it is to be a girl; is $p_B = 0.5$?
2. **Model 1: Sex Unconditional Model (Different probabilities).** Is the coin loaded; is $p_B \neq 0.5$?
3. **Model 2: Sex Conditional Model (Sex bias).** Do boys or girls run in families? That is, is there a tendency for families with more boys than girls to be more likely to produce another boy? Is the case the same for girls?
4. **Model 3: Stopping Rule Model (Waiting for a boy).** Is there evidence that couples stop having children once a boy is born?

Ultimately, our goal is to incorporate the family composition data represented as series of coin flips to find the "best" estimate for the probability of having a boy, p_B, and evaluate the assumptions built into these models. We will be using likelihood-based methods, not ordinary least squares, to fit and compare these models.

While the NLSY data is of interest, we start with a smaller, hypothetical data set of 30 families with a total of 50 children in order to illustrate concepts related to likelihoods (Table 2.1). The data are the frequencies of possible family gender compositions for one-, two-, and three-child families. The methods we develop on this small data set will then be applied to the one-, two- and three-family NLSY data. It is straightforward to include all of the family sizes up to the four- or five-child families in the NLSY data.

2.3 Model 0: Sex Unconditional, Equal Probabilities

For the Sex Unconditional models, having children is modeled using coin flips. With a coin flip model, the result of each flip is independent of results of other flips. With this version of the Sex Unconditional Model, the chance that a baby is a boy is specified to be $p_B = 0.5$. It makes no difference if the first and

2.4 Model 1: Sex Unconditional, Unequal Probabilities

TABLE 2.1: The gender composition of 30 families in the hypothetical data set of n=50 children.

Composition	Number of families	Number of children
B	6	6
G	7	7
BB	5	10
BG	4	8
GB	5	10
GGB	1	3
GBB	2	6
Total	30	50

third children are boys, the probability that the second child is a boy is 0.5; that is, the results for each child are **independent** of the others. Under this model you expect to see equal numbers of boys and girls.

2.4 Model 1: Sex Unconditional, Unequal Probabilities

You may want your model to allow for the probability of a boy, p_B, to be something different than 0.5. With this version of the Sex Unconditional model, $p_B > 0.5$ or $p_B < 0.5$ or $p_B = 0.5$, in which case you expect to see more boys than girls or fewer boys than girls or equal numbers of boys and girls, respectively. We would retain the assumption of independence; that is, the probability of a boy, p_B, is the same for each child. Seeing a boy for the first child will not lead you to change the probability that the second child is a boy; this would not imply that "sex runs in families."

2.4.1 What Is a Likelihood?

As is often the case in statistics, our objective is to find an estimate for a model parameter using our data; here, the parameter to estimate is the probability of a boy, p_B, and the data is the gender composition for each family. One way in which to interpret probability is to imagine repeatedly producing children. The probability of a boy will be the overall proportion of boys as the number of children increases. With likelihood methods, conceptually we consider different possible values for our parameter(s), p_B, and determine how likely we would be to see our observed data in each case, $\text{Lik}(p_B)$. We'll select as our estimate the value of p_B for which our data is most likely. A **likelihood** is a function that tells us how likely we are to observe our data for a given parameter value, p_B. For a single family which has a girl followed by two boys, GBB, the likelihood function looks like:

$$\text{Lik}(p_B) = P(G)P(B)P(B) = (1-p_B)p_B^2$$

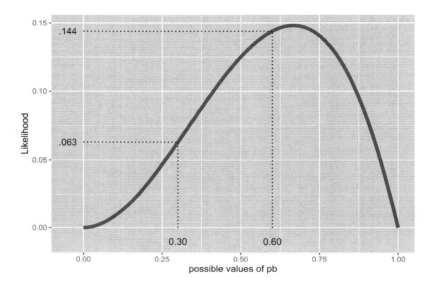

FIGURE 2.2: Likehood function for GBB.

From the likelihood in Figure 2.2, when $p_B = 0.3$ we see a family of a girl followed by two boys 6.3% ($0.7 \cdot 0.3^2$) of the time. However, it indicates that we are much more likely to see our data if $p_B = 0.6$ where the likelihood of GBB is $0.4 \cdot 0.6^2$ or 14.4%.

If the choice was between 0.3 and 0.6 for an estimate of p_B, we'd choose 0.6. The "best" estimate of p_B would be the value where we are most likely to see our data from all possible values between 0 and 1, which we refer to as the **maximum likelihood estimate** or MLE. We can approximate an MLE using graphical or numerical approaches. Graphically, here it looks like the MLE is just above 0.6. In many, but not all, circumstances, we can obtain an MLE exactly using calculus. In this simple example, the MLE is 2/3. This is consistent with our intuition since 2 out of the 3 children are boys.

Suppose another family consisting of three girls is added to our data set. We've already seen that the Sex Unconditional Model multiplies probabilities to construct a likelihood because children are independent of one another. Extending this idea, families can be assumed to be independent of one another

2.4 Model 1: Sex Unconditional, Unequal Probabilities

so that the likelihood for both families can be obtained by multiplication. With two families (GBB and GGG) our likelihood is now:

$$\begin{aligned}
\text{Lik}(p_B) &= P(GBB)P(GGG) \\
&= [(1-p_B)p_B^2][(1-p_B)^3] \\
&= (1-p_B)^4 p_B^2
\end{aligned}$$

A plot of this likelihood appears in Figure 2.3. It is right skewed with an MLE at approximately 0.3. Using calculus, we can show that the MLE is precisely 1/3 which is consistent with intuition given the 2 boys and 4 girls in our data.

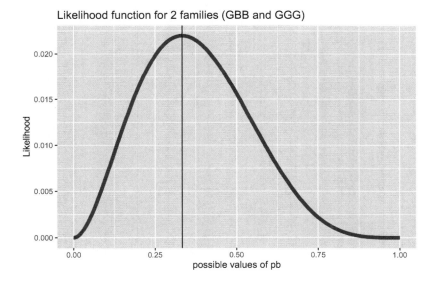

FIGURE 2.3: Likelihood function for the data of 2 families (GBB and GGG). The solid line is at the MLE, $p_B = 1/3$.

Turning now to our hypothetical data with 30 families who have a total of 50 children, we can create the likelihood contribution for each of the family compositions.

The likelihood function for the hypothetical data set can be found by taking the product of the entries in the last column of Table 2.2 and simplifying.

$$\begin{aligned}
\text{Lik}(p_B) &= p_B^6(1-p_B)^7 p_B^{10} \cdots \\
&= p_B^{30}(1-p_B)^{20}
\end{aligned} \quad (2.1)$$

It should be obvious that the likelihood for this Sex Unconditional Model (the coin flipping model) has the simple form:

TABLE 2.2: The likelihood factors for the hypothetical data set of n=50 children.

Composition	Likelihood contribution for one family	Number of families	Likelihood contribution for multiple families
B	p_B	6	p_B^6
G	$(1-p_B)$	7	$(1-p_B)^7$
BB	p_B^2	5	p_B^{10}
BG	$p_B(1-p_B)$	4	$p_B^4(1-p_B)^4$
GB	$(1-p_B)p_B$	5	$(1-p_B)^5 p_B^5$
GGB	$(1-p_B)^2 p_B$	1	$(1-p_B)^2 p_B$
GBB	$(1-p_B)p_B^2$	2	$(1-p_B)^2 p_B^4$
Total		30	

$$\text{Lik}(p_B) = p_B^{n_{\text{Boys}}}(1-p_B)^{n_{\text{Girls}}}$$

and as we asserted above, the MLE will be the (number of boys)/(number of kids) or 30/50 here. Now, more formally, we demonstrate how we use the likelihood principle to approximate the MLE or determine it exactly.

2.4.2 Finding MLEs

2.4.2.1 Graphically approximating an MLE

Figure 2.4(a) is the likelihood for the data set of 50 children. The height of each point is the likelihood and the possible values for p_B appear across the horizontal axis. It appears that our data is most likely when $p_B = 0.6$ as we would expect. Note that the log of the likelihood function in Figure 2.4(b) is maximized at the same spot: $p_B = 0.6$; we will see advantages of using log likelihoods a bit later. Figures 2.4(c) and (d) are also maximized at $p_B = 0.6$, but they illustrate less variability and a sharper peak since there is more data (although the same proportions of boys and girls).

2.4.2.2 Numerically approximating an MLE

Here a grid search is used with the software package R to find maximum likelihood estimates, something that can be done with most software. A grid search specifies a set of finite possible values for p_B and then the likelihood, $\text{Lik}(p_B)$, is computed for each of the possible values. First, we define a relatively coarse grid by specifying 50 values for p_B and then computing how likely we would see our data for each of these possible values. The second example uses a finer grid, 1,000 values for p_B, which allows us to determine a better (more precise) approximation of the MLE. In addition, most packages, like R, have

2.4 Model 1: Sex Unconditional, Unequal Probabilities

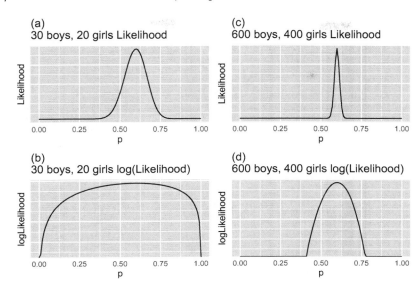

FIGURE 2.4: Likelihood and log-likelihood functions for 50 children (30 boys and 20 girls) and for 1000 children (600 boys and 400 girls).

an optimization function which can also be used to obtain MLEs. Both of these approaches are illustrated in the following code.

```
Lik.f <- function(nBoys,nGirls,nGrid){
    # possible values for prob a boy is born
    pb <- seq(0, 1, length = nGrid)
    lik <- pb^{nBoys} * (1 - pb)^{nGirls}
    # maximum likelihood over nGrid values of pb
    max(lik)
    # value of pb where likelihood maximized
    pb[lik==max(lik)]
}
# estimated maximum likelihood estimator for p_B
Lik.f(nBoys = 30, nGirls = 20, nGrid = 50)
```

```
## [1] 0.5918
```

```
# more precise MLE for p_B based on finer grid (more points)
Lik.f(nBoys = 30, nGirls = 20, nGrid = 1000)
```

```
## [1] 0.5996
```

```
## Another approach: using R's optimize function
##   Note that the log-likelihood is optimized here
oLik.f <- function(pb){
   return(30*log(pb) + 20*log(1-pb))
 }
optimize(oLik.f, interval=c(0,1), maximum=TRUE)
```

```
## $maximum
## [1] 0.6
##
## $objective
## [1] -33.65
```

2.4.2.3 MLEs using calculus (optional)

Calculus may provide another way to determine an MLE. Here, we can ascertain the value of p_B where the likelihood is a maximum by using the first derivative of the likelihood with respect to p_B. We obtain the first derivative using the Product Rule, set it to 0, solve for p_B, and verify that a maximum occurs there.

$$\frac{d}{dp_B} p_B^{30}(1-p_B)^{20} = 30 p_B^{29}(1-p_B)^{20} - p_B^{30} 20(1-p_B)^{19} = 0$$

This approach would produce $p_B = .60$ as we intuited earlier; however, we'll find that likelihoods can get a lot more complicated than this one and there is a simpler way to proceed. This simpler approach is based on the fact that the log of a likelihood is maximized at the same value of p_B as the likelihood. Likelihoods are typically products which require the application of the Product Rule when differentiating, whereas log-likelihoods are sums which are much easier to differentiate. In addition, likelihoods can become tiny with large data sets. So we can take the log of the likelihood, differentiate it, and find the value of p_B where the log-likelihood is a maximum and have the MLE for p_B. We observed this visually in Figure 2.4, where (a) and (b) are maximized at the same p_B (as are (c) and (d)). For the data set with 50 children:

$$\text{Lik}(p_B) = p_B^{30}(1-p_B)^{20}$$
$$\log(\text{Lik}(p_B)) = 30\log(p_B) + 20\log(1-p_B)$$
$$\frac{d}{dp_B}\log(\text{Lik}(p_B)) = \frac{30}{p_B} - \frac{20}{1-p_B} = 0$$

2.4 Model 1: Sex Unconditional, Unequal Probabilities

It is now straightforward to determine that the log-likelihood is maximized when $p_B = 3/5$. We say that the MLE is $\hat{p}_B = 0.6$. This is the exact estimate that was approximated above.

2.4.2.4 How does sample size affect the likelihood?

Consider two hypothetical cases under the Sex Unconditional Model:

Hypothetical Case 1: n = 50 children with 30 boys and 20 girls In previous sections, we found the MLE, $\hat{p}_B = 0.6$.

Hypothetical Case 2: n = 1000 children with 600 boys and 400 girls Our earlier work suggests that the MLE here is also $\hat{p}_B = 0.6$.

The graphs of the likelihoods and log-likelihoods for these two cases in Figure 2.4 give us an idea of how the increase in the sample size affects the precision of our estimates. The likelihoods and log-likelihoods for the two sample sizes have similar forms; however, the graphs with the larger sample size are much narrower, reflecting the greater precision we have with more data. With only 50 children there is a wide range of p_B values that lead to values of the log-likelihood near its maximum, so it's less clear what the optimal p_B is. As we have seen in statistics courses before, a larger sample size will result in less variation in our estimates, thereby affecting the power of hypothesis tests and the width of the confidence intervals.

2.4.3 Summary

Using likelihoods to find estimates of parameters is conceptually intuitive—select the estimate for your parameter where your data is most likely. Often MLEs make a lot of intuitive sense in the context of a problem as well; for example, here the MLE for the probability of a boy is the observed proportion of boys in the data. It may seem like a lot of work for such an obvious result, but MLEs have some nice, useful theoretical properties, and we'll see that many more complex models can be fit using the principle of maximum likelihood.

In summary, we constructed a likelihood that reflected features of our Sex Unconditional Model, then we approximated the parameter value for which our data is most likely using a graph or software, or we determined our optimal parameter value exactly using calculus. You may not be familiar with calculus, yet the concept is clear from the graphs: just find the value of p_B where the likelihood or log likelihood is a maximum. Our "best" estimate for p_B, the MLE, is where our data is most likely to be observed.

Work to understand the *idea* of a likelihood. Likelihoods are the foundation upon which estimates are obtained and models compared for most of this course. Do not be overly concerned with calculus and computation at this point.

2.4.4 Is a Likelihood a Probability Function? (optional)

No. Even though we use probabilities to construct likelihoods, a likelihood is not a probability function. A probability function fixes parameter values and takes as inputs possible outcomes, returning the probability of seeing different outcomes given the parameter value. For example, you flip a loaded coin which comes up heads 25% of the time. After 5 flips, you observe the outcome of three heads and two tails. A *probability function* provides the probability of observing (3H,2T) when $p_H = 0.25$. If you flip this same coin another 5 times and observe all tails (5T), the probability function provides the probability of (5T) when $p_H = 0.25$.

In contrast, a likelihood is constructed by fixing the data, say (3H,2T). It takes as input *possible parameter values* and returns the probability of seeing the fixed data for each parameter value. For example, the likelihood will provide the chance of seeing the data (3H,2T) if $p_H = 0.6$, the chance of seeing the data (3H,2T) if $p_H = 0.3$, and so on. With the likelihood we can find the value of p_H where we are most likely to see our data.

2.5 Model 2: Sex Conditional

Our first research question involves determining whether sex runs in the family. Do families who already have boys tend to have more additional boys than expected by chance, and do families who already have girls tend to have more additional girls than expected by chance? What do you think? And how could we use a statistical model to investigate this phenomenon? There are a number of different ways to construct a model for this question. Here's one possibility.

2.5.1 Model Specification

Unlike the previous model, the p_B in a Sex Conditional Model *depends* on existing family compositions. We introduce **conditional probabilities** and conditional notation to make the dependence explicit. One way to interpret the notation $P(A|B)$ is the "probability of A *given* B has occurred." Another way to read this notation is the "probability of A *conditional* on B." Here, let $p_{B|N}$ represent the probability the next child is a boy given that there are equal

2.5 Model 2: Sex Conditional

TABLE 2.3: Family contributions to the likelihood for a Sex Conditional Model using a hypothetical data set of n=50 children from 30 families.

Composition	Likelihood contribution	Prior Status	Number of families
B	$p_{B|N}$	neutral	6
G	$(1 - p_{B|N})$	neutral	7
BB	$(p_{B|N})(p_{B|B\ Bias})$	neutral, boy bias	5
BG	$(p_{B|N})(1 - p_{B|B\ Bias})$	neutral, boy bias	4
GB	$(1 - p_{B|N})(p_{B|G\ Bias})$	neutral, girl bias	5
GGB	$(1 - p_{B|N})(1 - p_{B|G\ Bias})(p_{B|G\ Bias})$	neutral, girl bias, girl bias	1
GBB	$(1 - p_{B|N})(p_{B|G\ Bias})(p_{B|N})$	neutral, girl bias, neutral	2
Total			30

numbers of boys and girls (sex-neutral) in the existing family. Let $p_{B|B\ Bias}$ represent the probability the next child is a boy if the family is boy-biased; i.e., there are more boys than girls prior to this child. Similarly, let $p_{B|G\ Bias}$ represent the probability the next child is a boy if the family is girl-biased; i.e., there are more girls than boys prior to this child.

Before we are mired in notation and calculus, let's think about how these conditional probabilities can be used to describe sex running in families. While we only had one parameter, p_B, to estimate in the Sex Unconditional Model, here we have three parameters: $p_{B|N}$, $p_{B|B\ Bias}$, and $p_{B|G\ Bias}$. Clearly if all three of these probabilities are equal, the probability a child is a boy does not depend upon the existing gender composition of the family and there is no evidence of sex running in families. A conditional probability $p_{B|B\ Bias}$ that is larger than $p_{B|N}$ suggests families with more boys are more likely to produce additional boys in contrast to families with equal boys and girls. This finding would support the theory of "boys run in families." An analogous argument holds for girls. In addition, comparisons of $p_{B|B\ Bias}$ and $p_{B|G\ Bias}$ to the parameter estimate p_B from the Sex Unconditional Model may be interesting and can be performed using likelihoods.

While it may seem that including families with a single child (singletons) would not be helpful for assessing whether there is a preponderance of one sex or another in families, in fact singleton families would be helpful in estimating $p_{B|N}$ because singletons join "neutral families."

2.5.2 Application to Hypothetical Data

Using the family composition data for 50 children in the 30 families that appears in Table 2.3, we construct a likelihood. The six singleton families with only one boy contribute $p_{B|N}^6$ to the likelihood and the seven families with only one girl contribute $p_{G|N}^7$ or $(1 - p_{B|N})^7$. [Why do we use $1 - p_{B|N}$

instead of $p_{G|N}$?] There are five families with two boys each with probability $(p_{B|N})(p_{B|\text{B Bias}})$ contributing:

$$[(p_{B|N})(p_{B|\text{B Bias}})]^5.$$

We construct the likelihood using data from all 30 families assuming families are independent to get:

$$\text{Lik}(p_{B|N},\ p_{B|\text{B Bias}},\ p_{B|\text{G Bias}}) = \big[(p_{B|N})^{17}(1-p_{B|N})^{15}(p_{B|\text{B Bias}})^5$$
$$(1-p_{B|\text{B Bias}})^4(p_{B|\text{G Bias}})^8(1-p_{B|\text{G Bias}})\big] \quad (2.2)$$

A couple of points are worth noting. First, there are 50 factors in the likelihood corresponding to the 50 children in these 30 families. Second, in the Sex Unconditional example we only had one parameter, p_B; here we have three parameters. This likelihood does not simplify like the Sex Unconditional Model to one that is a product of only two powers: one of p_B and the other of $1 - p_B$. Yet, the basic idea we discussed regarding using a likelihood to find parameter estimates is the same. To obtain the MLEs, we need to find the combination of values for our three parameters where the data is most likely to be observed. Conceptually, we are trying different combinations of possible values for these three parameters, one after another, until we find *the* combination where the likelihood is a maximum. It will not be as easy to graph this likelihood and we will need multivariable calculus to locate the optimal combination of parameter values where the likelihood is a maximum. In this text, we do not assume you know multivariable calculus, but we do want you to retain the concepts associated with maximum likelihood estimates. In practice, we use software to obtain MLEs.

With calculus, we can take partial derivatives of the likelihood with respect to each parameter assuming the other parameters are fixed. As we saw in Section 2.4.2.3, differentiating the log of the likelihood often makes things easier. This same approach is recommended here. Set each partial derivative to 0 and solve for all parameters simultaneously.

Knowing that it is easier to work with log-likelihoods, let's take the log of the likelihood we constructed in Equation (2.2).

$$\log(\text{Lik}(p_{B|N},\ p_{B|\text{B Bias}},\ p_{B|\text{G Bias}})) = 17\log(p_{B|N}) + 15\log(1-p_{B|N})$$
$$+ 5\log(p_{B|\text{B Bias}}) + 4\log(1-p_{B|\text{B Bias}}) + 8\log(p_{B|\text{G Bias}}) + 1\log(1-p_{B|\text{G Bias}})$$

Taking a partial derivative with respect to $p_{B|N}$

$$\frac{17}{p_{B|N}} - \frac{15}{1 - p_{B|N}} = 0$$

$$\hat{p}_{B|N} = \frac{17}{32}$$

$$= 0.53$$

This estimate follows naturally. First consider all of the children who enter into a family with an equal number of boys and girls. From Table 2.3, we can see there are 32 such children (30 are first kids and 2 are third kids in families with 1 boy and 1 girl). Of those children, 17 are boys. So, given that a child joins a sex-neutral family, the chance they are a boy is 17/32. Similar calculations for $p_{B|B \text{ Bias}}$ and $p_{B|G \text{ Bias}}$ yield:

$$\hat{p}_{B|N} = 17/32 = 0.53$$
$$\hat{p}_{B|B \text{ Bias}} = 5/9 = 0.56$$
$$\hat{p}_{B|G \text{ Bias}} = 8/9 = 0.89$$

If we anticipate any "sex running in families" effect, we would expect $p_{B|B \text{ Bias}}$ to be larger than the probability of a boy in the neutral setting, $p_{B|N}$. In our small hypothetical example, $\hat{p}_{B|B \text{ Bias}}$ is slightly greater than 0.53, providing light support for the "sex runs in families" theory when it comes to boys. What about girls? Do families with more girls than boys tend to have a greater probability of having a girl? We found that the MLE for the probability of a girl in a girl-biased setting is 1-0.89=0.11. [1] This data does not provide evidence that girls run in families since $\hat{p}_{G|G_{bias}} = 0.11 < \hat{p}_{G|N} = 0.47$; there is a markedly lower probability of a girl if the family is already girl biased. This data is, however, hypothetical. Let's take a look at some real data and see what we find.

2.6 Case Study: Analysis of the NLSY Data

2.6.1 Model Building Plan

You should now have a feel for using the Likelihood Principle to obtain estimates of parameters using family gender composition data. Next, these

---

[1] Note: A nice property of MLEs is demonstrated here. We have the MLE for $p_{B|G \text{ Bias}}$, and we want the MLE of $p_{G|G \text{ Bias}} = 1 - p_{B|G \text{ Bias}}$. We can get it by replacing $p_{B|G \text{ Bias}}$ with its MLE; i.e., $\hat{p}_{G|G \text{ Bias}} = 1 - \hat{p}_{B|G \text{ Bias}}$. In mathematical terms, you can get the MLE of a function by applying the function to the original MLE.

TABLE 2.4: Number of families and children in families with given composition in NLSY data. Sex ratio and proportion males are given by family size.

Family Composition	Number of families	Number of children	males : females	p_B
B	930	930	97:100	0.494
G	951	951		
BB	582	1164	104:100	0.511
BG	666	1332		
GB	666	1332		
GG	530	1060		
BBB	186	558	104:100	0.510
BBG	177	531		
BGG	173	519		
BGB	148	444		
GBB	151	453		
GGB	125	375		
GBG	182	546		
GGG	159	477		

ideas will be applied to the NLSY data summarized in Table 2 of [Rodgers and Doughty, 2001]. In addition to considering the Sex Unconditional and Conditional Models, we investigate some models that incorporate choices couples may make about when to stop having more children.

2.6.2 Exploratory Data Analysis

We begin by performing an exploratory data analysis (EDA) aimed at shedding some light on our research questions. We are looking for clues as to which of our models is most plausible. The first statistic of interest is the proportion of boys in the sample. There are 5,416 boys out of the 10,672 children or a proportion of .507 boys. While this proportion is very close to .500, it is worth noting that a difference of .007 could be meaningful in population terms.

Table 2.4 displays family composition data for the 5,626 families with one, two, or three children in the NLSY data set. This data set includes 10,672 children. Because our interest centers on the proportion of males, let's calculate sex ratios and proportions of males for each family size. For one-child families the male to female ratio is less than one (97 males:100 females), whereas both two- and three-child families have ratios of 104 boys to 100 girls, what we may expect in a population which favors males. While our research questions do not specifically call for these measures stratified by family size, it still provides us with an idea of gender imbalance in the data.

Table 2.5 provides insight into whether sex runs in families if the probability of a boy is 0.5. Simple probability suggests that the percentage of 2-child families

2.6 Case Study: Analysis of the NLSY Data

TABLE 2.5: Proportion of families in NLSY data with all the same sex by number of children in the family. Note that 1-child families are all homogeneous with respect to sex, so we look at 2- and 3-child families.

Number of children	Number of families	Number with all same sex	Percent with same sex
Two Children	2444	1112	45%
Three Children	1301	345	27%

TABLE 2.6: Proportion of families in NLSY data with only one boy who is born last.

Number of children	Number of families	Number with one boy last	Percent with boy last
One Child	1881	930	49.4%
Two Children	2444	666	27.2%
Three Children	1301	125	8.6%

with all the same sex would be 50% (BB or GG vs. BG or GB) but in our data we see only 45%. For 3-child families, we have 8 possible orderings of boys and girls and so we would expect 2 out of the 8 orderings (25%) to be of the same sex (BBB or GGG), but in fact 27% have the same sex among the 3-children families. These results do not provide overwhelming evidence of sex running in families. There are some potentially complicating factors: the probability of a boy may not be 0.5 or couples may be waiting for a boy or a girl or both.

Table 2.6 contains the number of families by size and the percentage of those which are families with one boy who is last. Some of these families may have "waited" for a boy and then quit childbearing after a boy was born. We see the proportion of one-child families with a boy is slightly less than the 50% expected. We'd expect one out of four, or 25%, of 2-child family configurations to have one boy last and there is 27% in our dataset. Only 8.6% of 3-child families have one boy last, but in theory we would expect one out of eight or 12.5% of 3-child families to have one boy last. So if, in fact, the probability of a boy is 50%, there does not appear to be evidence supporting the notion that families wait for a boy.

There are many other ways to formulate and explore the idea that sex runs in families or that couples wait for a boy (or a girl). See Rodgers and Doughty [2001] for other examples.

2.6.3 Likelihood for the Sex Unconditional Model

We construct a likelihood for the Sex Unconditional Model for the one-, two- and three-child families from the NLSY. See Table 2.4 for the frequencies of each gender composition.

TABLE 2.7: Contributions to the likelihood function for the Sex Unconditional Model for a sample of family compositions from the NLSY data.

Family composition	Likelihood contribution
G	$(1-p_B)^{951}$
GB	$(1-p_B)^{666} p_B^{666}$
BGB	$p_B^{2*148}(1-p_B)^{148}$

Families with different compositions will contribute different factors to the likelihood. For example, Table 2.7 shows a sample of contributions for a few family compositions, where coefficients come from Table 2.4.

Now we create the entire likelihood for our data under the Sex Unconditional Model.

$$\begin{aligned}\text{Lik}(p_B) &= p_B^{930} p_G^{951} p_{BB}^{582} \cdots p_{BBG}^{177} \cdots p_{GGG}^{159} \\ &= p_B^{930+2*582+666+666+\cdots+182}(1-p_B)^{951+666+666+2*530+\cdots+3*159} \\ &= p_B^{5416}(1-p_B)^{5256}\end{aligned}$$

This very simple likelihood implies that each child contributes a factor of the form p_B or $1-p_B$. Given that there are 10,672 children, what would be your best guess of the estimated probability of a boy for this model? We can determine the MLE for p_B using our previous work.

$$\begin{aligned}\hat{p_B} &= \frac{nBoys}{nBoys + nGirls} \\ &= \frac{5416}{5416 + 5256} \\ &= 0.507\end{aligned}$$

2.6.4 Likelihood for the Sex Conditional Model

The contribution to a Sex Conditional Model likelihood for the same family compositions we considered in the previous section appear in Table 2.8.

The products of the last three columns of Table 2.9 provide the likelihood contributions for the Sex Conditional Model for all of the one-, two- and three-child NLSY families. We write the likelihood as a function of the three parameters $p_{B|N}, p_{B|B \text{ Bias}},$ and $p_{B|G \text{ Bias}}$.

2.6 Case Study: Analysis of the NLSY Data

TABLE 2.8: Contributions to the likelihood function for the Sex Conditional Model for a sample of family compositions from the NLSY data.

Family composition	Likelihood contribution
G	$(1-p_{B|N})^{951}$
GB	$(1-p_{B|N})^{666} p_{B|G\ \text{Bias}}^{666}$
BGB	$p_{B|N}^{2*148}(1-p_{B|B\ \text{Bias}})^{148}$

TABLE 2.9: Likelihood contributions for NLSY families in Sex Unconditional and Sex Conditional Models.

			Sex Conditional Model		
Family composition	Num. families	Sex Unconditional	Child 1	Child 2	Child 3
B	930	p_B	$p_{B|N}$		
G	951	$(1-p_B)$	$1-p_{B|N}$		
BB	582	p_B^2	$p_{B|N}$	$p_{B|B\ \text{Bias}}$	
BG	666	$p_B(1-p_B)$	$p_{B|N}$	$1-p_{B|B\ \text{Bias}}$	
GB	666	$(1-p_B)p_B$	$1-p_{B|N}$	$p_{B|G\ \text{Bias}}$	
GG	530	$(1-p_B)^2$	$1-p_{B|N}$	$1-p_{B|G\ \text{Bias}}$	
BBB	186	p_B^3	$p_{B|N}$	$p_{B|B\ \text{Bias}}$	$p_{B|B\ \text{Bias}}$
BBG	177	$p_B^2(1-p_B)$	$p_{B|N}$	$p_{B|B\ \text{Bias}}$	$1-p_{B|B\ \text{Bias}}$
BGG	173	$p_B(1-p_B)^2$	$p_{B|N}$	$1-p_{B|B\ \text{Bias}}$	$1-p_{B|N}$
BGB	148	$p_B^2(1-p_B)$	$p_{B|N}$	$1-p_{B|B\ \text{Bias}}$	$p_{B|N}$
GBB	151	$p_B^2(1-p_B)$	$1-p_{B|N}$	$p_{B|G\ \text{Bias}}$	$p_{B|N}$
GGB	125	$p_B(1-p_B)^2$	$1-p_{B|N}$	$1-p_{B|G\ \text{Bias}}$	$p_{B|G\ \text{Bias}}$
GBG	182	$p_B(1-p_B)^2$	$1-p_{B|N}$	$p_{B|G\ \text{Bias}}$	$1-p_{B|N}$
GGG	159	$(1-p_B)^3$	$1-p_{B|N}$	$1-p_{B|G\ \text{Bias}}$	$1-p_{B|G\ \text{Bias}}$
log-likelihood		-7396.067		-7374.238	
AIC		14794.13		14751.48	
BIC		14810.68		14749.18	

$$\text{Lik}(p_{B|N}, p_{B|B\ \text{Bias}}, p_{B|G\ \text{Bias}})$$
$$= \left[p_{B|N}^{930}(1-p_{B|N})^{951}(p_{B|N}p_{B|B\ \text{Bias}})^{582}\ldots\right.$$
$$\left.((1-p_{B|N})(1-p_{B|G\ \text{Bias}})(1-p_{B|G\ \text{Bias}}))^{159}\right]$$
$$= \left[p_{B|N}^{3161}(1-p_{B|N})^{3119}p_{B|B\ \text{Bias}}^{1131}(1-p_{B|B\ \text{Bias}})^{1164}p_{B|G\ \text{Bias}}^{1124}(1-p_{B|G\ \text{Bias}})^{973}\right]$$

$$\log(\text{Lik}(p_{B|N}, p_{B|B\ \text{Bias}}, p_{B|G\ \text{Bias}})) =$$
$$3161\log(p_{B|N}) + 3119\log(1-p_{B|N}) + 1131\log(p_{B|B\ \text{Bias}})$$
$$+ 1164\log(1-p_{B|B\ \text{Bias}}) + 1124\log(p_{B|G\ \text{Bias}}) + 973\log(1-p_{B|G\ \text{Bias}}) \tag{2.3}$$

To use calculus to estimate the probability of a boy entering a neutral family (a family with equal boys and girls), $p_{B|N}$, we begin with the logarithm of the likelihood in equation (2.3). Differentiating the log-likelihood with respect to $p_{B|N}$ holding all other parameters constant yields an intuitive estimate.

$$\hat{p}_{B|N} = \frac{3161}{3161 + 3119}$$
$$= 0.5033$$

There are 6,280 times when a child is joining a neutral family and, of those times, 3,161 are boys. Thus the MLE of the probability of a boy joining a family where the numbers of boys and girls are equal (including when there are no children) is 0.5033.

Similarly, MLEs for $p_{B|B \text{ Bias}}$ and $p_{B|G \text{ Bias}}$ can be obtained:

$$\hat{p}_{B|B \text{ Bias}} = \frac{1131}{1131 + 1164}$$
$$= 0.4928$$

$$\hat{p}_{B|G \text{ Bias}} = \frac{1124}{1124 + 973}$$
$$= 0.5360$$

Are these results consistent with the notion that boys or girls run in families? We consider the Sex Conditional Model because we hypothesized there would be a higher probability of boys among children born into families with a boy bias. However, we found that, if there is a boy bias, the probability of a subsequent boy was estimated to be actually less (0.493) than the probability of a subsequent girl. Similarly, girls join families with more girls than boys approximately 46.4% of the time so that there is little support for the idea that either "girls or boys run in families."

Even though initial estimates don't support the idea, let's formally take a look as to whether prior gender composition affects the probability of a boy. To do so, we'll see if the Sex Conditional Model is statistically significantly better than the Sex Unconditional Model.

2.6.5 Model Comparisons

2.6.5.1 Nested models

Likelihoods are not only useful for fitting models, but they are also useful when comparing models. If the parameters for a reduced model are a subset of parameters for a larger model, we say the models are **nested** and the difference

2.6 Case Study: Analysis of the NLSY Data

between their likelihoods can be incorporated into a statistical test to help judge the benefit of including additional parameters. Another way in which to think of nesting is to consider whether parameters in the larger model can be equated to obtain the simpler model or whether some parameters in the larger model can be set to constants. Since $p_{B|B\ \text{Bias}}, p_{B|N}$ and $p_{B|G\ \text{Bias}}$ in the Sex Conditional Model can be set to p_B to obtain the Sex Unconditional Model, we can say the models are nested.

If the parameters are not nested, comparing models with the likelihood can still be useful but will take a different form. We'll see that the Akaike Information Criterion (AIC) and Bayesian Information Criterion (BIC) are functions of the log-likelihood that can be used to compare models even when the models are not nested. Either way we see that this notion of likelihood is pretty useful.

Hypotheses

$H_0 : p_{B|N} = p_{B|B\ \text{Bias}} = p_{B|G\ \text{Bias}} = p_B$ (Sex Unconditional Model) The probability of a boy does not depend on the prior family composition.

H_A : At least one parameter from $p_{B|N}, p_{B|B\ \text{Bias}}, p_{B|G\ \text{Bias}}$ differs from the others. (Sex Conditional Model) The probability of a boy does depend on the prior family composition.

We start with the idea of comparing the likelihoods or, equivalently, the log-likelihoods of each model at their maxima. To do so, we use the log-likelihoods to determine the MLEs, and then replace the parameters in the log-likelihood with their MLEs, thereby finding the maximum value for the log-likelihood of each model. Here we will refer to the first model, the Sex Unconditional Model, as the **reduced model**, noting that it has only a single parameter, p_B. The more complex model, the Sex Conditional Model, has three parameters and is referred to here as the **larger (full) model**. We'll use the MLEs derived earlier in Section 2.6.4.

The maximum of the log-likelihood for the reduced model can be found by replacing p_B in the log-likelihood with the MLE of p_B, 0.5075.

$$\log(\text{Lik}(0.5075)) = 5416 \log(.5075) + 5256 \log(1 - .5075)$$
$$= -7396.067$$

The maximum of the log-likelihood for the larger model can be found by replacing $p_{B|N}, p_{B|B\ \text{Bias}}, p_{B|G\ \text{Bias}}$ in the log-likelihood with 0.5033, 0.4928, and 0.5360, respectively.

$$\log(\text{Lik}(0.5033, 0.4928, 0.5360))) = 3161 \log(.5033) + 3119 \log(1 - .5033)$$
$$+ 1131 \log(.4928) + 1164 \log(1 - .4928)$$
$$+ 1124 \log(.5360) + 973 \log(1 - .5360)$$
$$= -7391.448$$

Take a look at the log-likelihoods—the maximum log-likelihood for the larger model is indeed larger (less negative). The maximum log-likelihood for the larger model is guaranteed to be at least as large as the maximum log-likelihood for the reduced model, so we'll be interested in whether this observed difference in maximum log-likelihoods, -7391.448 -(-7396.067) = 4.619, is significant.

A result from statistical theory states that, when the reduced model is the true model, twice the difference of the maximum log-likelihoods follows a χ^2 distribution with the degrees of freedom equal to the difference in the number of parameters between the two models. A difference of the maximum log-likelihoods can also be looked at as the log of the ratio of the likelihoods and for that reason the test is referred to as the **Likelihood Ratio Test (LRT)**.

Our test statistic is

$$\text{LRT} = 2[\max(\log(\text{Lik}(\text{larger model}))) - \max(\log(\text{Lik}(\text{reduced model})))]$$
$$= 2 \log \left(\frac{\max(\text{Lik}(\text{larger model}))}{\max(\text{Lik}(\text{reduced model}))} \right)$$

Intuitively, when the likelihood for the larger model is much greater than it is for the reduced model, we have evidence that the larger model is more closely aligned with the observed data. This isn't really a fair comparison on the face of it. We need to account for the fact that more parameters were estimated and used for the larger model. That is accomplished by taking into account the degrees of freedom for the χ^2 distribution. The expected value of the χ^2 distribution is its degrees of freedom. Thus when the difference in the number of parameters is large, the test statistic will need to be much larger to convince us that it is not simply chance variation with two identical models. Here, under the reduced model we'd expect our test statistic to be 2, when in fact it is over 9. The evidence favors our larger model. More precisely, the test statistic is $2(-7391.448 + 7396.073) = 9.238$ ($p = .0099$), where the p-value is the probability of obtaining a value above 9.238 from a χ^2 distribution with 2 degrees of freedom.

We have convincing evidence that the Sex Conditional Model provides a significant improvement over the Sex Unconditional Model. However, keep in mind that our point estimates for a probability of a boy were not what we had expected for "sex runs in families." It may be that this discrepancy stems from behavioral aspects of family formation. The next section on stopping rules explores how types of couples' decisions may affect the relative proportions of family compositions in the data.

Note: You may notice that the LRT is similar in spirit to the extra-sum-of-squares F-test used in linear regression. Recall that the extra-sum-of-squares F-test involves comparing two nested models. When the smaller model is true, the F-ratio follows an F-distribution which on average is 1.0. A large, unusual F-ratio provides evidence that the larger model provides a significant improvement.

Also note: It might have been more logical to start by using a Likelihood Ratio Test to determine whether the probability of having a boy differs significantly from 0.5. We leave this as an exercise.

2.7 Model 3: Stopping Rule Model (waiting for a boy)

Rodgers and Doughty [2001] offer one reason to explain the contradictory results: waiting for a male child. It has been noted by demographers that some parents are only interested in producing a male heir so that the appearance of a boy leads more often to the family ending childbearing. Stopping models investigate questions like: Are couples more likely to stop childbearing once they have a boy? Or are some parents waiting for a girl? Others might wish to have at least one boy and girl. The exploratory data analysis results in Table 2.6 provide some insight but cannot definitively settle the question about couples' stopping once they have a boy.

For stopping models, two probabilities are recorded for each child: the probability of the sex and the conditional probability of stopping after that child. As we have done in previous models, let p_B = probability the child is a boy. When conditioning, every possible condition must have a probability associated with it. Here the stopping conditions for Model 3 are: stop on first boy $(S|B1)$ or stopping on a child who is not the first boy $(S|N)$.

Additional parameters for the First Boy Stopping Model

- $p_{S|B1}$ = probability of stopping after the first boy
- $1 - p_{S|B1}$ = probability of not stopping after the first boy
- $p_{S|N}$ = probability of stopping after a child who is not the first boy
- $1 - p_{S|N}$ = probability of not stopping after a child who is not the first boy

With these additional parameters, likelihood contributions of the NLSY families are listed in Table 2.10. Our interest centers on whether the probability of

TABLE 2.10: Likelihood contributions for NLSY families in Model 3: Waiting for a boy.

Family Composition	Num. families	Likelihood Contribution	
B	930	p_B	$p_{S\mid B1}$
G	951	$(1-p_B)$	$p_{S\mid N}$
BB	582	p_B^2	$(1-p_{S\mid B1})p_{S\mid N}$
BG	666	$p_B(1-p_B)$	$(1-p_{S\mid B1})p_{S\mid N}$
GB	666	$(1-p_B)p_B$	$(1-p_{S\mid N})p_{S\mid B1}$
GG	530	$(1-p_B)^2$	$(1-p_{S\mid N})p_{S\mid N}$
BBB	186	p_B^3	$(1-p_{S\mid B1})(1-p_{S\mid N})p_{S\mid N}$
BBG	177	$p_B^2(1-p_B)$	$(1-p_{S\mid B1})(1-p_{S\mid N})p_{S\mid N}$
BGG	173	$p_B(1-p_B)^2$	$(1-p_{S\mid B1})(1-p_{S\mid N})p_{S\mid N}$
BGB	148	$p_B^2(1-p_B)$	$(1-p_{S\mid B1})(1-p_{S\mid N})p_{S\mid N}$
GBB	151	$p_B^2(1-p_B)$	$(1-p_{S\mid N})(1-p_{S\mid B1})p_{S\mid N}$
GGB	125	$p_B(1-p_B)^2$	$(1-p_{S\mid N})^2 p_{S\mid B1}$
GBG	182	$p_B(1-p_B)^2$	$(1-p_{S\mid N})(1-p_{S\mid B1})p_{S\mid N}$
GGG	159	$(1-p_B)^3$	$(1-p_{S\mid N})^2 p_{S\mid N}$

stopping after the first boy, $p_{S\mid B1}$ is greater than stopping when it is not a first boy, $p_{S\mid N}$.

Using calculus, the MLEs are derived to be $\hat{p}_B = 0.507$, $\hat{p}_{S\mid B1} = 0.432$, and $\hat{p}_{S\mid N} = 0.584$. These are consistent with intuition. The estimated proportion of boys for this model is the same as the estimate for the Sex Unconditional Model (Model 1). The estimates of the stopping parameters are consistent

TABLE 2.11: Patterns related to stopping decisions.

Child is...	total children	prop of all children	n.stops (n.families)	prop stopped after these children
a boy who is the only boy in the family up to that point	3,986	37.4%	1,721	43.2%
not an only boy in the family up to that point	6,686	62.2%	3,905	58.4%
a girl who is the only girl in the family up to that point	3,928	36.8%	1,794	45.7%
not an only girl in the family up to that point	3,832	63.2%	3,832	56.8%
	10,672		5,626	

2.7 Model 3: Stopping Rule Model (waiting for a boy)

with the fact that of the 3,986 first boys, parents stop 43.2% of the time and of the 6,686 children who are not first boys, childbearing stopped 58.4% of the time. See Table 2.11.

These results do, in fact, suggest that the probability a couple stops childbearing on the first boy is different than the probability of stopping at a child who is not the first boy, but the direction of the difference does not imply that couples "wait for a boy;" rather it appears that they are less likely to stop childbearing after the first boy in comparison to children who are not the first-born male.

Similarly, for girls, the MLEs are $\hat{p}_{S|G1} = 0.457$ and $\hat{p}_{S|N} = 0.568$. Once again, the estimates do not provide evidence of waiting for a girl.

2.7.1 Non-nested Models

How does the waiting for a boy model compare to the waiting for a girl model? Thus far we've seen how nested models can be compared. But these two models are not nested since one is not simply a reduced version of the other. Two measures referred to as information criteria, AIC and BIC, are useful when comparing non-nested models. Each measure can be calculated for a model using a function of the model's maximum log-likelihood. You can find the log-likelihood in the output from most modeling software packages.

- AIC = -2(maximum log-likelihood) $+ 2p$, where p represents the number of parameters in the fitted model. AIC stands for Akaike Information Criterion. Because smaller AICs imply better models, we can think of the second term as a penalty for model complexity—the more variables we use, the larger the AIC.
- BIC = -2(maximum log-likelihood) $+ p\log(n)$, where p is the number of parameters and n is the number of observations. BIC stands for Bayesian Information Criterion, also known as Schwarz's Bayesian criterion (SBC). Here we see that the penalty for the BIC differs from the AIC, where the log of the number of observations places a greater penalty on each extra predictor, especially for large data sets.

So which explanation of the data seems more plausible—waiting for a boy or waiting for a girl? These models are not nested (i.e., one is not a simplified version of the other), so it is not correct to perform a Likelihood Ratio Test, but we can legitimately compare these models using information criteria (Table 2.12).

Smaller AIC and BIC are preferred, so here the Waiting for a Boy Model is judged superior to the Waiting for a Girl Model, suggesting that couples waiting for a boy is a better explanation of the data than waiting for a girl. However, for either boys and girls, couples do not stop more frequently after the first occurrence.

TABLE 2.12: Measures of model performance with NLSY data: Waiting for a Boy vs. Waiting for a Girl Model.

	Waiting for a boy	Waiting for a girl
max log-likelihood	-14661	-14716
AIC	29324	29433
BIC	29332	29441

Other stopping rule models are possible. Another model could be that couples wait to stop until they have both a boy and a girl. We leave the consideration of this balance-preference model as an exercise.

2.8 Summary of Model Building

Using a Likelihood Ratio Test, we found statistical evidence that the Sex Conditional Model (Sex Bias) is preferred to the Sex Unconditional Models. However, the parameter estimates were not what we expected if we believe that sex runs in families. Quite to the contrary, the results suggested that if there were more of one sex in a family, the next child is likely to be of the other sex. The results may support the idea that gender composition tends to "even out" over time.

Using AICs and BICs to compare the non-nested models of waiting for a boy or waiting for a girl, we found that the model specifying stopping for a first boy was superior to the model for stopping for the first girl. Again, neither model suggested that couples were *more* likely to stop after the first male or female, rather it appeared just the opposite—couples were *less* likely to be stopping after the first boy or first girl.

These results may need to be considered conditional on the size of a family. In which case, a look at the exploratory data analysis results may be informative. The reported percentages in Table 2.5 could be compared to the percentages expected if the sex of the baby occurs randomly, P(all one sex|2-child family) = 1/2, and we observed 45%. For three-child families, P(all one sex|3-child family) = 1/4, and we observed 27%. There is very slight evidence for sex running in families for three-child families and none for two-child families.

Under a random model that assumes the probability of a boy is 50%, the percentage of one-, two- and three-child families with the first boy showing up last in the family is 50%, 25%, and 12.5%, respectively. Comparing these probabilities to what was observed in the data in Table 2.6, we find little support for the idea that couples are waiting for a boy.

We can perform a LRT to compare stopping at the first boy to a Random Stopping Model. The parameters for the first model (waiting for a boy) are $p_B, p_{S|B1}, p_{S|N}$ and the parameters for the second model (random stopping) are p_B and p_S. The results suggest that the Waiting for a Boy Model is significantly better than the Random Stopping Model. The Random Stopping Model takes into account that the odds of stopping after a child are not 50-50, but may be closer to the MLE for p_S of 52.7%. We leave the derivation of this result as an exercise.

2.9 Likelihood-Based Methods

With likelihood methods, we are no longer restricted to independent, identically distributed normal responses (iidN). Likelihood methods can accommodate non-normal responses and correlated data. Likelihood-based methods are useful for every model in this text, so that it is worth your time and effort to understand them.

Models that in the past you would fit using ordinary least squares can also be fit using the principle of maximum likelihood. It is pleasing to discover that under the right assumptions the maximum likelihood estimates (MLEs) for the intercept and slope in a linear regression are identical to ordinary least squares estimators (OLS) despite the fact that they are obtained in quite different ways.

Beyond the intuitively appealing aspects of MLEs, they also have some very desirable statistical properties. You learn more about these features in a statistical theory course. Here we briefly summarize the highlights in non-technical terms. MLEs are *consistent*; i.e., MLEs converge in probability to the true value of the parameter as the sample size increases. MLEs are *asymptotically normal*; as the sample size increases, the distribution of MLEs is closer to normal. MLEs are *efficient* because no consistent estimator has a lower mean squared error. Of all the estimators that produce unbiased estimates of the true parameter value, no estimator will have a smaller mean square error than the MLE. While likelihoods are powerful and flexible, there are times when likelihood-based methods fail: either MLEs do not exist, likelihoods cannot be written down, or MLEs cannot be written explicitly. It is also worth noting that other approaches to the likelihood, such as bootstrapping, can be employed.

2.10 Likelihoods and This Course

Rodgers and Doughty [2001] noted that

> Many factors have been identified that can potentially affect the human sex ratio at birth. A 1972 paper by Michael Teitelbaum accounted for around 30 such influences, including drinking water, coital rates, parental age, parental socioeconomic status, birth order, and even some societal-level influences like wars and environmental pathogens.

This chapter on likelihood ignored these complicating factors and was intentionally kept simple to impress you with the fact that likelihoods are conceptually straightforward. Likelihoods answer the sensible question of how likely you are to see your data in different settings. When the likelihood is simple as in this chapter, you can roughly determine an MLE by looking at a graph or you can be a little more precise by using calculus or, most conveniently, software. As we progress throughout the course, the likelihoods will become more complex and numerical methods may be required to obtain MLEs, yet the concept of an MLE will remain the same. Likelihoods will show up in parameter estimation, model performance assessment, and model comparisons.

One of the reasons many of the likelihoods will become complex is because of covariates. Here we estimated probabilities of having a boy in different settings, but we did not use any specific information about families other than sex composition. The problems in the remainder of the book will typically employ covariates. For example, suppose we had information on paternal age for each family. Consider the Sex Unconditional Model, and let

$$p_B = \frac{e^{\beta_0 + \beta_1 (\text{parental age})}}{1 + e^{\beta_0 + \beta_1 (\text{parental age})}}.$$

(We will give a good reason for this crazy-looking expression for p_B in later chapters.) The next step would be to replace p_B in the likelihood, $\text{Lik}(p_B)$, (Equation (2.1)), with the complicated expression for p_B. The result would be a function of β_0 and β_1. We could then use calculus to find the MLEs for β_0 and β_1.

Another compelling reason for likelihoods occurs when we encounter correlated

2.11 Exercises

data. For example, models with conditional probabilities do not conform to the independence assumption. The Sex Conditional Model is an example of such a model. We'll see that likelihoods can be useful when the data has structure such as multilevel that induces a correlation. A good portion of the book addresses this.

When the responses are not normal such as in generalized linear models, where we see binary responses and responses which are counts, we'll find it difficult to use the linear least squares regression models of the past and we'll find the flexibility of likelihood methods to be extremely useful. Likelihood methods will enable us to move *beyond multiple linear regression*!

2.11 Exercises

2.11.1 Conceptual Exercises

1. Suppose we plan to use data to estimate one parameter, p_B.
   - When using a likelihood to obtain an estimate for the parameter, which is preferred: a large or a small likelihood value? Why?
   - The height of a likelihood curve is the probability of the data for the given parameter. The horizontal axis represents different possible parameter values. Does the area under the likelihood curve for an interval from .25 to .75 equal the probability that the true probability of a boy is between 0.25 and 0.75?
2. Suppose the families with an "only child" were excluded for the Sex Conditional Model. How might the estimates for the three parameters be affected? Would it still be possible to perform a Likelihood Ratio Test to compare the Sex Unconditional and Sex Conditional Models? Why or why not?
3. Come up with an alternative model to investigate whether "sex runs in families."

2.11.2 Guided Exercises

1. Write out the likelihood for a model which assumes the probability of a girl equals the probability of a boy. Carry out a LRT to determine whether there is evidence that the two probabilities are not equal. Comment on the practical significance of this finding (there is not necessarily one correct answer).
2. **Case 3** In Case 1 we used hypothetical data with 30 boys and 20

girls. Case 2 was a much larger study with 600 boys and 400 girls. Consider Case 3, a hypothetical data set with 6000 boys and 4000 girls.

- Use the methods for Case 1 and Case 2 and determine the MLE for p_B for the independence model. Compare your result to the MLEs for Cases 1 and 2.
- Describe how the graph of the log-likelihood for Case 3 would compare to the log-likelihood graphs for Cases 1 and 2.
- Compute the log-likelihood for Case 3. Why is it incorrect to perform an LRT comparing Cases 1, 2, and 3?

3. Write out an expression for the likelihood of seeing our NLSY data (5,416 boys and 5,256 girls) if the true probability of a boy is:

(a) $p_B = 0.5$

(b) $p_B = 0.45$

(c) $p_B = 0.55$

(d) $p_B = 0.5075$

- Compute the value of the log-likelihood for each of the values of p_B above.
- Which of these four possibilities, $p_B = 0.45, p_B = 0.5, p_B = 0.55$, or $p_B = 0.5075$ would be the best estimate of p_B given what we observed (our data)?

4. Compare the Waiting for a Boy Model to a Random Stopping Model. The parameters for the first model (Waiting for a Boy) are p_B, $p_{S|B1}$, $p_{S|N}$ and the parameters for the second model (Random Stopping) are p_B and p_S. Use an intuitive approach to arrive at the MLEs for the parameters for each model. Perform a LRT to compare these two models.

2.11.3 Open-Ended Exercises

1. **Another stopping rule model: balance-preference.** Can you construct a model which suggests that childbearing is stopped when couples have a boy *and* a girl? Define the parameter(s) for balance-preference stopping combined with the sex conditional model and write out likelihood contributions for the same family compositions that appear in Table 2.3.

    - **Extra Credit** Obtain the MLEs for your balance-preference model parameters and interpret the results. Then, compare the

2.11 Exercises

balance-preference stopping rule model to the random stopping model using a LRT.

2. **The hot hand in basketball.** Gilovich et al. [1985] wrote a controversial but compelling article claiming that there is no such thing as "the hot hand" in basketball. That is, there is no empirical evidence that shooters have stretches where they are more likely to make consecutive shots, and basketball shots are essentially independent events. One of the many ways they tested for evidence of a "hot hand" was to record sequences of shots for players under game conditions and determine if players are more likely to make shots after made baskets than after misses. For instance, assume we recorded data from one player's first 5 three-point attempts over a 5-game period. We can assume games are independent, but we'll consider two models for shots within a game:

    - No Hot Hand (1 parameter): p_B = probability of making a basket (thus $1 - p_B$ = probability of not making a basket).
    - Hot Hand (2 parameters): p_B = probability of making a basket after a miss (or the first shot of a game); $p_{B|B}$ = probability of making a basket after making the previous shot.

    a. Fill out Table 2.13—write out the contribution of each game to the likelihood for both models along with the total likelihood for each model.

    b. Given that, for the No Hot Hand model, Lik$(p_B) = p_B^{10}(1-p_B)^{15}$ for the 5 games where we collected data, how do we know that 0.40 (the maximum likelihood estimator (MLE) of p_B) is a better estimate than, say, 0.30?

    c. Find the MLEs for the parameters in each model, and then use those MLEs to determine if there's significant evidence that the hot hand exists.

TABLE 2.13: Data for Open-ended Exercise 2. (B = made basket. M = missed basket.)

Game	First 5 shots	Likelihood (No Hot Hand)	Likelihood (Hot Hand)
1	BMMBB		
2	MBMBM		
3	MMBBB		
4	BMMMB		
5	MMMMM		
Total			

3
Distribution Theory

3.1 Learning Objectives

After finishing this chapter, you should be able to:

- Write definitions of non-normal random variables in the context of an application.
- Identify possible values for each random variable.
- Identify how changing values for a parameter affects the characteristics of the distribution.
- Recognize a form of the probability density function for each distribution.
- Identify the mean and variance for each distribution.
- Match the response for a study to a plausible random variable and provide reasons for ruling out other random variables.
- Match a histogram of sample data to plausible distributions.
- Create a mixture of distributions and evaluate the shape, mean, and variance.

```
# Packages required for Chapter 3
library(gridExtra)
library(knitr)
library(kableExtra)
library(tidyverse)
```

3.2 Introduction

What if it is not plausible that a response is normally distributed? You may want to construct a model to predict whether a prospective student will enroll at a school or model the lifetimes of patients following a particular surgery. In the first case you have a binary response (enrolls (1) or does not enroll (0)), and in the second case you are likely to have very skewed data with many similar values and a few hardy souls with extremely long survival. These responses are not expected to be normally distributed; other distributions will be needed to describe and model binary or lifetime data. Non-normal responses are encountered in a large number of situations. Luckily, there are quite a few possibilities for models. In this chapter we begin with some general definitions, terms, and notation for different types of distributions with some examples of applications. We then create new random variables using combinations of random variables (see Guided Exercises).

3.3 Discrete Random Variables

A discrete random variable has a countable number of possible values; for example, we may want to measure the number of people in a household or the number of crimes committed on a college campus. With discrete random variables, the associated probabilities can be calculated for each possible value using a **probability mass function** (pmf). A pmf is a function that calculates $P(Y = y)$, given each variable's parameters.

3.3.1 Binary Random Variable

Consider the event of flipping a (possibly unfair) coin. If the coin lands heads, let's consider this a success and record $Y = 1$. A series of these events is a **Bernoulli process**, independent trials that take on one of two values (e.g., 0 or 1). These values are often referred to as a failure and a success, and the probability of success is identical for each trial. Suppose we only flip the coin once, so we only have one parameter, the probability of flipping heads, p. If we know this value, we can express $P(Y = 1) = p$ and $P(Y = 0) = 1 - p$. In general, if we have a Bernoulli process with only one trial, we have a **binary distribution** (also called a **Bernoulli distribution**) where

$$P(Y = y) = p^y(1-p)^{1-y} \quad \text{for} \quad y = 0, 1. \tag{3.1}$$

If $Y \sim \text{Binary}(p)$, then Y has mean $E(Y) = p$ and standard deviation $SD(Y) = \sqrt{p(1-p)}$.

Example 1: Your playlist of 200 songs has 5 which you cannot stand. What is the probability that when you hit shuffle, a song you tolerate comes on?

3.3 Discrete Random Variables

Assuming all songs have equal odds of playing, we can calculate $p = \frac{200-5}{200} = 0.975$, so there is a 97.5% chance of a song you tolerate playing, since $P(Y = 1) = .975^1 * (1 - .975)^0$.

3.3.2 Binomial Random Variable

We can extend our knowledge of binary random variables. Suppose we flipped an unfair coin n times and recorded Y, the number of heads after n flips. If we consider a case where $p = 0.25$ and $n = 4$, then here $P(Y = 0)$ represents the probability of no successes in 4 trials, i.e., 4 consecutive failures. The probability of 4 consecutive failures is $P(Y = 0) = P(TTTT) = (1 - p)^4 = 0.75^4$. When we consider $P(Y = 1)$, we are interested in the probability of exactly 1 success *anywhere* among the 4 trials. There are $\binom{4}{1} = 4$ ways to have exactly 1 success in 4 trials, so $P(Y = 1) = \binom{4}{1}p^1(1-p)^{4-1} = (4)(0.25)(0.75)^3$. In general, if we carry out a sequence of n Bernoulli trials (with probability of success p) and record Y, the total number of successes, then Y follows a **binomial distribution**, where

$$P(Y = y) = \binom{n}{y}p^y(1-p)^{n-y} \quad \text{for} \quad y = 0, 1, \ldots, n. \quad (3.2)$$

If $Y \sim \text{Binomial}(n, p)$, then $\text{E}(Y) = np$ and $\text{SD}(Y) = \sqrt{np(1-p)}$. Typical shapes of a binomial distribution are found in Figure 3.1. On the left side n remains constant. We see that as p increases, the center of the distribution ($\text{E}(Y) = np$) shifts right. On the right, p is held constant. As n increases, the distribution becomes less skewed.

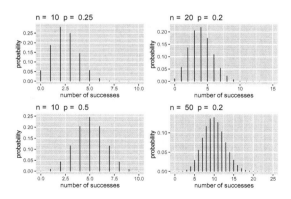

FIGURE 3.1: Binomial distributions with different values of n and p.

Note that if $n = 1$,

$$P(Y = y) = \binom{1}{y} p^y (1-p)^{1-y}$$
$$= p^y (1-p)^{1-y} \quad \text{for} \quad y = 0, 1,$$

a Bernoulli distribution! In fact, Bernoulli random variables are a special case of binomial random variables where $n = 1$.

In R we can use the function dbinom(y, n, p), which outputs the probability of y successes given n trials with probability p, i.e., $P(Y = y)$ for $Y \sim$ Binomial(n, p).

Example 2: While taking a multiple choice test, a student encountered 10 problems where she ended up completely guessing, randomly selecting one of the four options. What is the chance that she got exactly 2 of the 10 correct?

Knowing that the student randomly selected her answers, we assume she has a 25% chance of a correct response. Thus, $P(Y = 2) = \binom{10}{2}(.25)^2(.75)^8 = 0.282$. We can use R to verify this:

```
dbinom(2, size = 10, prob = .25)
```

[1] 0.2816

Therefore, there is a 28% chance of exactly 2 correct answers out of 10.

3.3.3 Geometric Random Variable

Suppose we are to perform independent, identical Bernoulli trials until the first success. If we wish to model Y, the number of failures before the first success, we can consider the following pmf:

$$P(Y = y) = (1-p)^y p \quad \text{for} \quad y = 0, 1, \ldots, \infty. \tag{3.3}$$

We can think about this function as modeling the probability of y failures, then 1 success. In this case, Y follows a **geometric distribution** with $E(Y) = \frac{1-p}{p}$ and $SD(Y) = \sqrt{\frac{1-p}{p^2}}$.

Typical shapes of geometric distributions are shown in Figure 3.2. Notice that as p increases, the range of plausible values decreases and means shift towards 0.

Once again, we can use R to aid our calculations. The function dgeom(y, p) will output the probability of y failures before the first success where $Y \sim \text{Geometric}(p)$.

3.3 Discrete Random Variables

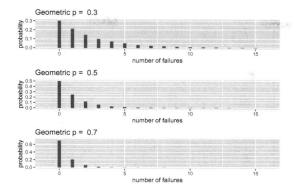

FIGURE 3.2: Geometric distributions with $p = 0.3$, 0.5 and 0.7.

Example 3: Consider rolling a fair, six-sided die until a five appears. What is the probability of rolling the first five on the third roll?

First note that $p = 1/6$. We are then interested in $P(Y = 2)$, as we would want 2 failures before our success. We know that $P(Y = 2) = (5/6)^2(1/6) = 0.116$. Verifying through R:

```
dgeom(2, prob = 1/6)
```

```
## [1] 0.1157
```

Thus, there is a 12% chance of rolling the first five on the third roll.

3.3.4 Negative Binomial Random Variable

What if we were to carry out multiple independent and identical Bernoulli trials until the rth success occurs? If we model Y, the number of failures before the rth success, then Y follows a **negative binomial distribution** where

$$P(Y = y) = \binom{y + r - 1}{r - 1}(1 - p)^y(p)^r \quad \text{for} \quad y = 0, 1, \ldots, \infty. \tag{3.4}$$

If $Y \sim$ Negative Binomial(r, p) then $\text{E}(Y) = \frac{r(1-p)}{p}$ and $\text{SD}(Y) = \sqrt{\frac{r(1-p)}{p^2}}$. Figure 3.3 displays three negative binomial distributions. Notice how centers shift right as r increases, and left as p increases.

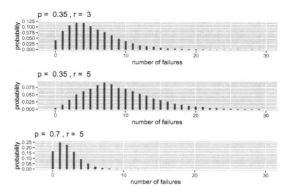

FIGURE 3.3: Negative binomial distributions with different values of p and r.

Note that if we set $r = 1$, then

$$P(Y = y) = \binom{y}{0}(1-p)^y p$$
$$= (1-p)^y p \quad \text{for} \quad y = 0, 1, \ldots, \infty,$$

which is the probability mass function of a geometric random variable! Thus, a geometric random variable is, in fact, a special case of a negative binomial random variable.

While negative binomial random variables typically are expressed as above using binomial coefficients (expressions such as $\binom{x}{y}$), we can generalize our definition to allow non-integer values of r. This will come in handy later when modeling. To do this, we need to first introduce the **gamma function**. The gamma function is defined as such

$$\Gamma(x) = \int_0^\infty t^{x-1} e^{-t} dt. \tag{3.5}$$

One important property of the gamma function is that for any integer n, $\Gamma(n) = (n-1)!$. Applying this, we can generalize the pmf of a negative binomial variable such that

$$P(Y = y) = \binom{y + r - 1}{r - 1}(1-p)^y (p)^r$$
$$= \frac{(y + r - 1)!}{(r - 1)! y!}(1-p)^y (p)^r$$
$$= \frac{\Gamma(y + r)}{\Gamma(r) y!}(1-p)^y (p)^r \quad \text{for} \quad y = 0, 1, \ldots, \infty.$$

3.3 Discrete Random Variables

With this formulation, r is no longer restricted to non-negative integers; rather r can be any non-negative real number.

In R we can use the function dnbinom(y, r, p) for the probability of y failures before the rth success given probability p.

Example 4: A contestant on a game show needs to answer 10 questions correctly to win the jackpot. However, if they get 3 incorrect answers, they are kicked off the show. Suppose one contestant consistently has a 90% chance of correctly responding to any question. What is the probability that she will correctly answer 10 questions before 3 incorrect responses?

Letting Y represent the number of incorrect responses, and setting $r = 10$, we want

$$P(Y < 3) = P(Y = 0) + P(Y = 1) + P(Y = 2)$$
$$= \binom{9}{9}(1 - 0.9)^0(0.9)^{10} + \binom{10}{9}(1 - 0.9)^1(0.9)^{10}$$
$$+ \binom{11}{9}(1 - 0.9)^2(0.9)^{10}$$
$$= 0.89$$

Using R:

```
# could also use pnbinom(2, 10, .9)
sum(dnbinom(0:2, size = 10, prob = .9))
```

```
## [1] 0.8891
```

Thus, there is a 89% chance that she gets 10 correct responses before missing 3.

3.3.5 Hypergeometric Random Variable

In all previous random variables, we considered a Bernoulli process, where the probability of a success remained constant across all trials. What if this probability is dynamic? The **hypergeometric random variable** helps us address some of these situations. Specifically, what if we wanted to select n items *without replacement* from a collection of N objects, m of which are considered successes? In that case, the probability of selecting a "success" depends on the previous selections. If we model Y, the number of successes after n selections, Y follows a **hypergeometric distribution** where

$$P(Y = y) = \frac{\binom{m}{y}\binom{N-m}{n-y}}{\binom{N}{n}} \quad \text{for} \quad y = 0, 1, \ldots, \min(m, n). \quad (3.6)$$

If Y follows a hypergeometric distribution and we define $p = m/N$, then $E(Y) = np$ and $SD(Y) = \sqrt{np(1-p)\frac{N-n}{N-1}}$. Figure 3.4 displays several hypergeometric distributions. On the left, N and n are held constant. As $m \to N/2$, the distribution becomes more and more symmetric. On the right, m and N are held constant. Both distributions are displayed on the same scale. We can see that as $n \to N$ (or $n \to 0$), the distribution becomes less variable.

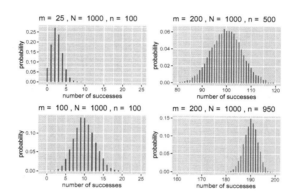

FIGURE 3.4: Hypergeometric distributions with different values of m, N, and n.

If we wish to calculate probabilities through R, `dhyper(y, m, N-m, n)` gives $P(Y = y)$ given n draws without replacement from m successes and $N - m$ failures.

Example 5: Suppose a deck of cards is randomly shuffled. What is the probability that all 4 queens are located within the first 10 cards?

We can model Y, the number of queens in the first 10 cards as a hypergeometric random variable where $n = 10$, $m = 4$, and $N = 52$. Then, $P(Y = 4) = \frac{\binom{4}{4}\binom{48}{6}}{\binom{52}{10}} = 0.0008$. We can avoid this calculation through R, of course:

```
dhyper(4, m = 4, n = 48, k = 10)
```

```
## [1] 0.0007757
```

So, there is a 0.08% chance of all 4 queens being within the first 10 cards of a randomly shuffled deck of cards.

3.3 Discrete Random Variables

3.3.6 Poisson Random Variable

Sometimes, random variables are based on a **Poisson process**. In a Poisson process, we are counting the number of events per unit of time or space and the number of events depends only on the length or size of the interval. We can then model Y, the number of events in one of these sections with the **Poisson distribution**, where

$$P(Y = y) = \frac{e^{-\lambda}\lambda^y}{y!} \quad \text{for} \quad y = 0, 1, \ldots, \infty, \tag{3.7}$$

where λ is the mean or expected count in the unit of time or space of interest. This probability mass function has $E(Y) = \lambda$ and $SD(Y) = \sqrt{\lambda}$. Three Poisson distributions are displayed in Figure 3.5. Notice how distributions become more symmetric as λ increases.

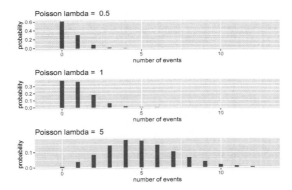

FIGURE 3.5: Poisson distributions with $\lambda = 0.5$, 1, and 5.

If we wish to use R, `dpois(y, lambda)` outputs the probability of y events given λ.

Example 6: A small town's police department issues 5 speeding tickets per month on average. Using a Poisson random variable, what is the likelihood that the police department issues 3 or fewer tickets in one month?

First, we note that here $P(Y \leq 3) = P(Y = 0) + P(Y = 1) + \cdots + P(Y = 3)$. Applying the probability mass function for a Poisson distribution with $\lambda = 5$, we find that

$$\begin{aligned}
P(Y \leq 3) &= P(Y = 0) + P(Y = 1) + P(Y = 2) + P(Y = 3) \\
&= \frac{e^{-5}5^0}{0!} + \frac{e^{-5}5^1}{1!} + \frac{e^{-5}5^2}{2!} + \frac{e^{-5}5^3}{3!} \\
&= 0.27.
\end{aligned}$$

We can verify through R:

```
sum(dpois(0:3, lambda = 5))    # or use ppois(3, 5)
```

```
## [1] 0.265
```

Therefore, there is a 27% chance of 3 or fewer tickets being issued within one month.

3.4 Continuous Random Variables

A continuous random variable can take on an uncountably infinite number of values. With continuous random variables, we define probabilities using **probability density functions** (pdfs). Probabilities are calculated by computing the area under the density curve over the interval of interest. So, given a pdf, $f(y)$, we can compute

$$P(a \leq Y \leq b) = \int_a^b f(y) dy.$$

This hints at a few properties of continuous random variables:

- $\int_{-\infty}^{\infty} f(y) dy = 1.$

- For any value y, $P(Y = y) = \int_y^y f(y) dy = 0.$

- Because of the above property, $P(y < Y) = P(y \leq Y)$. We will typically use the first notation rather than the second, but both are equally valid.

3.4.1 Exponential Random Variable

Suppose we have a Poisson process with rate λ, and we wish to model the wait time Y until the first event. We could model Y using an **exponential distribution**, where

$$f(y) = \lambda e^{-\lambda y} \quad \text{for} \quad y > 0, \tag{3.8}$$

where $\mathrm{E}(Y) = 1/\lambda$ and $\mathrm{SD}(Y) = 1/\lambda$. Figure 3.6 displays three exponential distributions with different λ values. As λ increases, $\mathrm{E}(Y)$ tends towards 0, and distributions "die off" quicker.

3.4 Continuous Random Variables

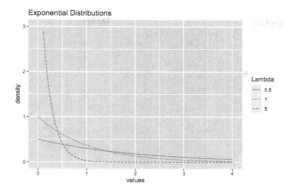

FIGURE 3.6: Exponential distributions with $\lambda = 0.5, 1$, and 5.

If we wish to use R, `pexp(y, lambda)` outputs the probability $P(Y < y)$ given λ.

Example 7: Refer to Example 6. What is the probability that 10 days or fewer elapse between two tickets being issued?

We know the town's police issue 5 tickets per month. For simplicity's sake, assume each month has 30 days. Then, the town issues $\frac{1}{6}$ tickets per day. That is $\lambda = \frac{1}{6}$, and the average wait time between tickets is $\frac{1}{1/6} = 6$ days. Therefore,

$$P(Y < 10) = \int_0^{10} \tfrac{1}{6} e^{-\tfrac{1}{6} y} dy = 0.81.$$

We can also use R:

```
pexp(10, rate = 1/6)
```

[1] 0.8111

Hence, there is a 81% chance of waiting fewer than 10 days between tickets.

3.4.2 Gamma Random Variable

Once again consider a Poisson process. When discussing exponential random variables, we modeled the wait time before one event occurred. If Y represents the wait time before r events occur in a Poisson process with rate λ, Y follows a **gamma distribution** where

$$f(y) = \frac{\lambda^r}{\Gamma(r)} y^{r-1} e^{-\lambda y} \quad \text{for} \quad y > 0. \tag{3.9}$$

If $Y \sim \text{Gamma}(r, \lambda)$ then $\text{E}(Y) = r/\lambda$ and $\text{SD}(Y) = \sqrt{r/\lambda^2}$. A few gamma distributions are displayed in Figure 3.7. Observe that means increase as r increases, but decrease as λ increases.

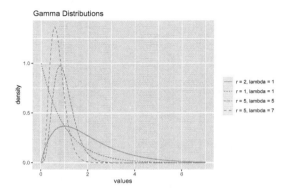

FIGURE 3.7: Gamma distributions with different values of r and λ.

Note that if we let $r = 1$, we have the following pdf,

$$f(y) = \frac{\lambda}{\Gamma(1)} y^{1-1} e^{-\lambda y}$$
$$= \lambda e^{-\lambda y} \quad \text{for} \quad y > 0,$$

an exponential distribution. Just as how the geometric distribution was a special case of the negative binomial, exponential distributions are in fact a special case of gamma distributions!

Just like negative binomial, the pdf of a gamma distribution is defined for all real, non-negative r.

In R, `pgamma(y, r, lambda)` outputs the probability $P(Y < y)$ given r and λ.

Example 8: Two friends are out fishing. On average they catch two fish per hour, and their goal is to catch 5 fish. What is the probability that they take less than 3 hours to reach their goal?

Using a gamma random variable, we set $r = 5$ and $\lambda = 2$. So,

$$P(Y < 3) = \int_0^3 \frac{2^4}{\Gamma(5)} y^4 e^{-2y} dy = 0.715.$$

Using R:

3.4 Continuous Random Variables

```
pgamma(3, shape = 5, rate = 2)
```

```
## [1] 0.7149
```

There is a 71.5% chance of catching 5 fish within the first 3 hours.

3.4.3 Normal (Gaussian) Random Variable

You have already at least informally seen normal random variables when evaluating LLSR assumptions. To recall, we required responses to be normally distributed at each level of X. Like any continuous random variable, normal (also called Gaussian) random variables have their own pdf, dependent on μ, the population mean of the variable of interest, and σ, the population standard deviation. We find that

$$f(y) = \frac{e^{-(y-\mu)^2/(2\sigma^2)}}{\sqrt{2\pi\sigma^2}} \quad \text{for} \quad -\infty < y < \infty. \tag{3.10}$$

As the parameter names suggest, $E(Y) = \mu$ and $SD(Y) = \sigma$. Often, normal distributions are referred to as $N(\mu, \sigma)$, implying a normal distribution with mean μ and standard deviation σ. The distribution $N(0, 1)$ is often referred to as the **standard normal distribution**. A few normal distributions are displayed in Figure 3.8.

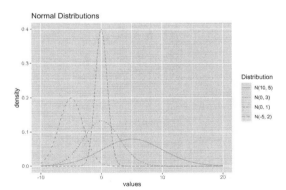

FIGURE 3.8: Normal distributions with different values of μ and σ.

In R, pnorm(y, mean, sd) outputs the probability $P(Y < y)$ given a mean and standard deviation.

Example 9: The weight of a box of Fruity Tootie cereal is approximately normally distributed with an average weight of 15 ounces and a standard deviation of 0.5 ounces. What is the probability that the weight of a randomly selected box is more than 15.5 ounces?

Using a normal distribution,

$$P(Y > 15.5) = \int_{15.5}^{\infty} \frac{e^{-(y-15)^2/(2 \cdot 0.5^2)}}{\sqrt{2\pi \cdot 0.5^2}} dy = 0.159$$

We can use R as well:

```
pnorm(15.5, mean = 15, sd = 0.5, lower.tail = FALSE)
```

```
## [1] 0.1587
```

There is a 16% chance of a randomly selected box weighing more than 15.5 ounces.

3.4.4 Beta Random Variable

So far, all of our continuous variables have had no upper bound. If we want to limit our possible values to a smaller interval, we may turn to a **beta random variable**. In fact, we often use beta random variables to model distributions of probabilities—bounded below by 0 and above by 1. The pdf is parameterized by two values, α and β ($\alpha, \beta > 0$). We can describe a beta random variable by the following pdf:

$$f(y) = \frac{\Gamma(\alpha+\beta)}{\Gamma(\alpha)\Gamma(\beta)} y^{\alpha-1}(1-y)^{\beta-1} \quad \text{for} \quad 0 < y < 1. \tag{3.11}$$

If $Y \sim \text{Beta}(\alpha, \beta)$, then $E(Y) = \alpha/(\alpha+\beta)$ and $SD(Y) = \sqrt{\frac{\alpha\beta}{(\alpha+\beta)^2(\alpha+\beta+1)}}$. Figure 3.9 displays several beta distributions. Note that when $\alpha = \beta$, distributions are symmetric. The distribution is left-skewed when $\alpha > \beta$ and right-skewed when $\beta > \alpha$.

If $\alpha = \beta = 1$, then

$$f(y) = \frac{\Gamma(1)}{\Gamma(1)\Gamma(1)} y^0 (1-y)^0$$
$$= 1 \quad \text{for} \quad 0 < y < 1.$$

This distribution is referred to as a **uniform distribution**.

3.5 Distributions Used in Testing

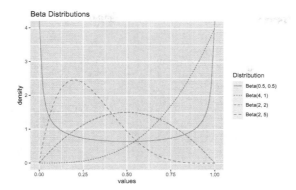

FIGURE 3.9: Beta distributions with different values of α and β.

In R, `pbeta(y, alpha, beta)` yields $P(Y < y)$ assuming $Y \sim \text{Beta}(\alpha, \beta)$.

Example 10: A private college in the Midwest models the probabilities of prospective students accepting an admission decision through a beta distribution with $\alpha = \frac{4}{3}$ and $\beta = 2$. What is the probability that a randomly selected student has probability of accepting greater than 80%?

Letting $Y \sim \text{Beta}(4/3, 2)$, we can calculate

$$P(Y > 0.8) = \int_{0.8}^{1} \frac{\Gamma(4/3+2)}{\Gamma(4/3)\Gamma(2)} y^{4/3-1}(1-y)^{2-1} dy = 0.06.$$

Alternatively, in R:

```
pbeta(0.8, shape1 = 4/3, shape2 = 2, lower.tail = FALSE)
```

```
## [1] 0.0593
```

Hence, there is a 6% chance that a randomly selected student has a probability of accepting an admission decision above 80%.

3.5 Distributions Used in Testing

We have spent most of this chapter discussing probability distributions that may come in handy when modeling. The following distributions, while rarely used in modeling, prove useful in hypothesis testing as certain commonly used test statistics follow these distributions.

3.5.1 χ^2 Distribution

You have probably already encountered χ^2 tests before. For example, χ^2 tests are used with two-way contingency tables to investigate the association between row and column variables. χ^2 tests are also used in goodness-of-fit testing such as comparing counts expected according to Mendelian ratios to observed data. In those situations, χ^2 tests compare observed counts to what would be expected under the null hypotheses and reject the null when these observed discrepancies are too large.

In this course, we encounter χ^2 distributions in several testing situations. In Section 2.6.5 we performed likelihood ratio tests (LRTs) to compare nested models. When a larger model provides no significant improvement over a reduced model, the LRT statistic (which is twice the difference in the log-likelihoods) follows a χ^2 distribution with the degrees of freedom equal to the difference in the number of parameters.

In general, χ^2 distributions with k degrees of freedom are right skewed with a mean k and standard deviation $\sqrt{2k}$. Figure 3.10 displays chi-square distributions with different values of k.

The χ^2 distribution is a special case of a gamma distribution. Specifically, a χ^2 distribution with k degrees of freedom can be expressed as a gamma distribution with $\lambda = 1/2$ and $r = k/2$.

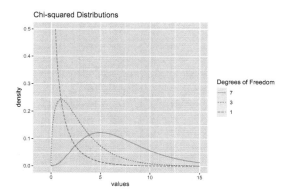

FIGURE 3.10: χ^2 distributions with 1, 3, and 7 degrees of freedom..

In R, `pchisq(y, df)` outputs $P(Y < y)$ given k degrees of freedom.

3.5.2 Student's t-Distribution

You likely have seen Student's t-distribution (developed by William Sealy Gosset under the penname *Student*) in a previous statistics course. You may have used it when drawing inferences about the means of normally distributed populations with unknown population standard deviations. t-distributions are parameterized by their degrees of freedom, k.

A t-distribution with k degrees of freedom has mean 0 and standard deviation $k/(k-2)$ (standard deviation is only defined for $k > 2$). As $k \to \infty$ the t-distribution approaches the standard normal distribution.

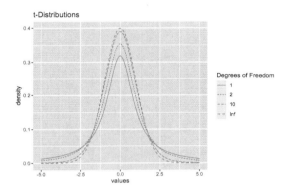

FIGURE 3.11: t-distributions with 1, 2, 10, and Infinite degrees of freedom.

Figure 3.11 displays some t-distributions, where a t-distribution with infinite degrees of freedom is equivalent to a standard normal distribution (with mean 0 and standard deviation 1). In R, `pt(y, df)` outputs $P(Y < y)$ given k degrees of freedom.

3.5.3 F-Distribution

F-distributions are also used when performing statistical tests. Like the χ^2 distribution, the values from an F-distribution are non-negative and the distribution is right skewed; in fact, an F-distribution can be derived as the ratio of two χ^2 random variables. R.A. Fisher (for whom the test is named) devised this test statistic to compare two different estimates of the same variance parameter, and it has a prominent role in Analysis of Variance (ANOVA). Model comparisons are often based on the comparison of variance estimates, e.g., the extra sums-of-squares F test. F-distributions are indexed by two degrees-of-freedom values, one for the numerator (k_1) and one for the denominator (k_2). The expected value for an F-distribution with k_1, k_2 degrees of freedom under the null hypothesis is $\frac{k_2}{k_2-2}$, which approaches 1 as $k_2 \to \infty$.

The standard deviation decreases as k_1 increases for fixed k_2, as seen in Figure 3.12, which illustrates several F-distributions.

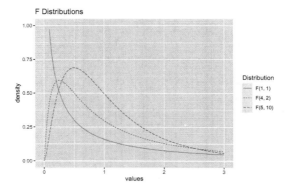

FIGURE 3.12: F-distributions with different degrees of freedom.

3.6 Additional Resources

Table 3.1 briefly details most of the random variables discussed in this chapter.

3.7 Exercises

3.7.1 Conceptual Exercises

1. At what value of p is the standard deviation of a binary random variable smallest? When is standard deviation largest?

2. How are hypergeometric and binomial random variables different? How are they similar?

3. How are exponential and Poisson random variables related?

4. How are geometric and exponential random variables similar? How are they different?

5. A university's college of sciences is electing a new board of 5 members. There are 35 applicants, 10 of which come from the math department.

3.7 Exercises

TABLE 3.1: Review of mentioned random variables.

Distribution Name	pmf / pdf	Parameters	Possible Y Values	Description
Binomial	$\binom{n}{y}p^y(1-p)^{n-y}$	p, n	$0, 1, \ldots, n$	Number of successes after n trials
Geometric	$(1-p)^y p$	p	$0, 1, \ldots, \infty$	Number of failures until the first success
Negative Binomial	$\binom{y+r-1}{r-1}(1-p)^y(p)^r$	p, r	$0, 1, \ldots, \infty$	Number of failures before r successes
Hypergeometric	$\binom{m}{y}\binom{N-m}{n-y}/\binom{N}{n}$	n, m, N	$0, 1, \ldots, \min(m,n)$	Number of successes after n trials without replacement
Poisson	$e^{-\lambda}\lambda^y/y!$	λ	$0, 1, \ldots, \infty$	Number of events in a fixed interval
Exponential	$\lambda e^{-\lambda y}$	λ	$(0, \infty)$	Wait time for one event in a Poisson process
Gamma	$\dfrac{\lambda^r}{\Gamma(r)}y^{r-1}e^{-\lambda y}$	λ, r	$(0, \infty)$	Wait time for r events in a Poisson process
Normal	$\dfrac{e^{-(y-\mu)^2/(2\sigma^2)}}{\sqrt{2\pi\sigma^2}}$	μ, σ	$(-\infty, \infty)$	Used to model many naturally occurring phenomena
Beta	$\dfrac{\Gamma(\alpha+\beta)}{\Gamma(\alpha)\Gamma(\beta)}y^{\alpha-1}(1-y)^{\beta-1}$	α, β	$(0, 1)$	Useful for modeling probabilities

What distribution could be helpful to model the probability of electing X board members from the math department?

6. Chapter 1 asked you to consider a scenario where *"The Minnesota Pollution Control Agency is interested in using traffic volume data to generate predictions of particulate distributions as measured in counts per cubic feet."* What distribution might be useful to model this count per cubic foot? Why?

7. Chapter 1 also asked you to consider a scenario where *"Researchers are attempting to see if socioeconomic status and parental stability are predictive of low birthweight. They classify a low birthweight as below 2500 g, hence our response is binary: 1 for low birthweight, and 0 when the birthweight is not low."* What distribution might be useful to model if a newborn has low birthweight?

8. Chapter 1 also asked you to consider a scenario where *"Researchers are interested in how elephant age affects mating patterns among males. In particular, do older elephants have greater mating success, and is there an optimal age for mating among males? Data collected includes, for each elephant, age and number of matings in a given year."* Which distribution would be useful to model the number of matings in a given year for these elephants? Why?

9. Describe a scenario which could be modeled using a gamma distribution.

3.7.2 Guided Exercises

1. **Beta-binomial distribution.** We can generate more distributions by mixing two random variables. Beta-binomial random variables are binomial random variables with fixed n whose parameter p follows a beta distribution with fixed parameters α, β. In more detail, we would first draw p_1 from our beta distribution, and then generate our first observation y_1, a random number of successes from a binomial (n, p_1) distribution. Then, we would generate a new p_2 from our beta distribution, and use a binomial distribution with parameters n, p_2 to generate our second observation y_2. We would continue this process until desired.

   Note that all of the observations y_i will be integer values from $0, 1, \ldots, n$. With this in mind, use `rbinom()` to simulate 1,000 observations from a plain old vanilla binomial random variable with $n = 10$ and $p = 0.8$. Plot a histogram of these binomial observations. Then, do the following to generate a beta-binomial distribution:

   a. Draw p_i from the beta distribution with $\alpha = 4$ and $\beta = 1$.
   b. Generate an observation y_i from a binomial distribution with $n = 10$ and $p = p_i$.
   c. Repeat (a) and (b) 1,000 times ($i = 1, \ldots, 1000$).
   d. Plot a histogram of these beta-binomial observations.

   Compare the histograms of the "plain old" binomial and beta-binomial distributions. How do their shapes, standard deviations, means, possible values, etc. compare?

2. **Gamma-Poisson mixture I.** Use the R function `rpois()` to generate 10,000 x_i from a plain old vanilla Poisson random variable, $X \sim \text{Poisson}(\lambda = 1.5)$. Plot a histogram of this distribution and note its mean and standard deviation. Next, let $Y \sim \text{Gamma}(r = 3, \lambda = 2)$ and use `rgamma()` to generate 10,000 random y_i from this distribution. Now, consider 10,000 different Poisson distributions where $\lambda_i = y_i$. Randomly generate one z_i from each Poisson distribution. Plot a histogram of these z_i and compare it to your original histogram of X (where $X \sim \text{Poisson}(1.5)$). How do the means and standard deviations compare?

3. **Gamma-Poisson mixture II.** A negative binomial distribution can actually be expressed as a gamma-Poisson mixture. In the previous problem's gamma-Poisson mixture $Z \sim \text{Poisson}(\lambda)$ where

$\lambda \sim \text{Gamma}(r = 3, \lambda' = 2)$. Find the parameters of a negative binomial distribution $X \sim \text{Negative Binomial}(r, p)$ such that X is equivalent to Z. As a hint, the means of both distributions must be the same, so $r(1 - p)/p = 3/2$. Show through histograms and summary statistics that your negative binomial distribution is equivalent to your gamma-Poisson mixture from Problem 2. Argue that if you want a NB(r, p) random variable, you can instead sample from a Poisson distribution, where the λ values are themselves sampled from a gamma distribution with parameters r and $\lambda' = \frac{p}{1-p}$.

4. **Mixture of two normal distributions** Sometimes, a value may be best modeled by a mixture of two normal distributions. We would have 5 parameters in this case— $\mu_1, \sigma_1, \mu_2, \sigma_2, \alpha$, where $0 < \alpha < 1$ is a mixing parameter determining the probability that an observation comes from the first distribution. We would then have $f(y) = \alpha\ f_1(y) + (1 - \alpha)\ f_2(y)$ (where $f_i(y)$ is the pdf of the normal distribution with μ_i, σ_i). One phenomenon which could be modeled this way would be the waiting times between eruptions of Old Faithful geyser in Yellowstone National Park. The data can be accessed in R through `faithful`, and a histogram of wait times can be found in Figure 3.13. The MLEs of our 5 parameters would be the combination of values that produces the maximum probability of our observed data. We will try to approximate MLEs by hand. Find a combination of $\mu_1, \sigma_1, \mu_2, \sigma_2, \alpha$ for this distribution such that the logged likelihood is above -1050. (The command `dnorm(x, mean, sd)`, which outputs $f(y)$ assuming $Y \sim N(\mu, \sigma)$, will be helpful in calculating likelihoods.)

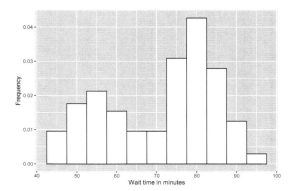

FIGURE 3.13: Waiting time between eruptions of Old Faithful.

4

Poisson Regression

4.1 Learning Objectives

After finishing this chapter, you should be able to:

- Describe why simple linear regression is not ideal for Poisson data.
- Write out a Poisson regression model and identify the assumptions for inference.
- Write out the likelihood for a Poisson regression and describe how it could be used to estimate coefficients for a model.
- Interpret estimated coefficients from a Poisson regression and construct confidence intervals for them.
- Use deviances for Poisson regression models to compare and assess models.
- Use an offset to account for varying effort in data collection.
- Fit and use a zero-inflated Poisson (ZIP) model.

```
# Packages required for Chapter 4
library(gridExtra)
library(knitr)
library(kableExtra)
library(mosaic)
library(xtable)
library(pscl)
library(multcomp)
library(pander)
library(MASS)
library(tidyverse)
```

4.2 Introduction to Poisson Regression

Consider the following questions:

1. Are the number of motorcycle deaths in a given year related to a state's helmet laws?
2. Does the number of employers conducting on-campus interviews during a year differ for public and private colleges?
3. Does the daily number of asthma-related visits to an Emergency Room differ depending on air pollution indices?
4. Has the number of deformed fish in randomly selected Minnesota lakes been affected by changes in trace minerals in the water over the last decade?

Each example involves predicting a response using one or more explanatory variables, although these examples have response variables that are counts per some unit of time or space. A Poisson random variable is often used to model counts; see Chapter 3 for properties of the Poisson distribution. Since a Poisson random variable is a count, its minimum value is zero and, in theory, the maximum is unbounded. We'd like to model our main parameter λ, the average number of occurrences per unit of time or space, as a function of one or more covariates. For example, in the first question above, λ_i represents the average number of motorcycle deaths in a year for state i, and we hope to show that state-to-state variability in λ_i can be explained by state helmet laws.

For a linear least squares regression model, the parameter of interest is the average response, μ_i, for subject i, and μ_i is modeled as a line in the case of one explanatory variable. By analogy, it might seem reasonable to try to model the Poisson parameter λ_i as a linear function of an explanatory variable, but there are some problems with this approach. In fact, a model like $\lambda_i = \beta_0 + \beta_1 x_i$ doesn't work well for Poisson data. A line is certain to yield negative values for certain x_i, but λ_i can only take on values from 0 to ∞. In addition, the equal variance assumption in linear regression inference is violated because as the mean rate for a Poisson variable increases, the variance also increases (recall from Chapter 3 that if Y is the observed count, then $E(Y) = Var(Y) = \lambda$).

One way to avoid these problems is to model $\log(\lambda_i)$ instead of λ_i as a function of the covariates. The $\log(\lambda_i)$ takes on values from $-\infty$ to ∞. We can also take into account the increase in the variance with an increasing mean using this approach. (Note that throughout *Beyond Multiple Linear Regression* we use log to represent the natural logarithm.) Thus, we will consider the **Poisson regression** model:

4.2 Introduction to Poisson Regression

$$log(\lambda_i) = \beta_0 + \beta_1 x_i$$

where the observed values $Y_i \sim$ Poisson with $\lambda = \lambda_i$ for a given x_i. For example, each state i can potentially have a different λ depending on its value of x_i, where x_i could represent presence or absence of a helmet law. Note that the Poisson regression model contains no separate error term like the ϵ we see in linear regression, because λ determines both the mean and the variance of a Poisson random variable.

4.2.1 Poisson Regression Assumptions

Much like linear least squares regression (LLSR), using Poisson regression to make inferences requires model assumptions.

1. **Poisson Response** The response variable is a count per unit of time or space, described by a Poisson distribution.
2. **Independence** The observations must be independent of one another.
3. **Mean=Variance** By definition, the mean of a Poisson random variable must be equal to its variance.
4. **Linearity** The log of the mean rate, $log(\lambda)$, must be a linear function of x.

4.2.2 A Graphical Look at Poisson Regression

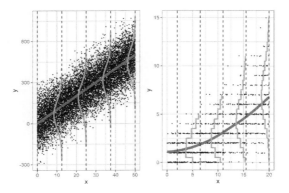

FIGURE 4.1: Regression models: Linear regression (left) and Poisson regression (right).

Figure 4.1 illustrates a comparison of the LLSR model for inference to Poisson regression using a log function of λ.

1. The graphic displaying the LLSR inferential model appears in the left panel of Figure 4.1. It shows that, for each level of X, the responses are approximately normal. The panel on the right side of Figure 4.1 depicts what a Poisson regression model looks like. For each level of X, the responses follow a Poisson distribution (Assumption 1). For Poisson regression, small values of λ are associated with a distribution that is noticeably skewed with lots of small values and only a few larger ones. As λ increases the distribution of the responses begins to look more and more like a normal distribution.
2. In the LLSR model, the variation in Y at each level of X, σ^2, is the same. For Poisson regression the responses at each level of X become more variable with increasing means, where variance=mean (Assumption 3).
3. In the case of LLSR, the mean responses for each level of X, $\mu_{Y|X}$, fall on a line. In the case of the Poisson model, the mean values of Y at each level of X, $\lambda_{Y|X}$, fall on a curve, not a line, although the logs of the means should follow a line (Assumption 4).

4.3 Case Studies Overview

We take a look at the Poisson regression model in the context of three case studies. Each case study is based on real data and real questions. Modeling household size in the Philippines introduces the idea of regression with a Poisson response along with its assumptions. A quadratic term is added to a model to determine an optimal size per household, and methods of model comparison are introduced. The campus crime case study introduces two big ideas in Poisson regression modeling: offsets, to account for sampling effort, and overdispersion, when actual variability exceeds what is expected by the model. Finally, the weekend drinking example uses a modification of a Poisson model to account for more zeros than would be expected for a Poisson random variable. These three case studies also provide context for some of the familiar concepts related to modeling such as exploratory data analysis (EDA), estimation, and residual plots.

4.4 Case Study: Household Size in the Philippines

How many other people live with you in your home? The number of people sharing a house differs from country to country and often from region to region. International agencies use household size when determining needs of populations, and the household sizes determine the magnitude of the household needs.

The Philippine Statistics Authority (PSA) spearheads the Family Income and Expenditure Survey (FIES) nationwide. The survey, which is undertaken every three years, is aimed at providing data on family income and expenditure, including levels of consumption by item of expenditure. Our data, from the 2015 FIES, is a subset of 1500 of the 40,000 observations [Philippine Statistics Authority, 2015]. Our data set focuses on five regions: Central Luzon, Metro Manila, Ilocos, Davao, and Visayas (see Figure 4.2).

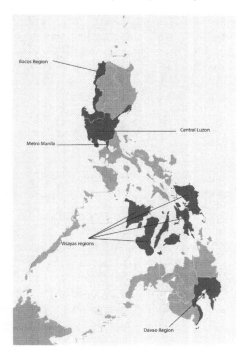

FIGURE 4.2: Regions of the Philippines.

At what age are heads of households in the Philippines most likely to find the largest number of people in their household? Is this association similar for poorer households (measured by the presence of a roof made from predomi-

TABLE 4.1: The first five observations from the Philippines Household case study.

X1	location	age	total	numLT5	roof
1	CentralLuzon	65	0	0	Predominantly Strong Material
2	MetroManila	75	3	0	Predominantly Strong Material
3	DavaoRegion	54	4	0	Predominantly Strong Material
4	Visayas	49	3	0	Predominantly Strong Material
5	MetroManila	74	3	0	Predominantly Strong Material

nantly light/salvaged materials)? We begin by explicitly defining our response, Y = number of household members other than the head of the household. We then define the explanatory variables: age of the head of the household, type of roof (predominantly light/salvaged material or predominantly strong material), and location (Central Luzon, Davao Region, Ilocos Region, Metro Manila, or Visayas). Note that predominantly light/salvaged materials are a combination of light material, mixed but predominantly light material, and mixed but predominantly salvaged material, and salvaged matrial. Our response is a count, so we consider a Poisson regression where the parameter of interest is λ, the average number of people, other than the head, per household. We will primarily examine the relationship between household size and age of the head of household, controlling for location and income.

4.4.1 Data Organization

The first five rows from our data set `fHH1.csv` are illustrated in Table 4.1. Each line of the data file refers to a household at the time of the survey:

- `location` = where the house is located (Central Luzon, Davao Region, Ilocos Region, Metro Manila, or Visayas)
- `age` = the age of the head of household
- `total` = the number of people in the household other than the head
- `numLT5` = the number in the household under 5 years of age
- `roof` = the type of roof in the household (either Predominantly Light/Salvaged Material, or Predominantly Strong Material, where stronger material can sometimes be used as a proxy for greater wealth)

4.4.2 Exploratory Data Analyses

For the rest of this case study, we will refer to the number of people in a household as the total number of people in that specific household *besides*

4.4 Case Study: Household Size in the Philippines

the head of household. The average number of people in a household is 3.68 (Var = 5.53), and there are anywhere from 0 to 16 people in the houses. Over 11.1% of these households are made from predominantly light and salvaged material. The mean number of people in a house for houses with a roof made from predominantly strong material is 3.69 (Var=5.55), whereas houses with a roof made from predominantly light/salvaged material average 3.64 people (Var=5.41). Of the various locations, Visayas has the largest household size, on average, with a mean of 3.90 in the household, and the Davao Region has the smallest with a mean of 3.39.

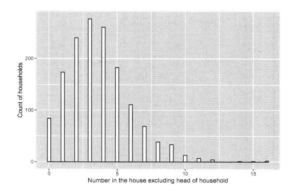

FIGURE 4.3: Distribution of household size in 5 Philippine regions.

Figure 4.3 reveals a fair amount of variability in the number in each house; responses range from 0 to 16 with many of the respondents reporting between 1 and 5 people in the house. Like many Poisson distributions, this graph is right skewed. It clearly does not suggest that the number of people in a household is a normally distributed response.

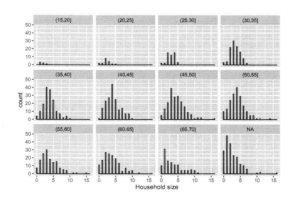

FIGURE 4.4: Distribution of household sizes by age group of the household head.

TABLE 4.2: Compare mean and variance of household size within each age group.

Age Groups	Mean	Variance	n
(15,20]	1.667	0.6667	6
(20,25]	2.167	1.5588	18
(25,30]	2.918	1.4099	49
(30,35]	3.444	2.1931	108
(35,40]	3.842	3.5735	158
(40,45]	4.234	4.4448	175
(45,50]	4.490	6.3963	194
(50,55]	4.011	5.2512	188
(55,60]	3.807	6.5319	145
(60,65]	3.706	6.1958	153
(65,70]	3.339	7.9980	115
NA	2.550	5.5436	191

Figure 4.4 further shows that responses can be reasonably modeled with a Poisson distribution when grouped by a key explanatory variable: age of the household head. These last two plots together suggest that Assumption 1 (Poisson Response) is satisfactory in this case study.

For Poisson random variables, the variance of Y (i.e., the square of the standard deviation of Y), is equal to its mean, where Y represents the size of an individual household. As the mean increases, the variance increases. So, if the response is a count and the mean and variance are approximately equal for each group of X, a Poisson regression model may be a good choice. In Table 4.2 we display age groups by 5-year increments, to check to see if the empirical means and variances of the number in the house are approximately equal for each age group. This provides us one way in which to check the Poisson Assumption 3 (mean = variance).

If there is a problem with this assumption, most often we see variances much larger than means. Here, as expected, we see more variability as age increases. However, it appears that the variance is smaller than the mean for lower ages, while the variance is greater than the mean for higher ages. Thus, there is some evidence of a violation of the mean=variance assumption (Assumption 3), although any violations are modest.

The Poisson regression model also implies that $\log(\lambda_i)$, not the mean household size λ_i, is a linear function of age; i.e., $log(\lambda_i) = \beta_0 + \beta_1 \text{age}_i$. Therefore, to check the linearity assumption (Assumption 4) for Poisson regression, we would like to plot $\log(\lambda_i)$ by age. Unfortunately, λ_i is unknown. Our best guess of

4.4 Case Study: Household Size in the Philippines

λ_i is the observed mean number in the household for each age (level of X). Because these means are computed for observed data, they are referred to as **empirical** means. Taking the logs of the empirical means and plotting by age provides a way to assess the linearity assumption. The smoothed curve added to Figure 4.5 suggests that there is a curvilinear relationship between age and the log of the mean household size, implying that adding a quadratic term should be considered. This finding is consistent with the researchers' hypothesis that there is an age at which a maximum household size occurs. It is worth noting that we are not modeling the log of the empirical means, rather it is the log of the *true* rate that is modeled. Looking at empirical means, however, does provide an idea of the form of the relationship between $\log(\lambda)$ and x_i.

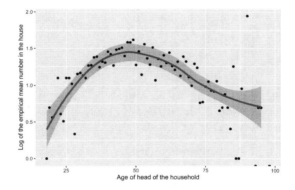

FIGURE 4.5: The log of the mean household sizes, besides the head of household, by age of the head of household, with loess smoother.

We can extend Figure 4.5 by fitting separate curves for each region (see Figure 4.6). This allows us to see if the relationship between mean household size and age is consistent across region. In this case, the relationships are pretty similar; if they weren't, we could consider adding an age-by-region interaction to our eventual Poisson regression model.

Finally, the independence assumption (Assumption 2) can be assessed using knowledge of the study design and the data collection process. In this case, we do not have enough information to assess the independence assumption with the information we are given. If each household was not selected individually in a random manner, but rather groups of households were selected from different regions with differing customs about living arrangements, the independence assumption would be violated. If this were the case, we could use a multilevel model like those discussed in later chapters with a village term.

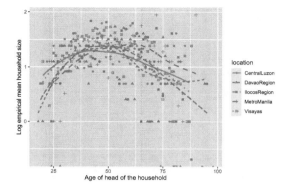

FIGURE 4.6: Empirical log of the mean household sizes vs. age of the head of household, with loess smoother by region.

4.4.3 Estimation and Inference

We first consider a model for which $\log(\lambda)$ is linear in age. We then will determine whether a model with a quadratic term in age provides a significant improvement based on trends we observed in the exploratory data analysis.

R reports an estimated regression equation for the linear Poisson model as:

$$log(\hat{\lambda}) = 1.55 - 0.0047 \text{age}$$

```
modela = glm(total ~ age, family = poisson, data = fHH1)
```

```
##                Estimate Std. Error  z value  Pr(>|z|)
## (Intercept)    1.549942  0.0502754  30.829   1.070e-208
## age           -0.004706  0.0009363  -5.026   5.013e-07

## Residual deviance =  2337   on   1498 df
## Dispersion parameter =  1
```

How can the coefficient estimates be interpreted in terms of this example? As done when interpreting slopes in the LLSR models, we consider how the estimated mean number in the house, λ, changes as the age of the household head increases by an additional year. But in place of looking at change in the mean number in the house, with a Poisson regression we consider the log of the mean number in the house and then convert back to original units. For example, consider a comparison of two models—one for a given age (x) and one after increasing age by 1 $(x+1)$:

4.4 Case Study: Household Size in the Philippines

$$\log(\lambda_X) = \beta_0 + \beta_1 X$$
$$\log(\lambda_{X+1}) = \beta_0 + \beta_1(X+1)$$
$$\log(\lambda_{X+1}) - \log(\lambda_X) = \beta_1$$
$$\log\left(\frac{\lambda_{X+1}}{\lambda_X}\right) = \beta_1 \quad (4.1)$$
$$\frac{\lambda_{X+1}}{\lambda_X} = e^{\beta_1}$$

These results suggest that by exponentiating the coefficient on age we obtain the multiplicative factor by which the mean count changes. In this case, the mean number in the house changes by a factor of $e^{-0.0047} = 0.995$ or decreases by 0.5% (since $1 - .995 = .005$) with each additional year older the household head is; or, we predict a 0.47% *increase* in mean household size for a 1-year *decrease* in age of the household head (since $1/.995 = 1.0047$). The quantity on the left-hand side of Equation (4.1) is referred to as a **rate ratio** or **relative risk**, and it represents a percent change in the response for a unit change in X. In fact, for regression models in general, whenever a variable (response or explanatory) is logged, we make interpretations about multiplicative effects on that variable, while with unlogged variables we can reach our usual interpretations about additive effects.

Typically, the standard errors for the estimated coefficients are included in Poisson regression output. Here the standard error for the estimated coefficient for age is 0.00094. We can use the standard error to construct a confidence interval for β_1. A 95% CI provides a range of plausible values for the **age** coefficient and can be constructed:

$$(\hat{\beta}_1 - Z^* \cdot SE(\hat{\beta}_1),\ \hat{\beta}_1 + Z^* \cdot SE(\hat{\beta}_1))$$
$$(-0.0047 - 1.96 * 0.00094,\ -0.0047 + 1.96 * 0.00094)$$
$$(-0.0065, -0.0029).$$

Exponentiating the endpoints yields a confidence interval for the relative risk; i.e., the percent change in household size for each additional year older. Thus $(e^{-0.0065}, e^{-0.0029}) = (0.993, 0.997)$ suggests that we are 95% confident that the mean number in the house decreases between 0.7% and 0.3% for each additional year older the head of household is. It is best to construct a confidence interval for the coefficient and then exponentiate the endpoints because the estimated coefficients more closely follow a normal distribution than the exponentiated coefficients. There are other approaches to constructing intervals in these circumstances, including profile likelihood, the delta method, and bootstrapping, and we will discuss some of those approaches later. In this case, for instance, the profile likelihood interval is nearly identical to the Wald-type (normal theory) confidence interval above.

```
# CI for betas using profile likelihood
confint(modela)
```

```
##                 2.5 %     97.5 %
## (Intercept)  1.451170  1.648249
## age         -0.006543 -0.002873
```

```
exp(confint(modela))
```

```
##              2.5 % 97.5 %
## (Intercept) 4.2681 5.1979
## age         0.9935 0.9971
```

If there is no association between age and household size, there is no change in household size for each additional year, so λ_X is equal to λ_{X+1} and the ratio λ_{X+1}/λ_X is 1. In other words, if there is no association between age and household size, then $\beta_1 = 0$ and $e^{\beta_1} = 1$. Note that our interval for e^{β_1}, (0.993,0.997), does not include 1, so the model with age is preferred to a model without age; i.e., age is significantly associated with household size. Note that we could have similarly confirmed that our interval for β_1 does not include 0 to show the significance of age as a predictor of household size.

Another way to test the significance of the age term is to calculate a **Wald-type statistic**. A Wald-type test statistic is the estimated coefficient divided by its standard error. When the true coefficient is 0, this test statistic follows a standard normal distribution for sufficiently large n. The estimated coefficient associated with the linear term in age is $\hat{\beta}_1 = -0.0047$ with standard error $SE(\hat{\beta}_1) = 0.00094$. The value for the Wald test statistic is then $Z = \hat{\beta}_1/SE(\hat{\beta}_1) = -5.026$, where Z follows a standard normal distribution if $\beta_1 = 0$. In this case, the two-sided p-value based on the standard normal distribution for testing $H_0 : \beta_1 = 0$ is almost 0 ($p = 0.000000501$). Therefore, we have statistically significant evidence (Z = -5.026, p < .001) that average household size decreases as age of the head of household increases.

4.4.4 Using Deviances to Compare Models

There is another way in which to assess how useful age is in our model. A **deviance** is a way in which to measure how the observed data deviates from the model predictions; it will be defined more precisely in Section 4.4.8, but it is similar to sum of squared errors (unexplained variability in the response) in

4.4 Case Study: Household Size in the Philippines

LLSR regression. Because we want models that minimize deviance, we calculate the **drop-in-deviance** when adding age to the model with no covariates (the **null model**). The deviances for the null model and the model with age can be found in the model output. A residual deviance for the model with age is reported as 2337.1 with 1498 df. The output also includes the deviance and degrees of freedom for the null model (2362.5 with 1499 df). The drop-in-deviance is 25.4 (2362.5 - 2337.1) with a difference of only 1 df, so that the addition of one extra term (age) reduced unexplained variability by 25.4. If the null model were true, we would expect the drop-in-deviance to follow a χ^2 distribution with 1 df. Therefore, the p-value for comparing the null model to the model with age is found by determining the probability that the value for a χ^2 random variable with one degree of freedom exceeds 25.4, which is essentially 0. Once again, we can conclude that we have statistically significant evidence ($\chi^2_{df=1} = 25.4$, $p < .001$) that average household size decreases as age of the head of household increases.

```
# model0 is the null/reduced model
model0 <- glm(total ~ 1, family = poisson, data = fHH1)
drop_in_dev <- anova(model0, modela, test = "Chisq")
```

	ResidDF	ResidDev	Deviance	Df	pval
1	1499	2362	NA	NA	NA
2	1498	2337	25.4	1	4.661e-07

More formally, we are testing:

$$\text{Null (reduced) Model} : \log(\lambda) = \beta_0 \text{ or } \beta_1 = 0$$

$$\text{Larger (full) Model} : \log(\lambda) = \beta_0 + \beta_1 \text{age or } \beta_1 \neq 0$$

In order to use the drop-in-deviance test, the models being compared must be **nested**; e.g., all the terms in the smaller model must appear in the larger model. Here the smaller model is the null model with the single term β_0 and the larger model has β_0 and β_1, so the two models are indeed nested. For nested models, we can compare the models' residual deviances to determine whether the larger model provides a significant improvement.

Here, then, is a summary of these two approaches to hypothesis testing about terms in Poisson regression models:

Drop-in-deviance test to compare models

- Compute the deviance for each model, then calculate: drop-in-deviance = residual deviance for reduced model − residual deviance for the larger model.

- When the reduced model is true, the drop-in-deviance $\sim \chi^2_d$ where d= the difference in the degrees of freedom associated with the two models (that is, the difference in the number of terms/coefficients).
- A large drop-in-deviance favors the larger model.

Wald test for a single coefficient

- Wald-type statistic = estimated coefficient / standard error
- When the true coefficient is 0, for sufficiently large n, the test statistic $\sim$ N(0,1).
- If the magnitude of the test statistic is large, there is evidence that the true coefficient is not 0.

The drop-in-deviance and the Wald-type tests usually provide consistent results; however, if there is a discrepancy, the drop-in-deviance is preferred. Not only does the drop-in-deviance test perform better in more cases, but it's also more flexible. If two models differ by one term, then the drop-in-deviance test essentially tests if a single coefficient is 0 like the Wald test does, while if two models differ by more than one term, the Wald test is no longer appropriate.

4.4.5 Using Likelihoods to Fit Models (optional)

Before continuing with model building, we take a short detour to see how coefficient estimates are determined in a Poisson regression model. The least squares approach requires a linear relationship between the parameter, λ_i (the expected or mean response for observation i), and x_i (the age for observation i). However, it is $\log(\lambda_i)$, not λ_i, that is linearly related to X with the Poisson model. The assumptions of equal variance and normality also do not hold for Poisson regression. Thus, the method of least squares will not be helpful for inference in Poisson Regression. Instead of least squares, we employ the likelihood principle to find estimates of our model coefficients. We look for those coefficient estimates for which the likelihood of our data is maximized; these are the **maximum likelihood estimates**.

The likelihood for n *independent* observations is the product of the probabilities. For example, if we observe five households with household sizes of 4, 2, 8, 6, and 1 person beyond the head, the likelihood is:

$$Likelihood = P(Y_1 = 4) * P(Y_2 = 2) * P(Y_3 = 8) * P(Y_4 = 6) * P(Y_5 = 1)$$

Recall that the probability of a Poisson response can be written

$$P(Y = y) = \frac{e^{-\lambda}\lambda^y}{y!}$$

for $y = 0, 1, 2, ...$ So, the likelihood can be written as

4.4 Case Study: Household Size in the Philippines

$$Likelihood = \frac{e^{-\lambda_1}\lambda_1^4}{4!} * \frac{e^{-\lambda_2}\lambda_2^2}{2!} * \frac{e^{-\lambda_3}\lambda_3^8}{8!} * \frac{e^{-\lambda_4}\lambda_4^6}{6!} * \frac{e^{-\lambda_5}\lambda_5^1}{1!}$$

where each λ_i can differ for each household depending on a particular x_i. As in Chapter 2, it will be easier to find a maximum if we take the log of the likelihood and ignore the constant term resulting from the sum of the factorials:

$$\begin{aligned}-logL \propto\ & \lambda_1 - 4log(\lambda_1) + \lambda_2 - 2log(\lambda_2) \\ & + \lambda_3 - 8log(\lambda_3) + \lambda_4 - 6log(\lambda_4) \\ & + \lambda_5 - log(\lambda_5) \end{aligned} \qquad (4.2)$$

Now if we had the age of the head of the household for each house (x_i), we consider the Poisson regression model:

$$log(\lambda_i) = \beta_0 + \beta_1 x_i$$

This implies that λ differs for each age and can be determined using

$$\lambda_i = e^{\beta_0 + \beta_1 x_i}.$$

If the ages are $X = c(32, 21, 55, 44, 28)$ years, our loglikelihood can be written:

$$\begin{aligned}logL \propto\ & [-e^{\beta_0+\beta_1 32} + 4(\beta_0 + \beta_1 32)] + [-e^{\beta_0+\beta_1 21} + 2(\beta_0 + \beta_1 21)] + \\ & [-e^{\beta_0+\beta_1 55} + 8(\beta_0 + \beta_1 55)] + [-e^{\beta_0+\beta_1 44} + 6(\beta_0 + \beta_1 44)] + \\ & [-e^{\beta_0+\beta_1 28} + (\beta_0 + \beta_1 28)] \end{aligned} \qquad (4.3)$$

To see this, match the terms in Equation (4.2) with those in Equation (4.3), noting that λ_i has been replaced with $e^{\beta_0+\beta_1 x_i}$. It is Equation (4.3) that will be used to estimate the coefficients β_0 and β_1. Although this looks a little more complicated than the loglikelihoods we saw in Chapter 2, the fundamental ideas are the same. In theory, we try out different possible values of β_0 and β_1 until we find the two for which the loglikelihood is largest. Most statistical software packages have automated search algorithms to find those values for β_0 and β_1 that maximize the loglikelihood.

4.4.6 Second Order Model

In Section 4.4.4, the Wald-type test and drop-in-deviance test both suggest that a linear term in age is useful. But our exploratory data analysis in Section 4.4.2 suggests that a quadratic model might be more appropriate. A quadratic model would allow us to see if there exists an age where the number in the house is, on average, a maximum. The output for a quadratic model appears below.

```
fHH1 <- fHH1 %>% mutate(age2 = age*age)
modela2 = glm(total ~ age + age2, family = poisson,
              data = fHH1)
```

```
##                 Estimate Std. Error z value  Pr(>|z|)
## (Intercept) -0.3325296  1.788e-01   -1.859  6.297e-02
## age          0.0708868  6.890e-03   10.288  8.007e-25
## age2        -0.0007083  6.406e-05  -11.058  2.009e-28

##  Residual deviance =  2201   on   1497 df
##  Dispersion parameter =  1
```

We can assess the importance of the quadratic term in two ways. First, the p-value for the Wald-type statistic for age^2 is statistically significant (Z = -11.058, p < 0.001). Another approach is to perform a drop-in-deviance test.

```
drop_in_dev <- anova(modela, modela2, test = "Chisq")
```

	ResidDF	ResidDev	Deviance	Df	pval
1	1498	2337	NA	NA	NA
2	1497	2201	136.1	1	1.854e-31

H_0: log(λ)=$\beta_0 + \beta_1$age (reduced model)

H_A : log(λ)=$\beta_0 + \beta_1$age $+ \beta_2$age^2 (larger model)

The first order model has a residual deviance of 2337.1 with 1498 df and the second order model, the quadratic model, has a residual deviance of 2200.9 with 1497 df. The drop-in-deviance by adding the quadratic term to the linear model is 2337.1 - 2200.9 = 136.2 which can be compared to a χ^2 distribution with one degree of freedom. The p-value is essentially 0, so the observed drop of 136.2 again provides significant support for including the quadratic term.

We now have an equation in age which yields the estimated log(mean number in the house).

$$\log(\text{mean numHouse}) = -0.333 + 0.071\text{age} - 0.00071\text{age}^2$$

4.4 Case Study: Household Size in the Philippines

As shown in the following, with calculus we can determine that the maximum estimated additional number in the house is $e^{1.441} = 4.225$ when the head of the household is 50.04 years old.

$$\log(\text{total}) = -0.333 + 0.071\text{age} - 0.00071\text{age}^2$$

$$\frac{d}{d\text{age}}\log(\text{total}) = 0 + 0.071 - 0.0014\text{age} = 0$$

$$\text{age} = 50.04$$

$$\max[\log(\text{total})] = -0.333 + 0.071 \times 50.04 - 0.00071 \times (50.04)^2 = 1.441$$

4.4.7 Adding a Covariate

We should consider other covariates that may be related to household size. By controlling for important covariates, we can obtain more precise estimates of the relationship between age and household size. In addition, we may discover that the relationship between age and household size may differ by levels of a covariate. One important covariate to consider is location. As described earlier in the case study, there are 5 different regions that are associated with the `Location` variable: Central Luzon, Metro Manila, Visayas, Davao Region, and Ilocos Region. Assessing the utility of including the covariate `Location` is, in essence, comparing two nested models; here the quadratic model is compared to the quadratic model plus terms for `Location`. Results from the fitted model appears below; note that Central Luzon is the reference region that all other regions are compared to.

```
modela2L = glm(total ~ age + age2 + location,
          family = poisson, data = fHH1)
```

```
##                      Estimate Std. Error  z value
## (Intercept)         -0.3843338  0.1820919  -2.1107
## age                  0.0703628  0.0069051  10.1900
## age2                -0.0007026  0.0000642 -10.9437
## locationDavaoRegion -0.0193872  0.0537827  -0.3605
## locationIlocosRegion 0.0609820  0.0526598   1.1580
## locationMetroManila  0.0544801  0.0472012   1.1542
## locationVisayas      0.1121092  0.0417496   2.6853
##                      Pr(>|z|)
## (Intercept)          3.480e-02
## age                  2.197e-24
## age2                 7.126e-28
## locationDavaoRegion  7.185e-01
## locationIlocosRegion 2.468e-01
```

```
## locationMetroManila   2.484e-01
## locationVisayas       7.247e-03

##  Residual deviance =  2188  on  1493 df
##  Dispersion parameter =  1
```

Our Poisson regression model now looks like:

$$\log(\text{total}) = -0.384 + 0.070 \cdot \text{age} - 0.00070 \cdot \text{age}^2 + 0.061 \cdot \text{IlocosRegion} +$$
$$0.054 \cdot \text{MetroManila} + 0.112 \cdot \text{Visayas} - 0.019 \cdot \text{DavaoRegion}$$

Notice that because there are 5 different locations, we must represent the effects of different locations through 4 indicator variables. For example, $\hat{\beta}_6 = -0.0194$ indicates that, after controlling for the age of the head of household, the log mean household size is 0.0194 lower for households in the Davao Region than for households in the reference location of Central Luzon. In more interpretable terms, mean household size is $e^{-0.0194} = 0.98$ times "higher" (i.e., 2% lower) in the Davao Region than in Central Luzon, when holding age constant.

```
drop_in_dev <- anova(modela2, modela2L, test = "Chisq")
```

	ResidDF	ResidDev	Deviance	Df	pval
1	1497	2201	NA	NA	NA
2	1493	2188	13.14	4	0.01059

To test if the mean household size significantly differs by location, we must use a drop-in-deviance test, rather than a Wald-type test, because four terms (instead of just one) are added when including the `location` variable. From the Analysis of Deviance table above, adding the four terms corresponding to location to the quadratic model with age produces a statistically significant improvement ($\chi^2 = 13.144, df = 4, p = 0.0106$), so there is significant evidence that mean household size differs by location, after controlling for age of the head of household. Further modeling (not shown) shows that after controlling for location and age of the head of household, mean household size did not differ between the two types of roofing material.

4.4.8 Residuals for Poisson Models (optional)

Residual plots may provide some insight into Poisson regression models, especially linearity and outliers, although the plots are not quite as useful here

4.4 Case Study: Household Size in the Philippines

as they are for linear least squares regression. There are a few options for computing residuals and predicted values. Residuals may have the form of residuals for LLSR models or the form of deviance residuals which, when squared, sum to the total deviance for the model. Predicted values can be estimates of the counts, $e^{\beta_0 + \beta_1 X}$, or log counts, $\beta_0 + \beta_1 X$. We will typically use the deviance residuals and predicted counts.

The residuals for linear least squares regression have the form:

$$\text{LLSR residual}_i = \text{obs}_i - \text{fit}_i$$
$$= Y_i - \hat{\mu}_i$$
$$= Y_i - (\hat{\beta}_0 + \hat{\beta}_1 X_i) \quad (4.4)$$

Residual sum of squares (RSS) are formed by squaring and adding these residuals, and we generally seek to minimize RSS in model building. We have several options for creating residuals for Poisson regression models. One is to create residuals in much the same way as we do in LLSR. For Poisson residuals, the predicted values are denoted by $\hat{\lambda}_i$ (in place of $\hat{\mu}_i$ in Equation (4.4)); they are then standardized by dividing by the standard error, $\sqrt{\hat{\lambda}_i}$. These kinds of residuals are referred to as **Pearson residuals**.

$$\text{Pearson residual}_i = \frac{Y_i - \hat{\lambda}_i}{\sqrt{\hat{\lambda}_i}}$$

Pearson residuals have the advantage that you are probably familiar with their meaning and the kinds of values you would expect. For example, after standardizing we expect most Pearson residuals to fall between -2 and 2. However, **deviance residuals** have some useful properties that make them a better choice for Poisson regression.

First, we define a **deviance residual** for an observation from a Poisson regression:

$$\text{deviance residual}_i = \text{sign}(Y_i - \hat{\lambda}_i)\sqrt{2\left[Y_i log\left(\frac{Y_i}{\hat{\lambda}_i}\right) - (Y_i - \hat{\lambda}_i)\right]}$$

where $\text{sign}(x)$ is defined such that:

$$\text{sign}(x) = \begin{cases} 1 & \text{if } x > 0 \\ -1 & \text{if } x < 0 \\ 0 & \text{if } x = 0 \end{cases}$$

As its name implies, a deviance residual describes how the observed data deviates from the fitted model. Squaring and summing the deviances for all observations produces the **residual deviance** $= \sum(\text{deviance residual})_i^2$.

Relatively speaking, observations for good fitting models will have small deviances; that is, the predicted values will deviate little from the observed. However, you can see that the deviance for an observation does not easily translate to a difference in observed and predicted responses as is the case with LLSR models.

A careful inspection of the deviance formula reveals several places where the deviance compares Y to $\hat{\lambda}$: the sign of the deviance is based on the difference between Y and $\hat{\lambda}$, and under the radical sign we see the ratio $Y/\hat{\lambda}$ and the difference $Y - \hat{\lambda}$. When $Y = \hat{\lambda}$, that is, when the model fits perfectly, the difference will be 0 and the ratio will be 1 (so that its log will be 0). So like the residuals in LLSR, an observation that fits perfectly will not contribute to the sum of the squared deviances. This definition of a deviance depends on the likelihood for Poisson models. Other models will have different forms for the deviance depending on their likelihood.

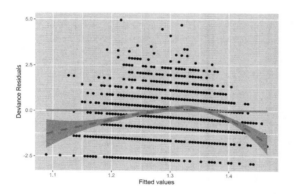

FIGURE 4.7: Residual plot for the Poisson model of household size by age of the household head.

A plot (Figure 4.7) of the deviance residuals versus predicted responses for the first order model exhibits curvature, supporting the idea that the model may improved by adding a quadratic term. Other details related to residual plots can be found in a variety of sources including McCullagh and Nelder [1989].

4.4.9 Goodness-of-Fit

The model residual deviance can be used to assess the degree to which the predicted values differ from the observed. When a model is true, we can expect the residual deviance to be distributed as a χ^2 random variable with degrees of freedom equal to the model's residual degrees of freedom. Our model thus far, the quadratic terms for age plus the indicators for location, has a residual deviance of 2187.8 with 1493 df. The probability of observing a deviance this large if the model fits is esentially 0, saying that there is significant evidence of lack-of-fit.

```
1-pchisq(modela2$deviance, modela2$df.residual)   # GOF test
```

[1] 0

There are several reasons why **lack-of-fit** may be observed. (1) We may be missing important covariates or interactions; a more comprehensive data set may be needed. (2) There may be extreme observations that may cause the deviance to be larger than expected; however, our residual plots did not reveal any unusual points. (3) Lastly, there may be a problem with the Poisson model. In particular, the Poisson model has only a single parameter, λ, for each combination of the levels of the predictors which must describe both the mean and the variance. This limitation can become manifest when the variance appears to be larger than the corresponding means. In that case, the response is more variable than the Poisson model would imply, and the response is considered to be **overdispersed**.

4.5 Linear Least Squares vs. Poisson Regression

Response
LLSR : Normal
PoissonRegression : Counts

Variance
LLSR : Equal for each level of X
PoissonRegression : Equal to the mean for each level of X

Model Fitting
LLSR : $\mu = \beta_0 + \beta_1 x$ using Least Squares
PoissonRegression : $log(\lambda) = \beta_0 + \beta_1 x$ using Maximum Likelihood

EDA

LLSR : Plot X vs. Y; add line

PoissonRegression : Find $log(\bar{y})$ for several subgroups; plot vs. X

Comparing Models

LLSR : Extra sum of squares F-tests; AIC/BIC

PoissonRegression : Drop in Deviance tests; AIC/BIC

Interpreting Coefficients

LLSR : $\beta_1 = $ change in μ_y for unit change in X

PoissonRegression : $e^{\beta_1} = $ percent change in λ for unit change in X

4.6 Case Study: Campus Crime

Students want to feel safe and secure when attending a college or university. In response to legislation, the US Department of Education seeks to provide data and reassurances to students and parents alike. All postsecondary institutions that participate in federal student aid programs are required by the Jeanne Clery Disclosure of Campus Security Policy and Campus Crime Statistics Act and the Higher Education Opportunity Act to collect and report data on crime occurring on campus to the Department of Education. In turn, this data is publicly available on the website of the Office of Postsecondary Education. We are interested in looking at whether there are regional differences in violent crime on campus, controlling for differences in the type of school.

4.6.1 Data Organization

Each row of `c_data.csv` contains crime information from a post secondary institution, either a college or university. The variables include:

- `Enrollment` = enrollment at the school
- `type` = college (C) or university (U)
- `nv` = the number of violent crimes for that institution for the given year
- `nvrate` = number of violent crimes per 1000 students
- `enroll1000` = enrollment at the school, in thousands
- `region` = region of the country (C = Central, MW = Midwest, NE = Northeast, SE = Southeast, SW = Southwest, and W = West)

4.6 Case Study: Campus Crime

```
# A tibble: 10 x 6
   Enrollment type     nv nvrate enroll1000 region
        <dbl> <chr> <dbl>  <dbl>      <dbl> <chr>
 1       5590 U        30 5.37         5.59 SE
 2        540 C         0 0            0.54 SE
 3      35747 U        23 0.643       35.7  W
 4      28176 C         1 0.0355      28.2  W
 5      10568 U         1 0.0946      10.6  SW
 6       3127 U         0 0            3.13 SW
 7      20675 U         7 0.339       20.7  W
 8      12548 C         0 0           12.5  W
 9      30063 U        19 0.632       30.1  C
10       4429 C         4 0.903        4.43 C
```

4.6.2 Exploratory Data Analysis

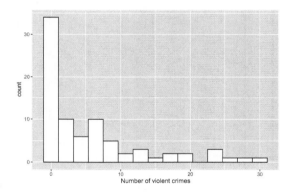

FIGURE 4.8: Histogram of number of violent crimes by institution.

A graph of the number of violent crimes, Figure 4.8, reveals the pattern often found with distributions of counts of rare events. Many schools reported no violent crimes or very few crimes. A few schools have a large number of crimes making for a distribution that appears to be far from normal. Therefore, Poisson regression should be used to model our data; Poisson random variables are often used to represent counts (e.g., number of violent crimes) per unit of time or space (e.g., one year).

Let's take a look at two covariates of interest for these schools: type of institution and region. In our data, the majority of institutions are universities (65% of the 81 schools) and only 35% are colleges. Interest centers on whether the different regions tend to have different crime rates. Table 4.3 contains the name of each region and each column represents the percentage of schools in

TABLE 4.3: Proportion of colleges and universities within region in the campus crime data set.

	C	MW	NE	SE	SW	W
C	0.294	0.3	0.381	0.4	0.2	0.5
U	0.706	0.7	0.619	0.6	0.8	0.5

TABLE 4.4: The mean and variance of the violent crime rate by region and type of institution.

region	type	MeanCount	VarCount	MeanRate	VarRate	n
C	C	1.6000	3.3000	0.3980	0.2781	5
C	U	4.7500	30.9318	0.2219	0.0349	12
MW	C	0.3333	0.3333	0.0163	0.0008	3
MW	U	8.7143	30.9048	0.4019	0.0621	7
NE	C	6.0000	32.8571	1.1250	1.1821	8
NE	U	5.9231	79.2436	0.4359	0.3850	13
S	C	1.1250	5.8393	0.1866	0.1047	8
S	U	8.6250	68.2500	0.5713	0.2778	16
W	C	0.5000	0.3333	0.0680	0.0129	4
W	U	12.5000	57.0000	0.4679	0.0247	4

that region which are colleges or universities. The proportion of colleges varies from a low of 20% in the Southwest (SW) to a high of 50% in the West (W).

While a Poisson regression model is a good first choice because the responses are counts per year, it is important to note that the counts are not directly comparable because they come from different size schools. This issue sometimes is referred to as the need to account for *sampling effort*; in other words, we expect schools with more students to have more reports of violent crime since there are more students who could be affected. We cannot directly compare the 30 violent crimes from the first school in the data set to no violent crimes for the second school when their enrollments are vastly different: 5,590 for school 1 versus 540 for school 2. We can take the differences in enrollments into account by including an **offset** in our model, which we will discuss in the next section. For the remainder of the EDA, we examine the violent crime counts in terms of the rate per 1,000 enrolled ($\frac{\text{number of violent crimes}}{\text{number enrolled}} \cdot 1000$).

Note that there is a noticeable outlier for a Southeastern school (5.4 violent crimes per 1000 students), and there is an observed rate of 0 for the Southwestern colleges which can lead to some computational issues. We therefore

4.6 Case Study: Campus Crime

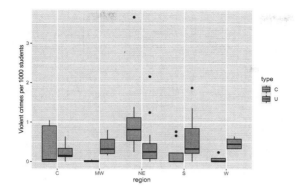

FIGURE 4.9: Boxplot of violent crime rate by region and type of institution (colleges (C) on the left, and universities (U) on the right).

combined the SW and SE to form a single category of the South, and we also removed the extreme observation from the data set.

Table 4.4 and Figure 4.9 display mean violent crime rates that are generally lower at the colleges within a region (with the exception of the Northeast). In addition, the regional pattern of rates at universities appears to differ from that of the colleges.

4.6.3 Accounting for Enrollment

Although working with the observed rates (per 1000 students) is useful during the exploratory data analysis, we do not use these rates explicitly in the model. The counts (per year) are the Poisson responses when modeling, so we must take into account the enrollment in a different way. Our approach is to include a term on the right side of the model called an **offset**, which is the log of the enrollment, in thousands. There is an intuitive heuristic for the form of the offset. If we think of λ as the mean number of violent crimes per year, then $\lambda/\text{enroll1000}$ represents the number per 1000 students, so that the yearly count is adjusted to be comparable across schools of different sizes. Adjusting the yearly count by enrollment is equivalent to adding $log(\text{enroll1000})$ to the right-hand side of the Poisson regression equation—essentially adding a predictor with a fixed coefficient of 1:

$$log(\frac{\lambda}{\text{enroll1000}}) = \beta_0 + \beta_1(\text{type})$$
$$log(\lambda) - log(\text{enroll1000}) = \beta_0 + \beta_1(\text{type})$$
$$log(\lambda) = \beta_0 + \beta_1(\text{type}) + log(\text{enroll1000})$$

While this heuristic is helpful, it is important to note that it is *not* $\frac{\lambda}{\text{enroll1000}}$ that we are modeling. We are still modeling $log(\lambda)$, but we're adding an offset to adjust for differing enrollments, where the offset has the unusual feature that the coefficient is fixed at 1.0. As a result, no estimated coefficient for `enroll1000` or $log(\text{enroll1000})$ will appear in the output. As this heuristic illustrates, modeling $log(\lambda)$ and adding an offset is equivalent to modeling rates, and coefficients can be interpreted that way.

4.7 Modeling Assumptions

In Table 4.4, we see that the variances are greatly higher than the mean counts in almost every group. Thus, we have reason to question the Poisson regression assumption of variability equal to the mean; we will have to return to this issue after some initial modeling. The fact that the variance of the rate of violent crimes per 1000 students tends to be on the same scale as the mean tells us that adjusting for enrollment may provide some help, although that may not completely solve our issues with excessive variance.

As far as other model assumptions, linearity with respect to $log(\lambda)$ is difficult to discern without continuous predictors, and it is not possible to assess independence without knowing how the schools were selected.

4.8 Initial Models

We are interested primarily in differences in violent crime between institutional types controlling for difference in regions, so we fit a model with region, institutional type, and our offset. Note that the central region is the reference level in our model.

```
modeltr <- glm(nv ~ type + region, family = poisson,
               offset = log(enroll1000), data = c.data)
```

```
##                Estimate Std. Error z value  Pr(>|z|)
## (Intercept)    -1.54780    0.1711  -9.0439  1.512e-19
## typeU           0.27956    0.1331   2.0997  3.576e-02
## regionMW        0.09912    0.1775   0.5583  5.766e-01
## regionNE        0.77813    0.1531   5.0836  3.703e-07
```

4.8 Initial Models

```
## regionS        0.58238        0.1490   3.9098 9.238e-05
## regionW        0.26275        0.1875   1.4011 1.612e-01

## Residual deviance = 348.7   on   74 df
## Dispersion parameter =   1
```

From our model, the Northeast and the South differ significantly from the Central region (p= 0.00000037 and p=0.0000924, respectively). The estimated coefficient of 0.778 means that the violent crime rate per 1,000 in the Northeast is nearly 2.2 ($e^{0.778}$) times that of the Central region controlling for the type of school. A Wald-type confidence interval for this factor can be constructed by first calculating a CI for the coefficient (0.778 $\pm$ 1.96 $\cdot$ 0.153) and then exponentiating (1.61 to 2.94).

4.8.1 Tukey's Honestly Significant Differences

Comparisons to regions other than the Central region can be accomplished by changing the reference region. If many comparisons are made, it would be best to adjust for multiple comparisons using a method such as **Tukey's Honestly Significant Differences**, which considers all pairwise comparisons among regions. This method helps control the large number of false positives that we would see if we ran multiple t-tests comparing groups. The honestly significant difference compares a standardized mean difference between two groups to a critical value from a studentized range distribution.

```
mult_comp <- summary(glht(modeltr, mcp(region="Tukey")))
```

```
## # A tibble: 10 x 5
##    comparison  estimate    SE z_value   p_value
##    <chr>          <dbl> <dbl>   <dbl>     <dbl>
## 1  MW - C        0.0991 0.178   0.558 0.980
## 2  NE - C        0.778  0.153   5.08  0.00000349
## 3  S  - C        0.582  0.149   3.91  0.000828
## 4  W  - C        0.263  0.188   1.40  0.621
## 5  NE - MW       0.679  0.155   4.37  0.000109
## 6  S  - MW       0.483  0.151   3.19  0.0121
## 7  W  - MW       0.164  0.189   0.864 0.908
## 8  S  - NE      -0.196  0.122  -1.61  0.486
## 9  W  - NE      -0.515  0.166  -3.11  0.0157
## 10 W  - S       -0.320  0.163  -1.96  0.280
```

In our case, Tukey's Honestly Significant Differences simultaneously evaluates all 10 mean differences between pairs of regions. We find that the Northeast has significantly higher rates of violent crimes than the Central, Midwest, and

Western regions, while the South has significantly higher rates of violent crimes than the Central and the Midwest, controlling for the type of institution. In the primary model, the University indicator is significant and, after exponentiating the coefficient, can be interpreted as an approximately ($e^{0.280}$) 32% increase in violent crime rate over colleges after controlling for region.

These results certainly suggest significant differences in regions and type of institution. However, the EDA findings suggest the effect of the type of institution may vary depending upon the region, so we consider a model with an interaction between region and type.

```
modeli <- glm(nv ~ type + region + region:type,
              family = poisson,
              offset = log(enroll1000), data = c.data)
```

```
##                  Estimate Std. Error z value  Pr(>|z|)
## (Intercept)       -1.4741     0.3536 -4.1694 3.054e-05
## typeU              0.1959     0.3775  0.5190 6.038e-01
## regionMW          -1.9765     1.0607 -1.8635 6.239e-02
## regionNE           1.5529     0.3819  4.0664 4.775e-05
## regionS           -0.1562     0.4859 -0.3216 7.478e-01
## regionW           -1.8337     0.7906 -2.3194 2.037e-02
## typeU:regionMW     2.1965     1.0765  2.0403 4.132e-02
## typeU:regionNE    -1.0698     0.4200 -2.5473 1.086e-02
## typeU:regionS      0.8121     0.5108  1.5899 1.118e-01
## typeU:regionW      2.4106     0.8140  2.9616 3.061e-03

##   Residual deviance =   276.7   on    70 df
##   Dispersion parameter =   1
```

These results provide convincing evidence of an interaction between the effect of region and the type of institution. A drop-in-deviance test like the one we carried out in the previous case study confirms the significance of the contribution of the interaction to this model. We have statistically significant evidence ($\chi^2 = 71.98, df = 4, p < .001$) that the difference between colleges and universities in violent crime rate differs by region. For example, our model estimates that violent crime rates are 13.6 ($e^{.196+2.411}$) times higher in universities in the West compared to colleges, while in the Northeast we estimate that violent crime rates are 2.4 ($\frac{1}{e^{.196-1.070}}$) times higher in colleges.

```
drop_in_dev <- anova(modeltr, modeli, test = "Chisq")
```

4.9 Overdispersion

```
  ResidDF ResidDev Deviance Df        pval
1      74    348.7       NA NA          NA
2      70    276.7    71.98  4   8.664e-15
```

The residual deviance (276.70 with 70 df) suggests significant lack-of-fit in the interaction model (p < .001). One possibility is that there are other important covariates that could be used to describe the differences in the violent crime rates. Without additional covariates to consider, we look for extreme observations, but we have already eliminated the most extreme of the observations.

In the absence of other covariates or extreme observations, we consider overdispersion as a possible explanation of the significant lack-of-fit.

4.9 Overdispersion

4.9.1 Dispersion Parameter Adjustment

Overdispersion suggests that there is more variation in the response than the model implies. Under a Poisson model, we would expect the means and variances of the response to be about the same in various groups. Without adjusting for overdispersion, we use incorrect, artificially small standard errors leading to artificially small p-values for model coefficients. We may also end up with artificially complex models.

We can take overdispersion into account in several different ways. The simplest is to use an estimated dispersion factor to inflate standard errors. Another way is to use a negative-binomial regression model. We begin with using an estimate of the dispersion parameter.

We can estimate a dispersion parameter, ϕ, by dividing the model deviance by its corresponding degrees of freedom; i.e., $\hat{\phi} = \frac{\sum (\text{Pearson residuals})^2}{n-p}$ where p is the number of model parameters. It follows from what we know about the χ^2 distribution that if there is no overdispersion, this estimate should be close to one. It will be larger than one in the presence of overdispersion. We inflate the standard errors by multiplying the variance by ϕ, so that the standard errors are larger than the likelihood approach would imply; i.e., $SE_Q(\hat{\beta}) = \sqrt{\hat{\phi}} * SE(\hat{\beta})$, where Q stands for "quasi-Poisson" since multiplying variances by ϕ is an ad-hoc solution. Our process for model building and comparison is called **quasilikelihood**—similar to likelihood but without exact underlying distributions. If we choose to use a dispersion parameter with our model, we refer to the approach as quasilikehood. The following output illustrates a quasi-Poisson approach to the interaction model:

```
modeliq <- glm(nv ~ type + region + region:type,
               family = quasipoisson,
               offset = log(enroll1000), data = c.data)
```

```
##                    Estimate Std. Error t value Pr(>|t|)
## (Intercept)         -1.4741     0.7455 -1.9773  0.05195
## typeU                0.1959     0.7961  0.2461  0.80631
## regionMW            -1.9765     2.2366 -0.8837  0.37987
## regionNE             1.5529     0.8053  1.9284  0.05786
## regionS             -0.1562     1.0246 -0.1525  0.87924
## regionW             -1.8337     1.6671 -1.0999  0.27513
## typeU:regionMW       2.1965     2.2701  0.9676  0.33659
## typeU:regionNE      -1.0698     0.8856 -1.2080  0.23111
## typeU:regionS        0.8121     1.0771  0.7540  0.45338
## typeU:regionW        2.4106     1.7164  1.4045  0.16460

##   Residual deviance =    276.7  on   70 df
##   Dispersion parameter =    4.447
```

In the absence of overdispersion, we expect the dispersion parameter estimate to be 1.0. The estimated dispersion parameter here is much larger than 1.0 (4.447) indicating overdispersion (extra variance) that should be accounted for. The larger estimated standard errors in the quasi-Poisson model reflect the adjustment. For example, the standard error for the West region term from a likelihood based approach is 0.7906, whereas the quasilikelihood standard error is $\sqrt{4.47}*0.7906$ or 1.6671. This term is no longer significant under the quasi-Poisson model. In fact, after adjusting for overdispersion (extra variation), none of the model coefficients in the quasi-Poisson model are significant at the .05 level! This is because standard errors were all increased by a factor of 2.1 ($\sqrt{\hat{\phi}} = \sqrt{4.447} = 2.1$), while estimated coefficients remain unchanged.

Note that tests for individual parameters are now based on the t-distribution rather than a standard normal distribution, with test statistic $t = \frac{\hat{\beta}}{SE_Q(\hat{\beta})}$ following an (approximate) t-distribution with $n - p$ degrees of freedom if the null hypothesis is true ($H_O : \beta = 0$). Drop-in-deviance tests can be similarly adjusted for overdispersion in the quasi-Poisson model. In this case, you can divide the test statistic (per degree of freedom) by the estimated dispersion parameter and compare the result to an F-distribution with the difference in the model degrees of freedom for the numerator and the degrees of freedom for the larger model in the denominator. That is, $F = \frac{\text{drop in deviance}}{\text{difference in df}} / \hat{\phi}$ follows an (approximate) F-distribution when the null hypothesis is true (H_0: reduced

4.9 Overdispersion

TABLE 4.5: Comparison of Poisson and quasi-Poisson inference.

	Poisson	quasi-Poisson
Estimate	$\hat{\beta}$	$\hat{\beta}$
Std error	$SE(\hat{\beta})$	$SE_Q(\hat{\beta}) = \sqrt{\hat{\phi}} SE(\hat{\beta})$
Wald-type test stat	$Z = \hat{\beta}/SE(\hat{\beta})$	$t = \hat{\beta}/SE_Q(\hat{\beta})$
Confidence interval	$\hat{\beta} \pm z' SE(\hat{\beta})$	$\hat{\beta} \pm t' SE_Q(\hat{\beta})$
Drop in deviance test	χ^2 = resid dev(reduced) - resid dev(full)	$F = (\chi^2/\text{difference in df})/\hat{\phi}$

model sufficient). The output below tests for an interaction between region and type of institution after adjusting for overdispersion (extra variance):

```
modeltrq <- glm(nv ~ type + region, family = quasipoisson,
                offset = log(enroll1000), data = c.data)
drop_in_dev <- anova(modeltrq, modeliq, test = "F")
```

```
  ResidDF ResidDev     F  Df      pval
1      74    348.7    NA  NA        NA
2      70    276.7 4.047   4  0.005213
```

Here, even after adjusting for overdispersion, we still have statistically significant evidence ($F = 4.05, p = .0052$) that the difference between colleges and universities in violent crime rate differs by region.

4.9.2 No Dispersion vs. Overdispersion

Table 4.5 summarizes the comparison between Poisson inference (tests and confidence intervals assuming no overdispersion) and quasi-Poisson inference (tests and confidence intervals after accounting for overdispersion).

4.9.3 Negative Binomial Modeling

Another approach to dealing with overdispersion is to model the response using a negative binomial instead of a Poisson distribution. An advantage of this approach is that it introduces another parameter in addition to λ, which gives the model more flexibility and, as opposed to the quasi-Poisson model, the negative binomial model assumes an explicit likelihood model. You may recall that negative binomial random variables take on non-negative integer values, which is consistent with modeling counts. This model posits selecting a λ for each institution and then generating a count using a Poisson random variable with the selected λ. With this approach, the counts will be more dispersed than would be expected for observations based on a single Poisson variable with rate λ. (See Guided Exercises on the Gamma-Poisson mixture in Chapter 3.)

Mathematically, you can think of the negative binomial model as a Poisson model where λ is also random, following a gamma distribution. Specifically, if $Y|\lambda \sim$ Poisson(λ) and $\lambda \sim$ gamma($r, \frac{1-p}{p}$), then $Y \sim$ NegBinom(r, p) where $E(Y) = \frac{pr}{1-p} = \mu$ and $Var(Y) = \frac{pr}{(1-p)^2} = \mu + \frac{\mu^2}{r}$. The overdispersion in this case is given by $\frac{\mu^2}{r}$, which approaches 0 as r increases (so smaller values of r indicate greater overdispersion).

Here is what happens if we apply a negative binomial regression model to the interaction model, which we've already established suffers from overdispersion issues under regular Poisson regression:

```
# Account for overdispersion with negative binomial model
modelinb <- glm.nb(nv ~ type + region + region:type,
            offset(log(enroll1000)), data = c.data2)
```

```
##                  Estimate Std. Error z value Pr(>|z|)
## (Intercept)       0.4904     0.4281   1.1455 0.252008
## typeU             1.2174     0.4608   2.6422 0.008237
## regionMW         -1.0953     0.8075  -1.3563 0.175004
## regionNE          1.3966     0.5053   2.7641 0.005709
## regionS           0.1461     0.5559   0.2627 0.792752
## regionW          -1.1858     0.6870  -1.7260 0.084347
## typeU:regionMW    1.6342     0.8498   1.9231 0.054469
## typeU:regionNE   -1.1259     0.5601  -2.0102 0.044411
## typeU:regionS     0.4513     0.5995   0.7527 0.451638
## typeU:regionW     2.0387     0.7527   2.7086 0.006758

##   Residual deviance =  199.6  on  69 df
##   Dispersion parameter (theta) =   1.313
```

These results differ from the quasi-Poisson model. Several effects are now statistically significant at the .05 level: the effect of type of institution for the Central region ($Z = 2.64, p = .008$), the difference between Northeast and Central regions for colleges ($Z = 2.76, p = .006$), the difference between Northeast and Central regions in type effect ($Z = -2.01, p = .044$), and the difference between West and Central regions in type effect ($Z = 2.71, p = .007$). In this case, compared to the quasi-Poisson model, negative binomial coefficient estimates are generally in the same direction and similar in size, but negative binomial standard errors are somewhat smaller.

In summary, we explored the possibility of differences in the violent crime rate between colleges and universities, controlling for region. Our initial efforts seemed to suggest that there are indeed differences between colleges and universities, and the pattern of those differences depends upon the region. However, this model exhibited significant lack-of-fit which remained after the removal of an extreme observation. In the absence of additional covariates, we accounted for the lack-of-fit by using a quasilikelihood approach and a negative binomial regression, which provided slightly different conclusions. We may want to look for additional covariates and/or more data.

4.10 Case Study: Weekend Drinking

Sometimes when analyzing Poisson data, you may see many more zeros in your data set than you would expect for a Poisson random variable. For example, an informal survey of students in an introductory statistics course included the question, "How many alcoholic drinks did you consume last weekend?". This survey was conducted on a dry campus where no alcohol is officially allowed, even among students of drinking age, so we expect that some portion of the respondents never drink. The non-drinkers would thus always report zero drinks. However, there will also be students who are drinkers reporting zero drinks because they just did not happen to drink during the past weekend. Our zeros, then, are a **mixture** of responses from non-drinkers and drinkers who abstained during the past weekend. Ideally, we'd like to sort out the non-drinkers and drinkers when performing our analysis.

4.10.1 Research Question

The purpose of this survey is to explore factors related to drinking behavior on a dry campus. What proportion of students on this dry campus never drink? What factors, such as off-campus living and sex, are related to whether students drink? Among those who do drink, to what extent is moving off campus associated with the number of drinks in a weekend? It is commonly assumed that males' alcohol consumption is greater than females'; is this true on this campus? Answering these questions would be a simple matter if we knew who was and was not a drinker in our sample. Unfortunately, the non-drinkers did not identify themselves as such, so we will need to use the data available with a model that allows us to estimate the proportion of drinkers and non-drinkers.

4.10.2 Data Organization

Each line of weekendDrinks.csv contains data provided by a student in an introductory statistics course. In this analysis, the response of interest is the respondent's report of the number of alcoholic drinks they consumed the previous weekend, whether the student lives off.campus, and sex. We will also consider whether a student is likely a firstYear student based on the dorm they live in. Here is a sample of observations from this data set:

```
head(zip.data[2:5])
```

	drinks	sex	off.campus	firstYear
1	0	f	0	TRUE
2	5	f	0	FALSE
3	10	m	0	FALSE
4	0	f	0	FALSE
5	0	m	0	FALSE
6	3	f	0	FALSE

4.10.3 Exploratory Data Analysis

As always we take stock of the amount of data; here there are 77 observations. Large sample sizes are preferred for the type of model we will consider, and n=77 is on the small side. We proceed with that in mind.

A premise of this analysis is that we believe that those responding zero drinks are coming from a mixture of non-drinkers and drinkers who abstained the weekend of the survey.

- **Non-drinkers**: respondents who never drink and would always reply with zero.
- **Drinkers**: obviously this includes those responding with one or more drinks, but it also includes people who are drinkers but did not happen to imbibe the past weekend. These people reply zero but are not considered non-drinkers.

Beginning the EDA with the response, number of drinks, we find that over 46% of the students reported no drinks during the past weekend. Figure 4.10a portrays the observed number of drinks reported by the students. The mean number of drinks reported the past weekend is 2.013. Our sample consists of 74% females and 26% males, only 9% of whom live off campus.

Because our response is a count, it is natural to consider a Poisson regression model. You may recall that a Poisson distribution has only one parameter, λ,

4.10 Case Study: Weekend Drinking

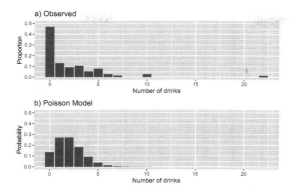

FIGURE 4.10: Observed (a) versus modeled (b) number of drinks.

for its mean and variance. Here we will include an additional parameter, α. We define α to be the true proportion of *non-drinkers* in the population.

The next step in the EDA is especially helpful if you suspect your data contains excess zeros. Figure 4.10b is what we might expect to see under a Poisson model. Bars represent the probabilities for a Poisson distribution (using the Poisson probability formula) with λ equal to the mean observed number of drinks, 2.013 drinks per weekend. Comparing this Poisson distribution to what we observed (Figure 4.10a), it is clear that many more zeros have been reported by the students than you would expect to see if the survey observations were coming from a Poisson distribution. This doesn't surprise us because we had expected a subset of the survey respondents to be non-drinkers; i.e., they would not be included in this Poisson process. This circumstance actually arises in many Poisson regression settings. We will define λ to be the mean number of drinks *among those who drink*, and α to be the proportion of *non-drinkers* ("true zeros"). Then, we will attempt to model λ and α (or functions of λ and α) simultaneously using covariates like sex, first-year status, and off-campus residence. This type of model is referred to as a **zero-inflated Poisson model** or **ZIP model**.

4.10.4 Modeling

We first fit a simple Poisson model with the covariates off.campus and sex.

```
pois.m1 <- glm(drinks ~ off.campus + sex, family = poisson,
               data = zip.data)
```

```
##              Estimate Std. Error z value Pr(>|z|)
```

```
## (Intercept)      0.1293      0.1241     1.041 2.976e-01
## off.campus       0.8976      0.2008     4.470 7.830e-06
## sexm             1.1154      0.1611     6.925 4.361e-12

##   Residual deviance =   230.5  on   74 df
##   Dispersion parameter =  1
```

```
# Exponentiated coefficients
exp(coef(pois.m1))
```

```
## (Intercept)   off.campus        sexm
##       1.138        2.454       3.051
```

```
# Goodness-of-fit test
gof.pvalue = 1 - pchisq(pois.m1$deviance, pois.m1$df.residual)
gof.pvalue
```

```
## [1] 0
```

Both covariates are statistically significant, but a goodness-of-fit test reveals that there remains significant lack-of-fit (residual deviance: 230.54 with only 74 df; p<.001 based on χ^2 test with 74 df). In the absence of important missing covariates or extreme observations, this lack-of-fit may be explained by the presence of a group of non-drinkers.

A zero-inflated Poisson regression model to take non-drinkers into account consists of two parts:

- One part models the association, among drinkers, between number of drinks and the predictors of sex and off-campus residence.
- The other part uses a predictor for first-year status to obtain an estimate of the proportion of non-drinkers based on the reported zeros.

The form for each part of the model follows. The first part looks like an ordinary Poisson regression model:

$$log(\lambda) = \beta_0 + \beta_1 \text{off.campus} + \beta_2 \text{sex}$$

where λ is the mean number of drinks in a weekend *among those who drink*. The second part has the form

4.10 Case Study: Weekend Drinking

$$logit(\alpha) = \beta_0 + \beta_1 \text{firstYear}$$

where α is the probability of being in the non-drinkers group and $logit(\alpha) = log(\alpha/(1-\alpha))$. We'll provide more detail on the logit in Chapter 6. There are many ways in which to structure this model; here we use different predictors in the two pieces, athough it would have been perfectly fine to use the same predictors for both pieces, or even no predictors for one of the pieces.

4.10.5 Fitting a ZIP Model

How is it possible to fit such a model? We cannot observe whether a respondent is a drinker or not (which probably would've been good to ask). The ZIP model is a special case of a more general type of statistical model referred to as a **latent variable model**. More specifically, it is a type of a **mixture model** where observations for one or more groups occur together and the group membership is unknown. Zero-inflated models are a particularly common example of a mixture model, but the response does not need to follow a Poisson distribution. Likelihood methods are at the core of this methodology, but fitting is an iterative process where it is necessary to start out with some guesses (or starting values). In general, it is important to know that models like ZIP exist, although we'll only explore interpretations and fitting options for a single case study here.

Here is the general idea of how ZIP models are fit. Imagine that the graph of the Poisson distribution in Figure 4.10b is removed from the observed data distribution in Figure 4.10a. Some zero responses will remain. These would correspond to non-drinkers, and the proportion of all observations these zeros constitute might make a reasonable estimate for α, the proportion of non-drinkers. The likelihood is used and some iterating in the fitting process is involved because the Poisson distribution in Figure 4.10b is based on the mean of the observed data, which means it is the average among all students, not only among drinkers. Furthermore, the likelihood incorporates the predictors, `sex` and `off.campus`. So there is a little more to it than computing the proportion of zeros, but this heuristic should provide you a general idea of how these kinds of models are fit. We will use the R function `zeroinfl` from the package `pscl` to fit a ZIP model.

```
zip.m2 <- zeroinfl(drinks ~ off.campus + sex | firstYear,
                   data = zip.data)
```

```
## $count
##              Estimate Std. Error z value Pr(>|z|)
## (Intercept)   0.7543     0.1440   5.238  1.624e-07
```

```
## off.campus       0.4159      0.2059   2.020 4.333e-02
## sexm             1.0209      0.1752   5.827 5.634e-09
##
## $zero
##                 Estimate Std. Error z value Pr(>|z|)
## (Intercept)      -0.6036     0.3114  -1.938  0.05261
## firstYearTRUE     1.1364     0.6095   1.864  0.06226

##  Log likelihood = -140.8
```

```
exp(coef(zip.m2))    # exponentiated coefficients
```

```
##   count_(Intercept)      count_off.campus
##              2.1261                1.5158
##         count_sexm        zero_(Intercept)
##              2.7757                0.5468
## zero_firstYearTRUE
##              3.1155
```

Our model uses `firstYear` to distinguish drinkers and non-drinkers ("Zero-inflation model coefficients") and `off.campus` and `sex` to help explain the differences in the number of drinks among drinkers ("Count model coefficients"). Again, we could have used the same covariates for the two pieces of a ZIP model, but neither `off.campus` nor `sex` proved to be a useful predictor of drinkers vs. non-drinkers after we accounted for first-year status.

We'll first consider the "Count model coefficients," which provide information on how the sex and off-campus status of a student who is a drinker are related to the number of drinks reported by that student over a weekend. As we have done with previous Poisson regression models, we exponentiate each coefficient for ease of interpretation. Thus, for those who drink, the average number of drinks for males is $e^{1.0209}$ or 2.76 times the number for females (Z = 5.827, p < 0.001) given that you are comparing people who live in comparable settings, i.e., either both on or both off campus. Among drinkers, the mean number of drinks for students living off campus is $e^{0.4159} = 1.52$ times that of students living on campus for those of the same sex (Z = 2.021, p = 0.0433).

The "Zero-inflation model coefficients" refer to separating drinkers from non-drinkers. An important consideration in separating drinkers from non-drinkers may be whether this is their first year, where `firstYear` is a 0/1 indicator variable.

We have
$$log(\alpha/(1-\alpha)) = -0.6036 + 1.1364 \text{firstYear}$$

However, we are interested in α, the proportion of non-drinkers. Exponentiating the coefficient for the first-year term for this model yields 3.12. Here it is interpreted as the odds ($\frac{\alpha}{1-\alpha}$) that a first-year student is a non-drinker is 3.12 times the odds that an upper-class student is a non-drinker. Furthermore, with a little algebra (solving the equation with $log(\alpha/(1-\alpha))$ for α), we have

$$\hat{\alpha} = \frac{e^{-0.6036+1.1364(\text{firstYear})}}{1+e^{-0.6036+1.1364(\text{firstYear})}}.$$

The estimated chance that a first-year student is a non-drinker is

$$\frac{e^{0.533}}{1+e^{0.533}} = 0.630$$

or 63.0%, while for non-first-year students, the estimated probability of being a non-drinker is 0.354. If you have seen logistic regression, you'll recognize that this transformation is what is used to estimate a probability. More on this in Chapter 6.

4.10.6 The Vuong Test (optional)

Moving from ordinary Poisson to zero-inflated Poisson has helped us address additional research questions: What proportion of students are non-drinkers, and what factors are associated with whether or not a student is a non-drinker? While a ZIP model seems more faithful to the nature and structure of this data, can we quantitatively show that a zero-inflated Poisson is better than an ordinary Poisson model?

We cannot use the drop-in-deviance test we discussed earlier because these two models are not nested within one another. Vuong [1989] devised a test to make this comparison for the special case of comparing a zero-inflated model and ordinary regression model. Essentially, the Vuong Test is able to compare predicted probabilities of **non-nested** models.

```
vuong(pois.m1, zip.m2)
```

```
Vuong Non-Nested Hypothesis Test-Statistic:
(test-statistic is asymptotically distributed N(0,1) under the
 null that the models are indistinguishible)
--------------------------------------------------------------
              Vuong z-statistic              H_A
Raw                       -2.689 model2 > model1
AIC-corrected             -2.534 model2 > model1
BIC-corrected             -2.353 model2 > model1
```

```
                p-value
Raw             0.0036
AIC-corrected   0.0056
BIC-corrected   0.0093
```

Here, we have structured the Vuong Test to compare Model 1: Ordinary Poisson Model to Model 2: Zero-inflation Model. If the two models do not differ, the test statistic for Vuong would be asymptotically standard Normal and the p-value would be relatively large. Here the first line of the output table indicates that the zero-inflation model is better ($Z = -2.69, p = .0036$). Note that the test depends upon sufficiently large n for the Normal approximation, so since our sample size (n=77) is somewhat small, we need to interpret this result with caution. More research is underway to address statistical issues related to these comparisons.

4.10.7 Residual Plot

Fitted values ($\hat{y}$) and residuals ($y - \hat{y}$) can be computed for zero-inflation models and plotted. Figure 4.11 reveals that one observation appears to be extreme (Y=22 drinks during the past weekend). Is this a legitimate observation or was there a transcribing error? Without the original respondents, we cannot settle this question. It might be worthwhile to get a sense of how influential this extreme observation is by removing Y=22 and refitting the model.

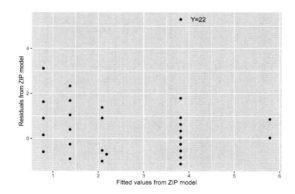

FIGURE 4.11: Residuals by fitted counts for ZIP model.

4.10.8 Limitations

Given that you have progressed this far in your statistical education, the weekend drinking survey question should raise some red flags. What time period constitutes the "weekend"? Will some students be thinking of only Saturday night, while others include Friday night or possibly Sunday evening? What constitutes a drink—a bottle of beer? How many drinks will a respondent report for a bottle of wine? Precise definitions would vastly improve the quality of this data. There is also an issue related to confidentiality. If the data is collected in class, will the teacher be able to identify the respondent? Will respondents worry that a particular response will affect their grade in the class or lead to repercussions on a dry campus?

In addition to these concerns, there are a number of other limitations that should be noted. Following the concern of whether this data represents a random sample of any population (it doesn't), we also must be concerned with the size of this data set. ZIP models are not appropriate for small samples and this data set is not impressively large.

At times, a mixture of zeros occurs naturally. It may not come about because of neglecting to ask a critical question on a survey, but the information about the subpopulation may simply not be ascertainable. For example, visitors from a state park were asked as they departed how many fish they caught, but those who report 0 could be either non-fishers or fishers who had bad luck. These kinds of circumstances occur often enough that ZIP models are becoming increasingly common.

Actually, applications which extend beyond ordinary Poisson regression applications—ZIPs and other Poisson modeling approaches such as hurdle models and quasi-Poisson applications—are becoming increasingly common. So it is worth taking a look at these variations of Poisson regression models. Here we have only skimmed the surface of zero-inflated models, but we want you to be aware of models of this type. ZIP models demonstrate that modeling can be flexible and creative—a theme we hope you will see throughout this book.

4.11 Exercises

4.11.1 Conceptual Exercises

Exercises 1-4 involve predicting a **response** using one or more **explanatory variables**, where these examples have response variables that are counts per some unit of time or space. List the response (both what is being counted and over what unit of time or space) and relevant explanatory variables.

1. Are the number of motorcycle deaths in a given year related to a state's helmet laws?

2. Does the number of employers conducting on-campus interviews during a year differ for public and private colleges?
3. Does the daily number of asthma-related visits to an Emergency Room differ depending on air pollution indices?
4. Has the number of deformed fish in randomly selected Minnesota lakes been affected by changes in trace minerals in the water over the last decade?
5. Models of the form $Y_i = \beta_0 + \beta_1 X_i + \epsilon_i, \epsilon_i \sim iidN(0, \sigma)$ are fit using the method of least squares. What method is used to fit Poisson regression models?
6. What should be done before adjusting for overdispersion?
7. Why are quasi-Poisson models used, and how do the results typically compare for corresponding models using regular Poisson regression?
8. Why is the log of mean counts, $\log(\bar{Y})$, not $\bar{Y}$, plotted against X when assessing the assumptions for Poisson regression?
9. How can the assumption of *mean=variance* be checked for Poisson regression? What if there are not many repeated observations at each level of X?
10. Is it possible that a predictor is significant for a model fit using Poisson regression, but not for a model for the same data fit using quasi-Poisson regression? Explain.

Complete (a)-(d) in the context of the study for Exercises 11-13.

a. Define the response.
b. What are the possible values for the response?
c. What does λ represent?
d. Would a zero-inflated model be considered here? If so, what would be a "true zero"?

11. **Fish (or, as they say in French, poisson).** A state wildlife biologist collected data from 250 park visitors as they left at the end of their stay. Each was asked to report the number of fish they caught during their one-week stay. On average, visitors caught 21.5 fish per week.

12. **Methadone program recidivism.** Program facilitators keep track of the number of times their program's patients relapse within five years of initial treatment. Data on 100 patients yielded a mean number of 2.8 relapses per patient within the five years of initial treatment.

13. **Clutch size.** Thirty nests were located and the number of eggs in each nest were counted at the start of a season. Later in the season following a particularly destructive storm, the mean clutch size of the 30 nests was only 1.7 eggs per nest.

TABLE 4.6: Sample data for Exercise 15.

Age	Time Online	Number of Dates Arranged Online
19	35	3
29	20	5
38	15	0
55	10	0

14. **Credit card use.** A survey of 1,000 consumers asked respondents how many credit cards they use. Interest centers on the relationship between credit card use and income, in $10,000. The estimated coefficient for income is 2.1.

    - Identify the predictor and interpret the estimated coefficient for the predictor in this context.
    - Describe how the assumption of linearity can be assessed in this example.
    - Suggest a way in which to assess the equal mean and variance assumption.

15. **Dating online.** Researchers are interested in the number of dates respondents arranged online and whether the rates differ by age group. Questions which elicit responses similar to this can be found in the Pew Survey concerning dating online and relationships [Smith and Duggan, 2013]. Each survey respondent was asked how many dates they have arranged online in the past 3 months as well as the typical amount of time, t, in hours, they spend online weekly. Some rows of data appear in Table 4.6.

    - Identify the response, predictor, and offset in this context. Does using an offset make sense?
    - Write out a model for this data. As part of your model description, define the parameter, λ.
    - Consider a zero-inflated Poisson model for this data. Describe what the 'true zeros' would be in this setting.

16. **Poisson approximation: rare events.** For rare diseases, the probability of a case occurring, p, in a very large population, n, is small. With a small p and large n, the random variable $Y =$ the number of cases out of n people can be approximated using a Poisson random variable with $\lambda = np$. If the count of those with the disease is observed in several different populations independently of one another, the Y_i represents the number of cases in the ith population and can

TABLE 4.7: Data from Scotto et al. (1974) on the number of cases of nonmelanoma skin cancer for women by age group in two metropolitan areas (Minneapolis-St. Paul and Dallas-Ft. Worth); the year is unknown.

Number of Cases	Population	Age Group	City
1	172675	15-24	1
16	123065	25-34	1
...	...	...	...
226	29007	75-84	2
65	7538	85+	2

The columns contain: number of cases, population size, age group, and city (1=Minneapolis-St. Paul, 2=Dallas-Ft. Worth).

be approximated using a Poisson random variable with $\lambda_i = n_i p_i$ where p_i is the probability of a case for the ith population. Poisson regression can take into account the differences in the population sizes, n_i, using as an offset $\log(n_i)$ as well as differences in a population characteristic like x_i. The coefficient of the offset is set at one; it is not estimated like the other coefficients. Thus the model statement has the form: $log(\lambda_i) = \beta_0 + \beta_1 x_i + log(n_i)$, where $Y_i \sim$ Poisson($\lambda_i = n_i p_i$). Note that λ_i depends on x_i which may differ for the different populations.

Scotto et al. [1974] wondered if skin cancer rates by age group differ by city. Based on their data in Table 4.7, identify and describe the following quantities which appear in the description of the Poisson approximation for rare events:

- A case,
- The population size, n_i,
- Probability, p_i,
- Poisson parameter, λ_i,
- Poisson random variables, Y_i, and
- The predictors, X_i.

4.11.2 Guided Exercises

1. **College burglaries.** We wish to build a regression model to describe the number of burglaries on a college campus in a year. Our population of interest will be U.S. liberal arts colleges.
    a. Describe why the response variable ($Y = \#$ burglaries on campus in a year) could be modeled by a Poisson distribution.

b. Describe explanatory variables which might explain differences in λ_i = mean number of burglaries per year on campus i.
c. Consider a campus with an average of 5 burglaries per year. Use `dpois()` to sketch a plot of the distribution of Y for this campus. Use `rpois()` to verify that both the mean and variance of Y are given by $\lambda = 5$.
d. Consider a campus with an average of 20 burglaries per year and repeat (c).

2. **Elephant mating.** How does age affect male elephant mating patterns? An article by Poole [1989] investigated whether mating success in male elephants increases with age and whether there is a peak age for mating success. To address this question, the research team followed 41 elephants for one year and recorded both their ages and their number of matings. The data [Ramsey and Schafer, 2002] is found in `elephant.csv`, and the variables are:
   - MATINGS = the number of matings in a given year
   - AGE = the age of the elephant in years.
   a. Create a histogram of MATINGS. Is there preliminary evidence that number of matings could be modeled as a Poisson response? Explain.
   b. Plot MATINGS by AGE. Add a least squares line. Is there evidence that modeling matings using a linear regression with age might not be appropriate? Explain. (Hints: fit a smoother; check residual plots).
   c. For each age, calculate the mean number of matings. Take the log of each mean and plot it by AGE.
      i. What assumption can be assessed with this plot?
      ii. Is there evidence of a quadratic trend on this plot?
   d. Fit a Poisson regression model with a linear term for AGE. Exponentiate and then interpret the coefficient for AGE.
   e. Construct a 95% confidence interval for the slope and interpret in context (you may want to exponentiate endpoints).
   f. Are the number of matings significantly related to age? Test with
      i. a Wald test and
      ii. a drop in deviance test.
   g. Add a quadratic term in AGE to determine whether there is a maximum age for the number of matings for elephants. Is a quadratic model preferred to a linear model? To investigate this question, use
      i. a Wald test and
      ii. a drop in deviance test.
   h. What can we say about the goodness-of-fit of the model with age as the sole predictor? Compare the residual deviance for

TABLE 4.8: A small subset of hypothetical data on Minnesota workplace rules on smoking.

Subject	X (location)	Y (cigarettes)
1	0	3
2	1	0
3	1	0
4	1	1
5	0	2
6	0	1

X is 0 for home and 1 for work.
Y is number of cigaretttes in a 2-hour period.

the linear model to a χ^2 distribution with the residual model degrees of freedom.
  i. Fit the linear model using quasi-Poisson regression. (Why?)
     i. How do the estimated coefficients change?
     ii. How do the standard errors change?
     iii. What is the estimated dispersion parameter?
     iv. An estimated dispersion parameter greater than 1 suggests overdispersion. When adjusting for overdispersion, are you more or less likely to obtain a significant result when testing coefficients? Why?

3. **Smoking at work and home.** An earlier study examined the effect of workplace rules in Minnesota which require smokers to smoke cigarettes outside. The number of cigarettes smoked by smokers in a 2-hour period was recorded, along with whether the smoker was at home or at work. A (very) small subset of the data appears in Table 4.8.

   - Model 1: Assume that $Y \sim \text{Poisson}(\lambda)$; there is no difference between home and work.
   - Model 2: Assume that $Y \sim \text{Poisson}(\lambda_W)$ when the smoker is at work, and $Y \sim \text{Poisson}(\lambda_H)$ when the smoker is at home.
   - Model 3: Assume that $Y \sim \text{Poisson}(\lambda)$ and $log(\lambda) = \beta_0 + \beta_1 X$.

   a. Write out the likelihood $L(\lambda)$ and the log-likelihood $logL(\lambda)$ in Model 1. Use the data values above, and simplify where possible.
   b. Intuitively, what would be a reasonable estimate for λ based on this data? Why?
   c. Find the maximum likelihood estimator for λ in Model 1 using

4.11 Exercises

an optimization routine in R (but not the `glm()` function). Use R to produce a plot of the likelihood function $L(\lambda)$.

d. Write out the log-likelihood function $logL(\lambda_W, \lambda_H)$ in Model 2. Use the data values above, and simplify where possible.

e. Intuitively, what would be reasonable estimates for λ_W and λ_H based on this data? Why?

f. Find the maximum likelihood estimators for λ_W and λ_H in Model 2 using an optimization routine in R (but not the `glm()` function).

4. **Smoking at work and home (continued).** We will use the same data set in this question as we used in Question 3.

   a. Write out the log-likelihood function $logL(\beta_0, \beta_1)$ in Model 3. Again, use the data values above, and simplify where possible.

   b. Find the maximum likelihood estimators for β_0 and β_1 in Model 3 using an optimization routine in R (but not the `glm()` function). Use R to produce a 3D plot of the log-likelihood function.

   c. Confirm your estimates for Model 1 and Model 3 using `glm()`. Then show that the MLEs for Model 3 agree with the MLEs for Model 2.

   For the remaining questions, we will focus exclusively on Model 3.

   d. State a (one-sided) hypothesis for β_1 in the context of the problem (i.e., explain how your hypothesis relates to smoking at home and at work). Note: we will nevertheless use two-sided tests and intervals in the following questions.

   e. Do we need to include an offset in our Poisson regression model? Why or why not?

   f. Give estimates of β_0 and β_1, and provide interpretations for both in the context of the problem.

   g. Provide and interpret a 95% confidence interval for β_1.

   h. Provide two *different* significance tests for β_1, in each case providing a test statistic and a p-value and a conclusion in the context of the problem.

   i. Provide a goodness-of-fit test for Model 3, again providing a test statistic, p-value, and conclusion in context.

   j. Can we generalize results of this study to all Minnesota smokers? Why or why not?

   k. Can we claim that rules restricting smoking in the workplace have caused lower levels of smoking at work? Explain.

   l. Give two ways in which this study might be improved (besides simply "bigger sample size").

5. **Campus crime.** The data set `campuscrime09.csv` contains the number of burglaries reported at a collection of 47 U.S. public

universities with over 10,000 students in the year 2009. In addition, covariates are included which may explain differences in crime rates, including total number of students, percentage of men, average SAT and ACT test scores, and tuition.

a. Perform an exploratory data analysis. Support your analysis with plots and summary statistics.
   i. Analyze whether number of burglaries could be reasonably modeled with a Poisson distribution.
   ii. Analyze which covariates you expect to be the best predictors of burglaries.
b. Consider a model with 4 predictors: act.comp + tuition + pct.male + total. Try fitting a linear regression with burg09 as the response. Are there any concerns with this linear regression model?
c. Run a Poisson regression model with the 4 predictors from (b). Interpret the coefficients for tuition and pct.male.
d. Replace tuition with tuition in thousands in your model from (c) – i.e., tuition.thous=tuition/1000. How does your new model compare to your model in (c)? Interpret the coefficient for tuition.thous.
e. We will consider the possibility of including the total number of students at a university as an offset.
   i. Explain why we might consider total as an offset.
   ii. Refit your model from (d) with total (actually, log(total)) as an offset rather than as a predictor. Does this new model appear to fit better or worse than the model from (d)?
   iii. Refit your model from (d) with log(total) rather than total – so log(total) is a predictor and not an offset. If total were a good candidate for an offset, what would we expect the coefficient of log(total) to be? Does a 95% confidence interval for that coefficient contain the value you expected?
f. Run the following model, then interpret the coefficients for tuition.thous and the interaction between tuition.thous and act.comp.

```
crime <- mutate(crime, total.thous = total/1000)
fit3 <- glm(burg09 ~ act.comp + tuition.thous +
         total.thous + act.comp:tuition.thous +
         act.comp:total.thous, family = poisson,
         data = crime)
```

6. **U.S. National Medical Expenditure Survey.** The data set

NMES1988 in the AER package contains a sample of individuals over 65 who are covered by Medicare in order to assess the demand for health care through physician office visits, outpatient visits, ER visits, hospital stays, etc. The data can be accessed by installing and loading the AER package and then running `data(NMES1988)`. More background information and references about the NMES1988 data can be found in help pages for the AER package.

   a. Show through graphical means that there are more respondents with 0 `visits` than might be expected under a Poisson model.
   b. Fit a ZIP model for the number of physician office `visits` using `chronic`, `health`, and `insurance` as predictors for the Poisson count, and `chronic` and `insurance` as the predictors for the binary part of the model. Then, provide interpretations in context for the following model parameters:

      • `chronic` in the Poisson part of the model
      • poor `health` in the Poisson part of the model
      • the Intercept in the logistic part of the model
      • `insurance` in the logistic part of the model

   c. Is there significant evidence that the ZIP model is an improvement over a simple Poisson regression model?

7. **Going vague: ambiguity in political issue statements.** In the following exercise, you will use a **hurdle model** to analyze the data. A hurdle model is similar to a zero-inflated Poisson model, but instead of assuming that "zeros" are comprised of two distinct groups—those who would always be 0 and those who happen to be 0 on this occasion (e.g., non-drinkers and drinkers who had zero drinks over the weekend in Case Study 4.10)—the hurdle model assumes that "zeros" are a single entity. Therefore, in a hurdle model, cases are classified as either "zeros" or "non-zeros", where "non-zeros" *hurdle* the 0 threshold—they must always have counts of 1 or above. We will use the `pscl` package and the `hurdle` function in it to analyze a hurdle model. Note that coefficients in the "zero hurdle model" section of the output relate predictors to the log-odds of being a *non-zero* (i.e., having at least one issue statement), which is opposite of the ZIP model.

In a 2018 study, Chapp et al. [2018] scraped every issue statement from webpages of candidates for the U.S. House of Representatives, counting the number of issues candidates commented on and scoring the level of ambiguity of each statement. We will focus on the issue counts, and determining which attributes (of both the district as a whole and the candidates themselves) are associated with candidate silence (commenting on 0 issues) and a willingness to comment on a greater number of issues. The data set `ambiguity.csv` contains the following variables:

- `name` : candidate name
- `distID` : unique identification number for Congressional district
- `ideology` : candidate left-right orientation
- `democrat` : 1 if Democrat, 0 if Republican
- `mismatch` : disagreement between candidate ideology and district voter ideology
- `incumbent` : 1 if incumbent, 0 if not
- `demHeterogeneity` : how much voters in a district differ according to race, education, occupation, etc.
- `attHeterogeneity` : how much voters in a district differ according to political ideology
- `distLean` : overall ideological lean in a district
- `totalIssuePages` : number of issues candidates commented on (response)

a. Create a frequency plot of `totalIssuePages`. Why might we consider using a hurdle model compared to a Poisson model? Why can't we use a zero-inflated Poisson model?
b. Create a plot of the empirical log odds of having at least one issue statement by ideology. You may want to group ideology values first. What can you conclude from this plot? (See Chapter 6 for more details.)
c. Create a scatterplot that shows the log of the mean number of issues vs. ideology group by party, among candidates with at least one issue statement. What can we conclude from this plot?
d. Create a hurdle model with `ideology` and `democrat` as predictors in both parts. Interpret `ideology` in both parts of the model.
e. Repeat (d), but include an interaction in both parts. Interpret the interaction in the zero hurdle part of the model.
f. Find the best model you can to determine `totalIssuePages`. Write a short paragraph discussing implications of your model.

4.11.3 Open-Ended Exercises

1. **Airbnb in NYC.** Awad et al. [2017] scraped 40628 Airbnb listings from New York City in March 2017 and put together the data set `NYCairbnb.csv`. Key variables include:
   - `id` = unique ID number for each unit
   - `last_scraped` = date when information scraped
   - `host_since` = date when host first listed the unit on Airbnb

4.11 Exercises

- `days` = `last_scraped` - `host_since` = number of days the unit has been listed
- `room_type` = Entire home/apt., Private room, or Shared room
- `bathrooms` = number of bathrooms
- `bedrooms` = number of bedrooms
- `price` = price per night (dollars)
- `number_of_reviews` = number of reviews for the unit on Airbnb
- `review_scores_cleanliness` = cleanliness score from reviews (1-10)
- `review_scores_location` = location score from reviews (1-10)
- `review_scores_value` = value score from reviews (1-10)
- `instant_bookable` = "t" if instantly bookable, "f" if not

Perform an EDA, build a model, and interpret model coefficients to describe variation in the number of reviews (a proxy for the number of rentals, which is not available) as a function of the variables provided. Don't forget to consider an offset, if needed.

2. **Crab satellites.** Brockmann [1996] carried out a study of nesting female horseshoe crabs. Female horseshoe crabs often have male crabs attached to a female's nest known as *satellites*. One objective of the study was to determine which characteristics of the female were associated with the number of satellites. Of particular interest is the relationship between the width of the female carapace and satellites.

   The data can be found in `crab.csv`. It includes:

   - `NumSat` = number of satellites
   - `Width` = carapace width (cm)
   - `Wt` = weight (kg)
   - `Sp` = spine condition (1 = both good, 2 = one worn or broken, 3 = both worn or broken)
   - `C` = color (1 = light medium, 2 = medium, 3 = dark medium, 4 = dark)

   Use Poisson regression to investigate the research question. Be sure you work to obtain an appropriate model before considering overdispersion. Should a hurdle model be considered here? If so, fit a hurdle model and interpret in context.

3. **Doctor visits.** Data was collected on doctor visits from a sample of 5,190 people in the 1977/1978 Australian Health Survey. Cameron and Trivedi [1986] sought to explain the variation in doctor visits using one or more explanatory variables. The data can be found in an R data set from `library(AER)` accessible with the command

data("DoctorVisits"). Variable descriptions can be found under help("DoctorVisits").

Explore the use of a zero-inflated model for this data. Begin with a histogram of the number of visits, complete an EDA, and then fit several models. Summarize your results.

4. **More fish.** The number of fish caught (count), persons in the party (persons), the number of children in the party (child), whether or not they brought a camper into the park (camper), and the length of stay (LOS) were recorded for 250 camping parties. The data can be found in fish2.csv (source: UCLA Statistical Consulting Group [2018]). Create and assess a model for the number of fish caught.

5

Generalized Linear Models: A Unifying Theory

5.1 Learning Objectives

- Determine if a probability distribution can be expressed in one-parameter exponential family form.
- Identify canonical links for distributions of one-parameter exponential family form.

```
# Packages required for Chapter 5
library(knitr)
```

5.2 One-Parameter Exponential Families

Thus far, we have expanded our repertoire of models from linear least squares regression to include Poisson regression. But in the early 1970s, Nelder and Wedderburn [1972] identified a broader class of models that generalizes the multiple linear regression we considered in the introductory chapter and are referred to as **generalized linear models (GLMs)**. All GLMs have similar forms for their likelihoods, MLEs, and variances. This makes it easier to find model estimates and their corresponding uncertainty. To determine whether a model based on a single parameter θ is a GLM, we consider the following properties. When a probability formula can be written in the form below

$$f(y; \theta) = e^{[a(y)b(\theta) + c(\theta) + d(y)]} \tag{5.1}$$

and if the support (the set of possible input values) does not depend upon θ, it is said to have a **one-parameter exponential family form**. We demonstrate that the Poisson distribution is a member of the one-parameter exponential family by writing its probability mass function (pmf) in the form of Equation (5.1) and assessing its support.

5.2.1 One-Parameter Exponential Family: Poisson

Recall we begin with

$$P(Y = y) = \frac{e^{-\lambda} \lambda^y}{y!} \quad \text{where} \quad y = 0, 1, 2 \ldots \infty$$

and consider the following useful identities for establishing exponential form:

$$a = e^{\log(a)}$$
$$a^x = e^{x \log(a)}$$
$$\log(ab) = \log(a) + \log(b)$$
$$\log\left(\frac{a}{b}\right) = \log(a) - \log(b)$$

Determining whether the Poisson model is a member of the one-parameter exponential family is a matter of writing the Poisson pmf in the form of Equation (5.1) and checking that the support does not depend upon λ. First, consider the condition concerning the support of the distribution. The set of possible values for any Poisson random variable is $y = 0, 1, 2 \ldots \infty$ which does not depend on λ. The support condition is met. Now we see if we can rewrite the probability mass function in one-parameter exponential family form.

$$P(Y = y) = e^{-\lambda} e^{y \log \lambda} e^{-\log(y!)}$$
$$= e^{y \log \lambda - \lambda - \log(y!)}$$

The first term in the exponent for Equation (5.1) must be the product of two factors, one solely a function of y, $a(y)$, and another, $b(\lambda)$, a function of λ only. The middle term in the exponent must be a function of λ only; no $y's$ should appear. The last term has only $y's$ and no λ. Since this appears to be the case here, we can identify the different functions in this form:

$$a(y) = y$$
$$b(\lambda) = \log(\lambda)$$
$$c(\lambda) = -\lambda$$
$$d(y) = -\log(y!)$$

These functions have useful interpretations in statistical theory. We won't be going into this in detail, but we will note that function $b(\lambda)$, or more generally $b(\theta)$, will be particularly helpful in GLMs. The function $b(\theta)$ is referred to as the **canonical link**. The canonical link is often a good choice to model as a linear function of the explanatory variables. That is, Poisson regression should be set up as $\log(\lambda) = \beta_0 + \beta_1 x_1 + \beta_2 x_2 + \cdots$. In fact, there is a distinct advantage to modeling the canonical link as opposed to other functions of θ, but it is also worth noting that other choices are possible, and at times preferred, depending upon the context of the application.

5.2 One-Parameter Exponential Families

There are other benefits of identifying a response as being from a one-parameter exponential family. For example, by creating a unifying theory for regression modeling, Nelder and Wedderburn made possible a common and efficient method for finding estimates of model parameters using iteratively reweighted least squares (IWLS). In addition, we can use the one-parameter exponential family form to determine the expected value and standard deviation of Y. With statistical theory you can show that

$$E(Y) = -\frac{c'(\theta)}{b'(\theta)} \quad \text{and} \quad \text{Var}(Y) = \frac{b''(\theta)c'(\theta) - c''(\theta)b'(\theta)}{[b'(\theta)]^3}$$

where differentiation is with respect to θ. Verifying these results for the Poisson response:

$$E(Y) = -\frac{-1}{1/\lambda} = \lambda \quad \text{and} \quad \text{Var}(Y) = \frac{1/\lambda^2}{(1/\lambda^3)} = \lambda$$

We'll find that other distributions are members of the one-parameter exponential family by writing their pdf or pmf in this manner and verifying the support condition. For example, we'll see that the binomial distribution meets these conditions, so it is also a member of the one-parameter exponential family. The normal distribution is a special case where we have two parameters, a mean μ and standard deviation σ. If we assume, however, that one of the parameters is known, then we can show that a normal random variable is also from a one-parameter exponential family.

5.2.2 One-Parameter Exponential Family: Normal

Here we determine whether a normal distribution is a one-parameter exponential family member. First we will need to assume that σ is known. Next, possible values for a normal random variable range from $-\infty$ to ∞, so the support does not depend on μ. Finally, we'll need to write the probability density function (pdf) in the one-parameter exponential family form. We start with the familiar form:

$$f(y) = \frac{1}{\sqrt{2\pi\sigma^2}} e^{-(y-\mu)^2/(2\sigma^2)}$$

Even writing $1/\sqrt{2\pi\sigma^2}$ as $e^{-\log\sigma - \log(2\pi)/2}$ we still do not have the pdf written in one-parameter exponential family form. We will first need to expand the exponent so that we have

$$f(y) = e^{[-\log\sigma - \log(2\pi)/2]} e^{[-(y^2 - 2y\mu + \mu^2)/(2\sigma^2)]}$$

Without loss of generality, we can assume $\sigma = 1$, so that

$$f(y) \propto e^{y\mu - \frac{1}{2}\mu^2 - \frac{1}{2}y^2}$$

and $a(y) = y$, $b(\mu) = \mu$, $c(\mu) = -\frac{1}{2}\mu^2$, and $d(y) = -\frac{1}{2}y^2$.

From this result, we can see that the canonical link for a normal response is μ which is consistent with what we've been doing with LLSR, since the simple linear regression model has the form:

$$\mu_{Y|X} = \beta_0 + \beta_1 X.$$

5.3 Generalized Linear Modeling

GLM theory suggests that the canonical link can be modeled as a linear combination of the explanatory variable(s). This approach unifies a number of modeling results used throughout the text. For example, likelihoods can be used to compare models in the same way for any member of the one-parameter exponential family.

We have now **generalized** our modeling to handle non-normal responses. In addition to normally distributed responses, we are able to handle Poisson responses, binomial responses, and more. Writing a pmf or pdf for a response in one-parameter exponential family form reveals the canonical link which can be modeled as a linear function of the predictors. This linear function of the predictors is the last piece of the puzzle for performing generalized linear modeling. But, in fact, it is really nothing new. We already use linear combinations and the canonical link when modeling normally distributed data.

Three components of a GLM

1. Distribution of Y (e.g., Poisson)
2. Link Function (a function of the parameter, e.g., $\log(\lambda)$ for Poisson)
3. Linear Predictor (choice of predictors, e.g., $\beta_0 + \beta_1 x_1 + \beta_2 x_2 + \cdots$)

Completing Table 5.1 is left as an exercise.

In the chapter on Poisson modeling, we provided heuristic rationale for using the log() function as our link. That is, counts would be non-negative but a linear function inevitably goes negative. By taking the logarithm of our parameter λ, we could use a linear predictor and not worry that it can take on negative values. Now we have theoretical justification for this choice, as the log is the canonical link for Poisson data. In the next chapter we encounter yet another type of response, a binary response, which calls for a different link function. Our work here suggests that we will model $\text{logit}(p) = \log\left(\frac{p}{1-p}\right)$ using a linear predictor.

TABLE 5.1: One-parameter exponential family form and canonical links.

Distribution	One-Parameter Exponential Family Form	Canonical Link
Binary		
Binomial		$\text{logit}(p)$
Poisson	$P(Y=y) = e^{y\log\lambda - \lambda - y!}$	$\log(\lambda)$
Normal	$f(y) \propto e^{y\mu - \frac{1}{2}\mu^2 - \frac{1}{2}y^2}$	μ
Exponential		
Gamma		
Geometric		

5.4 Exercises

1. For each distribution below,
    - Write the pmf or pdf in one-parameter exponential form, if possible.
    - Describe an example of a setting where this random variable might be used.
    - Identify the canonical link function, and
    - Compute $\mu = -\frac{c'(\theta)}{b'(\theta)}$ and $\sigma^2 = \frac{b''(\theta)c'(\theta) - c''(\theta)b'(\theta)}{[b'(\theta)]^3}$ and compare with known $E(Y)$ and $\text{Var}(Y)$.

   a) Binary: $Y = 1$ for a success, 0 for a failure
   $$P(Y=y;p) = p^y(1-p)^{(1-y)}$$

   b) Binomial (for fixed n): Y = number of successes in n independent, identical trials
   $$P(Y=y;p) = \binom{n}{y} p^y (1-p)^{(n-y)}$$

   c) Poisson: Y = number of events occurring in a given time (or space) when the average event rate is λ per unit of time (or space)
   $$P(Y=y;\lambda) = \frac{e^{-\lambda}\lambda^y}{y!}$$

d) Normal (with fixed σ – could set $\sigma = 1$ without loss of generality)

$$f(y;\mu) = \frac{1}{\sqrt{2\pi\sigma^2}} e^{-(y-\mu)^2/(2\sigma^2)}$$

e) Normal (with fixed μ – could set $\mu = 0$ without loss of generality)

$$f(y;\sigma) = \frac{1}{\sqrt{2\pi\sigma^2}} e^{-(y-\mu)^2/(2\sigma^2)}$$

f) Exponential: Y = time spent waiting for the first event in a Poisson process with an average rate of λ events per unit of time

$$f(y;\lambda) = \lambda e^{-\lambda y}$$

g) Gamma (for fixed r): Y = time spent waiting for the rth event in a Poisson process with an average rate of λ events per unit of time

$$f(y;\lambda) = \frac{\lambda^r}{\Gamma(r)} y^{r-1} e^{-\lambda y}$$

h) Geometric: Y = number of failures before the first success in a Bernoulli process

$$P(Y = y; p) = (1-p)^y p$$

i) Negative Binomial (for fixed r): Y = number of failures prior to the rth success in a Bernoulli process

$$P(Y = y; p) = \binom{y+r-1}{r-1} (1-p)^y p^r$$
$$= \frac{\Gamma(y+r)}{\Gamma(r) y!} (1-p)^y p^r$$

j) Pareto (for fixed k):

$$f(y;\theta) = \frac{\theta k^\theta}{y^{(\theta+1)}} \quad \text{for} \quad y \geq k; \theta \geq 1$$

2. Complete Table 5.1 containing your results of the preceding exercises.

6

Logistic Regression

6.1 Learning Objectives

- Identify a binomial random variable and assess the validity of the binomial assumptions.
- Write a generalized linear model for binomial responses in two forms, one as a function of the logit and one as a function of p.
- Explain how fitting a logistic regression differs from fitting a linear least squares regression (LLSR) model.
- Interpret estimated coefficients in logistic regression.
- Differentiate between logistic regression models with binary and binomial responses.
- Use the residual deviance to compare models, to test for lack-of-fit when appropriate, and to check for unusual observations or needed transformations.

```
# Packages required for Chapter 6
library(gridExtra)
library(mnormt)
library(lme4)
library(knitr)
library(pander)
library(tidyverse)
```

6.2 Introduction to Logistic Regression

Logistic regression is characterized by research questions with binary (yes/no or success/failure) or binomial (number of yesses or successes in n trials) responses:

a. Are students with poor grades more likely to binge drink?
b. Is exposure to a particular chemical associated with a cancer diagnosis?
c. Are the number of votes for a congressional candidate associated with the amount of campaign contributions?

Binary Responses: Recall from Section 3.3.1 that binary responses take on only two values: success (Y=1) or failure (Y=0), Yes (Y=1) or No (Y=0), etc. Binary responses are ubiquitous; they are one of the most common types of data that statisticians encounter. We are often interested in modeling the probability of success p based on a set of covariates, although sometimes we wish to use those covariates to classify a future observation as a success or a failure.

Examples (a) and (b) above would be considered to have binary responses (Does a student binge drink? Was a patient diagnosed with cancer?), assuming that we have a unique set of covariates for each individual student or patient.

Binomial Responses: Also recall from Section 3.3.2 that binomial responses are the number of successes in n identical, independent trials with constant probability p of success. A sequence of independent trials like this with the same probability of success is called a **Bernoulli process**. As with binary responses, our objective in modeling binomial responses is to quantify how the probability of success, p, is associated with relevant covariates.

Example (c) above would be considered to have a binomial response, assuming we have vote totals at the congressional district level rather than information on individual voters.

6.2.1 Logistic Regression Assumptions

Much like ordinary least squares (OLS), using **logistic regression** to make inferences requires model assumptions.

1. **Binary Response** The response variable is dichotomous (two possible responses) or the sum of dichotomous responses.
2. **Independence** The observations must be independent of one another.
3. **Variance Structure** By definition, the variance of a binomial random variable is $np(1-p)$, so that variability is highest when $p = .5$.
4. **Linearity** The log of the odds ratio, $\log(\frac{p}{1-p})$, must be a linear function of x. This will be explained further in the context of the first case study.

6.2.2 A Graphical Look at Logistic Regression

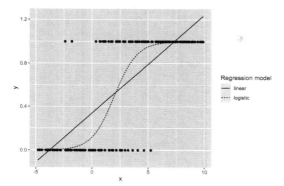

FIGURE 6.1: Linear vs. logistic regression models for binary response data.

Figure 6.1 illustrates a data set with a binary (0 or 1) response (Y) and a single continuous predictor (X). The solid line is a linear regression fit with least squares to model the probability of a success (Y=1) for a given value of X. With a binary response, the line doesn't fit the data well, and it produces predicted probabilities below 0 and above 1. On the other hand, the logistic regression fit (dashed curve) with its typical "S" shape follows the data closely and always produces predicted probabilities between 0 and 1. For these and several other reasons detailed in this chapter, we will focus on the following model for logistic regression with binary or binomial responses:

$$log(\frac{p_i}{1-p_i}) = \beta_0 + \beta_1 x_i$$

where the observed values $Y_i \sim$ binomial with $p = p_i$ for a given x_i and $n = 1$ for binary responses.

6.3 Case Studies Overview

We consider three case studies in this chapter. The first two involve binomial responses (Soccer Goalkeepers and Reconstructing Alabama), while the last case uses a binary response (Trying to Lose Weight). Even though binary responses are much more common, their models have a very similar form to binomial responses, so the first two case studies will illustrate important principles that also apply to the binary case. Here are the statistical concepts you will encounter for each case study.

TABLE 6.1: Soccer goalkeepers' penalty kick saves when their team is and is not behind.

	Saves	Scores	Total
Behind	2	22	24
Not Behind	39	141	180
Total	41	163	204

(Source: Roskes et al. 2011.)

The soccer goalkeeper data can be written in the form of a 2 × 2 table. This example is used to describe some of the underlying theory for logistic regression. We demonstrate how binomial probability mass functions (pmfs) can be written in one-parameter exponential family form, from which we can identify the canonical link as in Chapter 5. Using the canonical link, we write a Generalized Linear Model for binomial counts and determine corresponding MLEs for model coefficients. Interpretation of the estimated parameters involves a fundamental concept, the odds ratio.

The Reconstructing Alabama case study is another binomial example which introduces the notion of deviances, which are used to compare and assess models. Thus, we will investigate hypothesis tests and confidence intervals, including issues of interaction terms, overdispersion, and lack-of-fit. We will also check the assumptions of logistic regression using empirical logit plots and deviance residuals.

The last case study addresses why teens try to lose weight. Here the response is a binary variable which allows us to analyze individual level data. The analysis builds on concepts from the previous sections in the context of a random sample from CDC's Youth Risk Behavior Survey (YRBS).

6.4 Case Study: Soccer Goalkeepers

Does the probability of a save in a soccer match depend upon whether the goalkeeper's team is behind or not? Roskes et al. [2011] looked at penalty kicks in the men's World Cup soccer championships from 1982 to 2010, and they assembled data on 204 penalty kicks during shootouts. The data for this study is summarized in Table 6.1.

6.4 Case Study: Soccer Goalkeepers

6.4.1 Modeling Odds

Odds are one way to quantify a goalkeeper's performance. Here the odds that a goalkeeper makes a save when his team is behind is 2 to 22 or 0.09 to 1. Or equivalently, the odds that a goal is scored on a penalty kick is 22 to 2 or 11 to 1. An odds of 11 to 1 tells you that a shooter whose team is ahead will score 11 times for every 1 shot that the goalkeeper saves. When the goalkeeper's team is not behind the odds a goal is scored is 141 to 39 or 3.61 to 1. We see that the odds of a goal scored on a penalty kick are better when the goalkeeper's team is behind than when it is not behind (i.e., better odds of scoring for the shooter when the shooter's team is ahead). We can compare these odds by calculating the **odds ratio** (OR), 11/3.61 or 3.05, which tells us that the *odds* of a successful penalty kick are 3.05 times higher when the shooter's team is leading.

In our example, it is also possible to estimate the probability of a goal, p, for either circumstance. When the goalkeeper's team is behind, the probability of a successful penalty kick is $p = 22/24$ or 0.833. We can see that the ratio of the probability of a goal scored divided by the probability of no goal is $(22/24)/(2/24) = 22/2$ or 11, the odds we had calculated above. The same calculation can be made when the goalkeeper's team is not behind. In general, we now have several ways of finding the odds of success under certain circumstances:

$$\text{Odds} = \frac{\#\text{successes}}{\#\text{failures}} = \frac{\#\text{successes}/n}{\#\text{failures}/n} = \frac{p}{1-p}.$$

6.4.2 Logistic Regression Models for Binomial Responses

We would like to model the odds of success; however, odds are strictly positive. Therefore, similar to modeling $\log(\lambda)$ in Poisson regression, which allowed the response to take on values from $-\infty$ to ∞, we will model the log(odds), the **logit**, in logistic regression. Logits will be suitable for modeling with a linear function of the predictors:

$$\log\left(\frac{p}{1-p}\right) = \beta_0 + \beta_1 X$$

Models of this form are referred to as **binomial regression models**, or more generally as **logistic regression models**. Here we provide intuition for using and interpreting logistic regression models, and then in the short optional section that follows, we present rationale for these models using GLM theory.

In our example we could define $X = 0$ for not behind and $X = 1$ for behind and fit the model:

$$\log\left(\frac{p_X}{1-p_X}\right) = \beta_0 + \beta_1 X \tag{6.1}$$

where p_X is the probability of a successful penalty kick given X.

So, based on this model, the log odds of a successful penalty kick when the goalkeeper's team is not behind is:

$$\log\left(\frac{p_0}{1-p_0}\right) = \beta_0,$$

and the log odds when the team is behind is:

$$\log\left(\frac{p_1}{1-p_1}\right) = \beta_0 + \beta_1.$$

We can see that β_1 is the difference between the log odds of a successful penalty kick between games when the goalkeeper's team is behind and games when the team is not behind. Using rules of logs:

$$\beta_1 = (\beta_0 + \beta_1) - \beta_0 = \log\left(\frac{p_1}{1-p_1}\right) - \log\left(\frac{p_0}{1-p_0}\right) = \log\left(\frac{p_1/(1-p_1)}{p_0/(1-p_0)}\right).$$

Thus e^{β_1} is the ratio of the odds of scoring when the goalkeeper's team is not behind compared to scoring when the team is behind. In general, *exponentiated coefficients in logistic regression are odds ratios (OR)*. A general interpretation of an OR is the odds of success for group A compared to the odds of success for group B—how many times greater the odds of success are in group A compared to group B.

The logit model (Equation (6.1)) can also be re-written in a **probability form**:

$$p_X = \frac{e^{\beta_0 + \beta_1 X}}{1 + e^{\beta_0 + \beta_1 X}}$$

which can be re-written for games when the goalkeeper's team is behind as:

$$p_1 = \frac{e^{\beta_0 + \beta_1}}{1 + e^{\beta_0 + \beta_1}} \tag{6.2}$$

and for games when the goalkeeper's team is not behind as:

$$p_0 = \frac{e^{\beta_0}}{1 + e^{\beta_0}} \tag{6.3}$$

We use likelihood methods to estimate β_0 and β_1. As we had done in Chapter 2, we can write the likelihood for this example in the following form:

6.4 Case Study: Soccer Goalkeepers

$$\text{Lik}(p_1, p_0) = \binom{24}{22} p_1^{22}(1-p_1)^2 \binom{180}{141} p_0^{141}(1-p_0)^{39}$$

Our interest centers on estimating $\hat{\beta}_0$ and $\hat{\beta}_1$, not p_1 or p_0. So we replace p_1 in the likelihood with an expression for p_1 in terms of β_0 and β_1 as in Equation (6.2). Similarly, p_0 in Equation (6.3) involves only β_0. After removing constants, the new likelihood looks like:

$$\text{Lik}(\beta_0, \beta_1) \propto$$

$$\left(\frac{e^{\beta_0+\beta_1}}{1+e^{\beta_0+\beta_1}}\right)^{22} \left(1 - \frac{e^{\beta_0+\beta_1}}{1+e^{\beta_0+\beta_1}}\right)^2 \left(\frac{e^{\beta_0}}{1+e^{\beta_0}}\right)^{141} \left(1 - \frac{e^{\beta_0}}{1+e^{\beta_0}}\right)^{39}$$

Now what? Fitting the model means finding estimates of β_0 and β_1, but familiar methods from calculus for maximizing the likelihood don't work here. Instead, we consider all possible combinations of β_0 and β_1. That is, we will pick that pair of values for β_0 and β_1 that yield the largest likelihood for our data. Trial and error to find the best pair is tedious at best, but more efficient numerical methods are available. The MLEs for the coefficients in the soccer goalkeeper study are $\hat{\beta}_0 = 1.2852$ and $\hat{\beta}_1 = 1.1127$.

Exponentiating $\hat{\beta}_1$ provides an estimate of the odds ratio (the odds of scoring when the goalkeeper's team is behind, compared to the odds of scoring when the team is not behind) of 3.04, which is consistent with our calculations using the 2 × 2 table. We estimate that the odds of scoring when the goalkeeper's team is behind is over 3 times that of when the team is not behind or, in other words, the odds a shooter is successful in a penalty kick shootout are 3.04 times higher when his team is leading.

Time out for study discussion (optional).

- Discuss the following quote from the study abstract: "Because penalty takers shot at the two sides of the goal equally often, the goalkeepers' right-oriented bias was dysfunctional, allowing more goals to be scored."

- Construct an argument for why the greater success observed when the goalkeeper's team was behind might be better explained from the shooter's perspective.

Before we go on, you may be curious as to why there is *no error term* in our model statements for logistic or Poisson regression. One way to look at it is to consider that all models describe how observed values are generated. With the logistic model we assume that the observations are generated as binomial random variables. Each observation or realization of Y = number of successes in n independent and identical trials with a probability of success on any one trial of p is produced by $Y \sim \text{Binomial}(n, p)$. So the randomness in this model is not introduced by an added error term, but rather by appealing to a binomial probability distribution, where variability depends only on n and p through $\text{Var}(Y) = np(1-p)$, and where n is usually considered fixed and p the parameter of interest.

6.4.3 Theoretical Rationale (optional)

Recall from Chapter 5 that generalized linear models (GLMs) are a way in which to model a variety of different types of responses. In this chapter, we apply the general results of GLMs to the specific application of binomial responses. Let Y = the number scored out of n penalty kicks. The parameter, p, is the probability of a score on a single penalty kick. Recall that the theory of GLMs is based on the unifying notion of the one-parameter exponential family form:

$$f(y;\theta) = e^{[a(y)b(\theta)+c(\theta)+d(y)]}$$

To see that we can apply the general approach of GLMs to binomial responses, we first write an expression for the probability of a binomial response and then use a little algebra to rewrite it until we can demonstrate that it, too, can be written in one-parameter exponential family form with $\theta = p$. This will provide a way in which to specify the canonical link and the form for the model. Additional theory allows us to deduce the mean, standard deviation, and more from this form.

If Y follows a binomial distribution with n trials and probability of success p, we can write:

$$P(Y=y) = \binom{n}{y} p^y (1-p)^{(n-y)}$$
$$= e^{y \log(p) + (n-y)\log(1-p) + \log\binom{n}{y}}$$

However, this probability mass function is not quite in one-parameter exponential family form. Note that there are two terms in the exponent which consist of a product of functions of y and p. So more simplification is in order:

$$P(Y=y) = e^{y \log\left(\frac{p}{1-p}\right) + n \log(1-p) + \log\binom{n}{y}}$$

Don't forget to consider the support; we must make sure that the set of possible values for this response is not dependent upon p. For fixed n and any value of p, $0 < p < 1$, all integer values from 0 to n are possible, so the support is indeed independent of p.

The one-parameter exponential family form for binomial responses shows that the canonical link is $\log\left(\frac{p}{1-p}\right)$. Thus, GLM theory suggests that constructing a model using the logit, the log odds of a score, as a linear function of covariates is a reasonable approach.

6.5 Case Study: Reconstructing Alabama

This case study demonstrates how wide-ranging applications of statistics can be. Many would not associate statistics with historical research, but this case study shows that it can be done. U.S. Census data from 1870 helped historian Michael Fitzgerald of St. Olaf College gain insight into important questions about how railroads were supported during the Reconstruction Era.

In a paper entitled "Reconstructing Alabama: Reconstruction Era Demographic and Statistical Research," Ben Bayer performs an analysis of data from 1870 to explain factors that influence voting on referendums related to railroad subsidies [Bayer and Fitzgerald, 2011]. Positive votes are hypothesized to be inversely proportional to the distance a voter is from the proposed railroad, but the racial composition of a community (as measured by the percentage of blacks) is hypothesized to be associated with voting behavior as well. Separate analyses of three counties in Alabama—Hale, Clarke, and Dallas—were performed; we discuss Hale County here. This example differs from the soccer example in that it includes continuous covariates. Was voting on railroad referenda related to distance from the proposed railroad line and the racial composition of a community?

6.5.1 Data Organization

The unit of observation for this data is a community in Hale County. We will focus on the following variables from RR_Data_Hale.csv collected for each community (see Table 6.2):

- pctBlack = the percentage of blacks in the community

- distance = the distance, in miles, the proposed railroad is from the community

- YesVotes = the number of "Yes" votes in favor of the proposed railroad line (our primary response variable)

- NumVotes = total number of votes cast in the election

TABLE 6.2: Sample of the data for the Hale County, Alabama, railroad subsidy vote.

community	pctBlack	distance	YesVotes	NumVotes
Carthage	58.4	17	61	110
Cederville	92.4	7	0	15
Greensboro	59.4	0	1790	1804
Havana	58.4	12	16	68

6.5.2 Exploratory Analyses

We first look at a coded scatterplot to see our data. Figure 6.2 portrays the relationship between `distance` and `pctBlack` coded by the `InFavor` status (whether a community supported the referendum with over 50% Yes votes). From this scatterplot, we can see that all of the communities in favor of the railroad referendum are over 55% black, and all of those opposed are 7 miles or farther from the proposed line. The overall percentage of voters in Hale County in favor of the railroad is 87.9%.

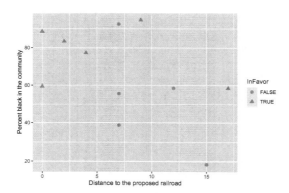

FIGURE 6.2: Scatterplot of distance from a proposed rail line and percent black in the community coded by whether the community was in favor of the referendum or not.

Recall that a model with two covariates has the form:

$$\log(\text{odds}) = \log\left(\frac{p}{1-p}\right) = \beta_0 + \beta_1 X_1 + \beta_2 X_2.$$

where p is the proportion of Yes votes in a community. In logistic regression, we expect the logits to be a linear function of X, the predictors. To assess the linearity assumption, we construct **empirical logit plots**, where "empirical"

6.5 Case Study: Reconstructing Alabama

means "based on sample data." Empirical logits are computed for each community by taking log $\left(\frac{\text{number of successes}}{\text{number of failures}}\right)$. In Figure 6.3, we see that the plot of empirical logits versus distance produces a plot that looks linear, as needed for the logistic regression assumption. In contrast, the empirical logits by percent black reveal that Greensboro deviates quite a bit from the otherwise linear pattern; this suggests that Greensboro is an outlier and possibly an influential point. Greensboro has 99.2% voting yes, with only 59.4% black.

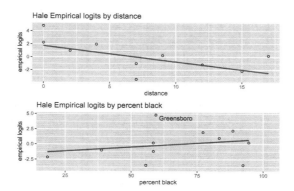

FIGURE 6.3: Empirical logit plots for the Railroad Referendum data.

In addition to examining how the response correlates with the predictors, it is a good idea to determine whether the predictors correlate with one another. Here, the correlation between distance and percent black is negative and moderately strong with $r = -0.49$. We'll watch to see if the correlation affects the stability of our odds ratio estimates.

6.5.3 Initial Models

The first model includes only one covariate, distance.

```
# Model with just distance
model.HaleD <- glm(cbind(YesVotes, NumVotes - YesVotes) ~
    distance, family = binomial, data = rrHale.df)
# alternative expression
model.HaleD.alt <- glm(YesVotes / NumVotes ~ distance,
    weights = NumVotes, family = binomial, data = rrHale.df)
```

```
##              Estimate Std. Error z value  Pr(>|z|)
## (Intercept)    3.3093    0.11313   29.25 4.268e-188
## distance      -0.2876    0.01302  -22.08 4.447e-108
```

```
## Residual deviance =    318.4   on   9 df
## Dispersion parameter =    1
```

Our estimated binomial regression model is:

$$\log\left(\frac{\hat{p}_i}{1-\hat{p}_i}\right) = 3.309 - 0.288 \text{distance}_i$$

where $\hat{p}_i$ is the estimated proportion of Yes votes in community i. The estimated odds ratio for distance, that is the exponentiated coefficient for distance, in this model is $e^{-0.288} = 0.750$. It can be interpreted as follows: for each additional mile from the proposed railroad, the support (odds of a Yes vote) declines by 25.0%.

The covariate `pctBlack` is then added to the first model.

```
model.HaleBD <- glm(cbind(YesVotes, NumVotes - YesVotes) ~
  distance + pctBlack, family = binomial, data = rrHale.df)
```

```
##              Estimate Std. Error z value    Pr(>|z|)
## (Intercept)  4.22202    0.296963  14.217   7.155e-46
## distance    -0.29173    0.013100 -22.270   7.236e-110
## pctBlack    -0.01323    0.003897  -3.394   6.881e-04

## Residual deviance =    307.2   on   8 df
## Dispersion parameter =    1
```

Despite the somewhat strong negative correlation between percent black and distance, the estimated odds ratio for distance remains approximately the same in this new model (OR $= e^{-0.29} = 0.747$); controlling for percent black does little to change our estimate of the effect of distance. For each additional mile from the proposed railroad, odds of a Yes vote declines by 25.3% after adjusting for the racial composition of a community. We also see that, for a fixed distance from the proposed railroad, the odds of a Yes vote declines by 1.3% (OR $= e^{-.0132} = .987$) for each additional percent black in the community.

6.5.4 Tests for Significance of Model Coefficients

Do we have statistically significant evidence that support for the railroad referendum decreases with higher proportions of black residents in a community, after accounting for the distance a community is from the railroad line? As discussed in Section 4.4 with Poisson regression, there are two primary approaches to testing significance of model coefficients: **Drop-in-deviance test to compare models** and **Wald test for a single coefficient**.

With our larger model given by $\log\left(\frac{p_i}{1-p_i}\right) = \beta_0 + \beta_1 \text{distance}_i + \beta_2 \text{pctBlack}_i$,

6.5 Case Study: Reconstructing Alabama

the Wald test produces a highly significant p-value ($Z = \frac{-0.0132}{0.0039} = -3.394$, $p = .00069$) indicating significant evidence that support for the railroad referendum decreases with higher proportions of black residents in a community, after adjusting for the distance a community is from the railroad line.

The drop-in-deviance test would compare the larger model above to the reduced model $\log\left(\frac{p_i}{1-p_i}\right) = \beta_0 + \beta_1 \text{distance}_i$ by comparing residual deviances from the two models.

```
drop_in_dev <- anova(model.HaleD, model.HaleBD, test = "Chisq")
```

```
  ResidDF ResidDev Deviance Df       pval
1       9    318.4       NA NA         NA
2       8    307.2    11.22  1  0.0008083
```

The drop-in-deviance test statistic is $318.44 - 307.22 = 11.22$ on $9 - 8 = 1$ df, producing a p-value of $.00081$, in close agreement with the Wald test.

A third approach to determining significance of β_2 would be to generate a 95% confidence interval and then checking if 0 falls within the interval or, equivalently, if 1 falls within a 95% confidence interval for e^{β_2}. The next section describes two approaches to producing a confidence interval for coefficients in logistic regression models.

6.5.5 Confidence Intervals for Model Coefficients

Since the Wald statistic follows a normal distribution with n large, we could generate a Wald-type (normal-based) confidence interval for β_2 using:

$$\hat{\beta}_2 \pm 1.96 \cdot \text{SE}(\hat{\beta}_2)$$

and then exponentiating endpoints if we prefer a confidence interval for the odds ratio e^{β_2}. In this case,

$$\begin{aligned}
95\% \text{ CI for } \beta_2 &= \hat{\beta}_2 \pm 1.96 \cdot \text{SE}(\hat{\beta}_2) \\
&= -0.0132 \pm 1.96 \cdot 0.0039 \\
&= -0.0132 \pm 0.00764 \\
&= (-0.0208, -0.0056) \\
95\% \text{ CI for } e^{\beta_2} &= (e^{-0.0208}, e^{-0.0056}) \\
&= (.979, .994) \\
95\% \text{ CI for } e^{10\beta_2} &= (e^{-0.208}, e^{-0.056}) \\
&= (.812, .946)
\end{aligned}$$

Thus, we can be 95% confident that a 10% increase in the proportion of black residents is associated with a 5.4% to 18.8% decrease in the odds of a Yes vote for the railroad referendum after controlling for distance. This same relationship could be expressed as (a) between a 0.6% and a 2.1% decrease in odds for each 1% increase in the black population, or (b) between a 5.7% ($1/e^{-.056}$) and a 23.1% ($1/e^{-.208}$) increase in odds for each 10% decrease in the black population, after adjusting for distance. Of course, with $n = 11$, we should be cautious about relying on a Wald-type interval in this example.

Another approach available in R is the **profile likelihood method**, similar to Section 4.4.

```
exp(confint(model.HaleBD))
```

```
               2.5 %    97.5 %
(Intercept) 38.2285  122.6116
distance     0.7276    0.7660
pctBlack     0.9794    0.9945
```

In the model with `distance` and `pctBlack`, the profile likelihood 95% confidence interval for e^{β_2} is (.979, .994), which is approximately equal to the Wald-based interval despite the small sample size. We can also confirm the statistically significant association between percent black and odds of voting Yes (after controlling for distance), because 1 is not a plausible value of e^{β_2} (where an odds ratio of 1 would imply that the odds of voting Yes do not change with percent black).

6.5.6 Testing for Goodness-of-Fit

As in Section 4.4.9, we can evaluate the goodness-of-fit for our model by comparing the residual deviance (307.22) to a χ^2 distribution with $n - p$ (8) degrees of freedom.

```
1-pchisq(307.2173, 8)   # Goodness-of-fit test
```

[1] 0

The model with `pctBlack` and `distance` has statistically significant evidence of lack-of-fit ($p < .001$).

6.5 Case Study: Reconstructing Alabama

Similar to the Poisson regression models, this lack-of-fit could result from (a) missing covariates, (b) outliers, or (c) overdispersion. We will first attempt to address (a) by fitting a model with an interaction between distance and percent black, to determine whether the effect of racial composition differs based on how far a community is from the proposed railroad.

```
model.HaleBxD <- glm(cbind(YesVotes, NumVotes - YesVotes) ~
  distance + pctBlack + distance:pctBlack,
  family = binomial, data = rrHale.df)
```

```
##                    Estimate Std. Error z value
## (Intercept)        7.550902  0.6383697  11.828
## distance          -0.614005  0.0573808 -10.701
## pctBlack          -0.064731  0.0091723  -7.057
## distance:pctBlack  0.005367  0.0008984   5.974
##                    Pr(>|z|)
## (Intercept)        2.783e-32
## distance           1.012e-26
## pctBlack           1.698e-12
## distance:pctBlack  2.321e-09

##  Residual deviance =  274.2  on  7 df
##  Dispersion parameter =  1
```

```
drop_in_dev <- anova(model.HaleBD, model.HaleBxD,
                     test = "Chisq")
```

```
  ResidDF ResidDev Deviance Df      pval
1       8    307.2       NA NA        NA
2       7    274.2    32.98  1 9.294e-09
```

We have statistically significant evidence (Wald test: $Z = 5.974, p < .001$; Drop-in-deviance test: $\chi^2 = 32.984, p < .001$) that the effect of the proportion of the community that is black on the odds of voting Yes depends on the distance of the community from the proposed railroad.

To interpret the interaction coefficient in context, we will compare two cases: one where a community is right on the proposed railroad (`distance = 0`), and the other where the community is 15 miles away (`distance = 15`). The significant interaction implies that the effect of `pctBlack` should differ in these two cases. In the first case, the coefficient for `pctBlack` is -0.0647, while in the second case, the relevant coefficient is $-0.0647 + 15(.00537) = 0.0158$. Thus, for a community right on the proposed railroad, a 1% increase in percent black is associated with a 6.3% ($e^{-.0647} = .937$) decrease in the odds of voting Yes, while for a community 15 miles away, a 1% increase in percent black is associated with a ($e^{.0158} = 1.016$) 1.6% *increase* in the odds of voting Yes. A significant interaction term doesn't always imply a change in the direction of the association, but it does here.

Because our interaction model still exhibits lack-of-fit (residual deviance of 274.23 on just 7 df), and because we have used the covariates at our disposal, we will assess this model for potential outliers and overdispersion by examining the model's residuals.

6.5.7 Residuals for Binomial Regression

With LLSR, residuals were used to assess model assumptions and identify outliers. For binomial regression, two different types of residuals are typically used. One residual, the **Pearson residual**, has a form similar to that used with LLSR. Specifically, the Pearson residual is calculated using:

$$\text{Pearson residual}_i = \frac{\text{actual count} - \text{predicted count}}{\text{SD of count}} = \frac{Y_i - m_i \hat{p}_i}{\sqrt{m_i \hat{p}_i (1 - \hat{p}_i)}}$$

where m_i is the number of trials for the ith observation and $\hat{p}_i$ is the estimated probability of success for that same observation.

A **deviance residual** is an alternative residual for binomial regression based on the discrepancy between the observed values and those estimated using the likelihood. A deviance residual can be calculated for each observation using:

$$d_i = \text{sign}(Y_i - m_i \hat{p}_i) \sqrt{2[Y_i \log\left(\frac{Y_i}{m_i \hat{p}_i}\right) + (m_i - Y_i) \log\left(\frac{m_i - Y_i}{m_i - m_i \hat{p}_i}\right)]}$$

When the number of trials is large for all of the observations and the models are appropriate, both sets of residuals should follow a standard normal distribution.

The sum of the individual deviance residuals is referred to as the **deviance** or **residual deviance**. The residual deviance is used to assess the model. As the name suggests, a model with a small deviance is preferred. In the case of binomial regression, when the denominators, m_i, are large and a model fits, the residual deviance follows a χ^2 distribution with $n - p$ degrees of freedom (the residual degrees of freedom). Thus for a good fitting model the residual deviance should be approximately equal to its corresponding degrees of freedom. When binomial data meets these conditions, the deviance can be used for a

6.5 Case Study: Reconstructing Alabama

goodness-of-fit test. The p-value for lack-of-fit is the proportion of values from a χ^2_{n-p} distribution that are greater than the observed residual deviance.

We begin a residual analysis of our interaction model by plotting the residuals against the fitted values in Figure 6.4. This kind of plot for binomial regression would produce two linear trends with similar negative slopes if there were equal sample sizes m_i for each observation.

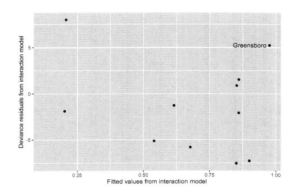

FIGURE 6.4: Fitted values by residuals for the interaction model for the Railroad Referendum data.

From this residual plot, Greensboro does not stand out as an outlier. If it did, we could remove Greensboro and refit our interaction model, checking to see if model coefficients changed in a noticeable way. Instead, we will continue to include Greensboro in our modeling efforts. Because the large residual deviance cannot be explained by outliers, and given we have included all of the covariates at hand as well as an interaction term, the observed binomial counts are likely overdispersed. This means that they exhibit more variation than the model would suggest, and we must consider ways to handle this overdispersion.

6.5.8 Overdispersion

Similar to Poisson regression, we can adjust for overdispersion in binomial regression. With overdispersion there is **extra-binomial variation**, so the actual variance will be greater than the variance of a binomial variable, $np(1-p)$. One way to adjust for overdispersion is to estimate a multiplier (dispersion parameter), $\hat{\phi}$, for the variance that will inflate it and reflect the reduction in the amount of information we would otherwise have with independent observations. We used a similar approach to adjust for overdispersion in a Poisson regression model in Section 4.9, and we will use the same estimate here: $\hat{\phi} = \dfrac{\sum (\text{Pearson residuals})^2}{n-p}$.

When overdispersion is adjusted for in this way, we can no longer use maximum likelihood to fit our regression model; instead we use a quasilikelihood approach. Quasilikelihood is similar to likelihood-based inference, but because the model uses the dispersion parameter, it is not a binomial model with a true likelihood (we call it **quasibinomial**). R offers quasilikelihood as an option when model fitting. The quasilikelihood approach will yield the same coefficient point estimates as maximum likelihood; however, the variances will be larger in the presence of overdispersion (assuming $\phi > 1$). We will see other ways in which to deal with overdispersion and clusters in the remaining chapters in the book, but the following describes how overdispersion is accounted for using $\hat{\phi}$:

Summary: accounting for overdispersion

- Use the dispersion parameter $\hat{\phi} = \frac{\sum (\text{Pearson residuals})^2}{n-p}$ to inflate standard errors of model coefficients.
- Wald test statistics: multiply the standard errors by $\sqrt{\hat{\phi}}$ so that $\text{SE}_Q(\hat{\beta}) = \sqrt{\hat{\phi}} \cdot \text{SE}(\hat{\beta})$ and conduct tests using the t-distribution.
- Confidence intervals use the adjusted standard errors and multiplier based on t, so they are thereby wider: $\hat{\beta} \pm t_{n-p} \cdot \text{SE}_Q(\hat{\beta})$.
- Drop-in-deviance test statistic comparing Model 1 (larger model with p parameters) to Model 2 (smaller model with $q < p$ parameters) is $F = \frac{1}{\hat{\phi}} \cdot \frac{D_2 - D_1}{p-q}$ where D_1 and D_2 are the residual deviances for models 1 and 2, respectively, and $p - q$ is the difference in the number of parameters for the two models. Note that both $D_2 - D_1$ and $p - q$ are positive. This test statistic is compared to an F-distribution with $p - q$ and $n - p$ degrees of freedom.

Output for a model which adjusts our interaction model for overdispersion appears below, where $\hat{\phi} = 51.6$ is used to adjust the standard errors for the coefficients and the drop-in-deviance tests during model building. Standard errors will be inflated by a factor of $\sqrt{51.6} = 7.2$. As a result, there are no significant terms in the adjusted interaction model below.

```
model.HaleBxDq <- glm(cbind(YesVotes, NumVotes - YesVotes) ~
  distance + pctBlack + distance:pctBlack,
  family = quasibinomial, data = rrHale.df)
```

```
##                     Estimate Std. Error  t value
## (Intercept)         7.550902   4.585464   1.6467
## distance           -0.614005   0.412171  -1.4897
## pctBlack           -0.064731   0.065885  -0.9825
## distance:pctBlack   0.005367   0.006453   0.8316
##                     Pr(>|t|)
```

6.5 Case Study: Reconstructing Alabama

```
## (Intercept)          0.1436
## distance             0.1799
## pctBlack             0.3586
## distance:pctBlack    0.4331

##   Residual deviance =   274.2  on   7 df
##   Dispersion parameter =   51.6
```

We therefore remove the interaction term and refit the model, adjusting for the extra-binomial variation that still exists.

```
model.HaleBDq <- glm(cbind(YesVotes, NumVotes - YesVotes) ~
  distance + pctBlack,
  family = quasibinomial, data = rrHale.df)
```

```
##                Estimate Std. Error t value Pr(>|t|)
## (Intercept)    4.22202    1.99031   2.1213  0.06669
## distance      -0.29173    0.08780  -3.3228  0.01050
## pctBlack      -0.01323    0.02612  -0.5064  0.62620

##   Residual deviance =   307.2  on   8 df
##   Dispersion parameter =   44.92
```

By removing the interaction term and using the overdispersion parameter, we see that distance is significantly associated with support, but percent black is no longer significant after adjusting for distance.

Because quasilikelihood methods do not change estimated coefficients, we still estimate a 25% decline $(1 - e^{-0.292})$ in support for each additional mile from the proposed railroad (odds ratio of .75).

```
exp(confint(model.HaleBDq))
```

```
                2.5 %    97.5 %
(Intercept)    1.3609  5006.722
distance       0.6091     0.871
pctBlack       0.9366     1.044
```

While we previously found a 95% confidence interval for the odds ratio associated with distance of (.728, .766), our confidence interval is now much wider: (.609, .871). Appropriately accounting for overdispersion has changed both the significance of certain terms and the precision of our coefficient estimates.

6.5.9 Summary

We began by fitting a logistic regression model with `distance` alone. Then we added the covariate `pctBlack`, and the Wald-type test and the drop-in-deviance test both provided strong support for the addition of `pctBlack` to the model. The model with `distance` and `pctBlack` had a large residual deviance suggesting an ill-fitted model. When we looked at the residuals, we saw that Greensboro is an extreme observation. Models without Greensboro were fitted and compared to our initial models. Seeing no appreciable improvement or differences with Greensboro removed, we left it in the model. There remained a large residual deviance so we attempted to account for it by using an estimated dispersion parameter similar to Section 4.9 with Poisson regression. The final model included distance and percent black, although percent black was no longer significant after adjusting for overdispersion.

6.6 Linear Least Squares vs. Binomial Regression

<u>Response</u>
LLSR : normal
Binomial Regression : number of successes in n trials

<u>Variance</u>
LLSR : equal for each level of X
Binomial Regression : $np(1-p)$ for each level of X

<u>Model Fitting</u>
LLSR : $\mu = \beta_0 + \beta_1 x$ using Least Squares
Binomial Regression : $\log\left(\dfrac{p}{1-p}\right) = \beta_0 + \beta_1 x$ using Maximum Likelihood

<u>EDA</u>
LLSR : plot X vs. Y; add line
Binomial Regression : find log(odds) for several subgroups; plot vs. X

Comparing Models
LLSR : extra sum of squares F-tests; AIC/BIC
Binomial Regression : drop-in-deviance tests; AIC/BIC

Interpreting Coefficients
LLSR : $\beta_1 =$ change in mean response for unit change in X
Binomial Regression : $e^{\beta_1} =$ percent change in odds for unit change in X

6.7 Case Study: Trying to Lose Weight

The final case study uses individual-specific information so that our response, rather than the number of successes out of some number of trials, is simply a binary variable taking on values of 0 or 1 (for failure/success, no/yes, etc.). This type of problem—**binary logistic regression**—is exceedingly common in practice. Here we examine characteristics of young people who are trying to lose weight. The prevalence of obesity among U.S. youth suggests that wanting to lose weight is sensible and desirable for some young people such as those with a high body mass index (BMI). On the flip side, there are young people who do not need to lose weight but make ill-advised attempts to do so nonetheless. A multitude of studies on weight loss focus specifically on youth and propose a variety of motivations for the young wanting to lose weight; athletics and the media are two commonly cited sources of motivation for losing weight for young people.

Sports have been implicated as a reason for young people wanting to shed pounds, but not all studies are consistent with this idea. For example, a study by Martinsen et al. [2009] reported that, despite preconceptions to the contrary, there was a higher rate of self-reported eating disorders among controls (non-elite athletes) as opposed to elite athletes. Interestingly, the kind of sport was not found to be a factor, as participants in leanness sports (for example, distance running, swimming, gymnastics, dance, and diving) did not differ in the proportion with eating disorders when compared to those in non-leanness sports. So, in our analysis, we will not make a distinction between different sports.

Other studies suggest that mass media is the culprit. They argue that students' exposure to unrealistically thin celebrities may provide unhealthy motivation for some, particularly young women, to try to slim down. An examination and analysis of a large number of related studies (referred to as a **meta-analysis**)

[Grabe et al., 2008] found a strong relationship between exposure to mass media and the amount of time that adolescents spend talking about what they see in the media, deciphering what it means, and figuring out how they can be more like the celebrities.

We are interested in the following questions: Are the odds that young females report trying to lose weight greater than the odds that males do? Is increasing BMI associated with an interest in losing weight, regardless of sex? Does sports participation increase the desire to lose weight? Is media exposure associated with more interest in losing weight?

We have a sample of 500 teens from data collected in 2009 through the U.S. Youth Risk Behavior Surveillance System (YRBSS) [Centers for Disease Control and Prevention, 2009]. The YRBSS is an annual national school-based survey conducted by the Centers for Disease Control and Prevention (CDC) and state, territorial, and local education and health agencies and tribal governments. More information on this survey can be found here[1].

6.7.1 Data Organization

Here are the three questions from the YRBSS we use for our investigation:

Q66. Which of the following are you trying to do about your weight?

- A. Lose weight
- B. Gain weight
- C. Stay the same weight
- D. I am not trying to do anything about my weight

Q81. On an average school day, how many hours do you watch TV?

- A. I do not watch TV on an average school day
- B. Less than 1 hour per day
- C. 1 hour per day
- D. 2 hours per day
- E. 3 hours per day
- F. 4 hours per day
- G. 5 or more hours per day

Q84. During the past 12 months, on how many sports teams did you play? (Include any teams run by your school or community groups.)

- A. 0 teams

---

[1]http://www.cdc.gov/HealthyYouth/yrbs/index.htm

6.7 Case Study: Trying to Lose Weight

TABLE 6.3: Mean BMI percentile by sex and desire to lose weight.

Sex	Weight loss status	mean BMI percentile	SD	n
Female	No weight loss	43.2	25.8	89
	Lose weight	72.4	23.0	125
Male	No weight loss	58.8	28.2	157
	Lose weight	85.7	18.0	74

- B. 1 team
- C. 2 teams
- D. 3 or more teams

Answers to Q66 are used to define our response variable: $Y = 1$ corresponds to "(A) trying to lose weight", while $Y = 0$ corresponds to the other non-missing values. Q84 provides information on students' sports participation and is treated as numerical, 0 through 3, with 3 representing 3 or more. As a proxy for media exposure, we use answers to Q81 as numerical values 0, 0.5, 1, 2, 3, 4, and 5, with 5 representing 5 or more. Media exposure and sports participation are also considered as categorical factors, that is, as variables with distinct levels which can be denoted by indicator variables as opposed to their numerical values.

BMI is included in this study as the percentile for a given BMI for members of the same sex. This facilitates comparisons when modeling with males and females. We will use the terms *BMI* and *BMI percentile* interchangeably with the understanding that we are always referring to the percentile.

With our sample, we use only the cases that include all of the data for these four questions. This is referred to as a **complete case analysis**. That brings our sample of 500 to 445. There are limitations of complete case analyses that we address in the Discussion.

6.7.2 Exploratory Data Analysis

Nearly half (44.7%) of our sample of 445 youths report that they are trying to lose weight, 48.1% of the sample are females, and 59.3% play on one or more sports teams. Also, 8.8% report that they do not watch any TV on school days, whereas another 13.0% watched 5 or more hours each day. Interestingly, the median BMI percentile for our 445 youths is 68. The most dramatic difference in the proportions of those who are trying to lose weight is by sex; 58% of the females want to lose weight in contrast to only 32% of the males (see Figure 6.5). This provides strong support for the inclusion of a sex term in every model considered.

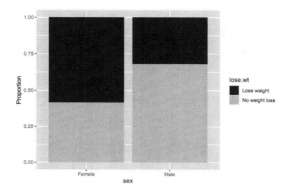

FIGURE 6.5: Weight loss plans vs. sex.

Table 6.3 displays the mean BMI of those wanting and not wanting to lose weight for males and females. The mean BMI is greater for those trying to lose weight compared to those not trying to lose weight, regardless of sex. The size of the difference is remarkably similar for the two sexes.

If we consider including a BMI term in our model(s), the logit should be linearly related to BMI. We can investigate this assumption by constructing an empirical logit plot. In order to calculate empirical logits, we first divide our data by sex. Within each sex, we generate 10 groups of equal sizes, the first holding the bottom 10% in BMI percentile for that sex, the second holding the next lowest 10%, etc. Within each group, we calculate the proportion, $\hat{p}$ that reported wanting to lose weight, and then the empirical log odds, $log(\frac{\hat{p}}{1-\hat{p}})$, that a young person in that group wants to lose weight.

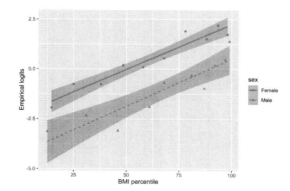

FIGURE 6.6: Empirical logits of trying to lose weight by BMI and sex.

Figure 6.6 presents the empirical logits for the BMI intervals by sex. Both males

6.7 Case Study: Trying to Lose Weight

and females exhibit an increasing linear trend on the logit scale indicating that increasing BMI is associated with a greater desire to lose weight and that modeling log odds as a linear function of BMI is reasonable. The slope for the females appears to be similar to the slope for males, so we do not need to consider an interaction term between BMI and sex in the model.

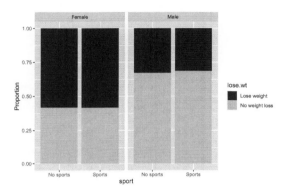

FIGURE 6.7: Weight loss plans vs. sex and sports participation.

Out of those who play sports, 44% want to lose weight, whereas 46% want to lose weight among those who do not play sports. Figure 6.7 compares the proportion of respondents who want to lose weight by their sex and sport participation. The data suggest that sports participation is associated with the same or even a slightly lower desire to lose weight, contrary to what had originally been hypothesized. While the overall levels of those wanting to lose weight differ considerably between the sexes, the differences between those in and out of sports within sex appear to be very small. A term for sports participation or number of teams will be considered, but there is not compelling evidence that an interaction term will be needed.

It was posited that increased exposure to media, here measured as hours of TV daily, is associated with increased desire to lose weight, particularly for females. Overall, the percentage who want to lose weight ranges from 38% of those watching 5 hours of TV per day to 55% among those watching 2 hours daily. There is minimal variation in the proportion wanting to lose weight with both sexes combined. However, we are more interested in differences between the sexes (see Figure 6.8). We create empirical logits using the proportion of students trying to lose weight for each level of hours spent watching TV daily and look at the trends in the logits separately for males and females. From Figure 6.9, there does not appear to be a linear relationship for males or females.

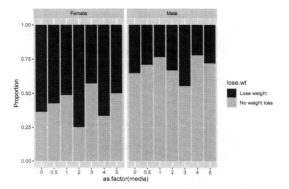

FIGURE 6.8: Weight loss plans vs. daily hours of TV and sex.

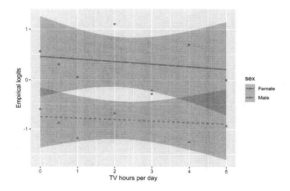

FIGURE 6.9: Empirical logits for the odds of trying to lose weight by TV watching and sex.

6.7.3 Initial Models

Our strategy for modeling is to use our questions of interest and what we have learned in the exploratory data analysis. For each model we interpret the coefficient of interest, look at the corresponding Wald test and, as a final step, compare the deviances for the different models we considered.

We first use a model where sex is our only predictor.

```
model1 <- glm(lose.wt.01 ~ female, family = binomial,
              data = risk2009)
```

```
##              Estimate Std. Error z value Pr(>|z|)
## (Intercept)   -0.7522     0.1410  -5.334 9.588e-08
```

6.7 Case Study: Trying to Lose Weight

```
## female              1.0919      0.1978    5.520 3.382e-08

## Residual deviance =   580.3   on  443 df
```

Our estimated binomial regression model is:

$$\log\left(\frac{\hat{p}}{1-\hat{p}}\right) = -0.75 + 1.09\text{female}$$

where $\hat{p}$ is the estimated proportion of youth wanting to lose weight. We can interpret the coefficient on `female` by exponentiating $e^{1.0919} = 2.98$ (95% CI = $(2.03, 4.41)$) indicating that the odds of a female trying to lose weight is nearly three times the odds of a male trying to lose weight ($Z = 5.520$, $p = 3.38e-08$). We retain sex in the model and consider adding the BMI percentile:

```
model2 <- glm(lose.wt.01 ~ female + bmipct,
              family = binomial, data = risk2009)
```

```
##               Estimate Std. Error z value  Pr(>|z|)
## (Intercept)  -4.25914    0.44927  -9.480  2.541e-21
## female        1.86067    0.25896   7.185  6.714e-13
## bmipct        0.04715    0.00524   8.997  2.313e-19

## Residual deviance =  463  on  442 df
```

We see that there is statistically significant evidence ($Z = 8.997, p < .001$) that BMI is positively associated with the odds of trying to lose weight, after controlling for sex. Clearly BMI percentile belongs in the model with sex.

Our estimated binomial regression model is:

$$\log\left(\frac{\hat{p}}{1-\hat{p}}\right) = -4.26 + 1.86\text{female} + 0.047\text{bmipct}$$

To interpret the coefficient on `bmipct`, we will consider a 10-unit increase in `bmipct`. Because $e^{10*0.047} = 1.602$, then there is an estimated 60.2% increase in the odds of wanting to lose weight for each additional 10 percentile points of BMI for members of the same sex. Just as we had done in other multiple regression models, we need to interpret our coefficient *given that the other variables remain constant*. An interaction term for BMI by sex was tested (not shown) and it was not significant ($Z = -0.70, p = 0.485$), so the effect of BMI does not differ by sex.

We next add sport to our model. Sports participation was considered for inclusion in the model in three ways: an indicator of sports participation (0 = no teams, 1 = one or more teams), treating the number of teams (0, 1, 2, or 3) as numeric, and treating the number of teams as a factor. The models below treat sports participation using an indicator variable, but all three models produced similar results.

```
model3 <- glm(lose.wt.01 ~ female + bmipct + sport,
              family = binomial, data = risk2009)
```

```
##                Estimate Std. Error z value  Pr(>|z|)
## (Intercept)    -4.17138   0.468463 -8.9044  5.367e-19
## female          1.84951   0.259514  7.1268  1.027e-12
## bmipct          0.04728   0.005251  9.0032  2.193e-19
## sportSports    -0.14767   0.235101 -0.6281  5.299e-01
```

```
##   Residual deviance =  462.6   on   441 df
```

```
model3int <- glm(lose.wt.01 ~ female + bmipct + sport +
                 female:sport + bmipct:sport,
                 family = binomial, data = risk2009)
```

```
##                       Estimate Std. Error z value
## (Intercept)           -3.643635   0.604821  -6.024
## female                 1.451017   0.378547   3.833
## bmipct                 0.042530   0.007211   5.898
## sportSports           -1.187199   0.893057  -1.329
## female:sportSports     0.731516   0.523566   1.397
## bmipct:sportSports     0.009908   0.010463   0.947
##                       Pr(>|z|)
## (Intercept)           1.698e-09
## female                1.265e-04
## bmipct                3.684e-09
## sportSports           1.837e-01
## female:sportSports    1.624e-01
## bmipct:sportSports    3.436e-01
```

```
##   Residual deviance =  460.5   on   439 df
```

6.7 Case Study: Trying to Lose Weight

Sports teams were not significant in any of these models, nor were interaction terms (sex by sports and bmipct by sports). As a result, sports participation was no longer considered for inclusion in the model.

We last look at adding `media` to our model.

```
model4 <- glm(lose.wt.01 ~ female + bmipct + media,
              family = binomial, data = risk2009)
```

```
##              Estimate Std. Error z value  Pr(>|z|)
## (Intercept) -4.08892    0.462947  -8.832  1.025e-18
## female       1.84776    0.259636   7.117  1.105e-12
## bmipct       0.04783    0.005287   9.046  1.485e-19
## media       -0.09938    0.072464  -1.371  1.702e-01

##   Residual deviance =  461.1  on  441 df
```

Media is not a statistically significant term ($Z = -1.371$, $p = 0.170$). However, because our interest centers on how media may affect attempts to lose weight and how its effect might be different for females and males, we fit a model with a media term and a sex by media interaction term (not shown). Neither term was statistically significant, so we have no support in our data that media exposure as measured by hours spent watching TV is associated with the odds a teen is trying to lose weight after accounting for sex and BMI.

6.7.4 Drop-in-Deviance Tests

```
drop_in_dev <- anova(model1, model2, model3, model4,
                     test="Chisq")
```

```
  ResidDF ResidDev Deviance Df        pval
1     443    580.3       NA NA          NA
2     442    463.0 117.3301  1   2.431e-27
3     441    462.6   0.3947  1   5.298e-01
4     441    461.1   1.5007  0          NA

        df   AIC
model1   2 584.3
model2   3 469.0
model3   4 470.6
model4   4 469.1
```

Comparing models using differences in deviances requires that the models be **nested**, meaning each smaller model is a simplified version of the larger model. In our case, Models 1, 2, and 4 are nested, as are Models 1, 2, and 3, but Models 3 and 4 cannot be compared using a drop-in-deviance test.

There is a large drop-in-deviance adding BMI to the model with sex (Model 1 to Model 2, 117.3), which is clearly statistically significant when compared to a χ^2 distribution with 1 df. The drop-in-deviance for adding an indicator variable for sports to the model with sex and BMI is only 462.99 - 462.59 = 0.40. There is a difference of a single parameter, so the drop-in-deviance would be compared to a χ^2 distribution with 1 df. The resulting p-value is very large (.53) suggesting that adding an indicator for sports is not helpful once we've already accounted for BMI and sex. For comparing Models 3 and 4, one approach is to look at the AIC. In this case, the AIC is (barely) smaller for the model with media, providing evidence that the latter model is slightly preferable.

6.7.5 Model Discussion and Summary

We found that the odds of wanting to lose weight are considerably greater for females compared to males. In addition, respondents with greater BMI percentiles express a greater desire to lose weight for members of the same sex. Regardless of sex or BMI percentile, sports participation and TV watching are not associated with different odds for wanting to lose weight.

A limitation of this analysis is that we used complete cases in place of a method of imputing responses or modeling missingness. This reduced our sample from 500 to 445, and it may have introduced bias. For example, if respondents who watch a lot of TV were unwilling to reveal as much, and if they differed with respect to their desire to lose weight from those respondents who reported watching little TV, our inferences regarding the relationship between lots of TV and desire to lose weight may be biased.

Other limitations may result from definitions. Trying to lose weight is self-reported and may not correlate with any action undertaken to do so. The number of sports teams may not accurately reflect sports-related pressures to lose weight. For example, elite athletes may focus on a single sport and be subject to greater pressures, whereas athletes who casually participate in three sports may not feel any pressure to lose weight. Hours spent watching TV are not likely to encompass the totality of media exposure, particularly because exposure to celebrities occurs often online. Furthermore, this analysis does not explore in any detail maladaptions—inappropriate motivations for wanting to lose weight. For example, we did not focus our study on subsets of respondents with low BMI who are attempting to lose weight.

6.8 Exercises

It would be instructive to use data science methodologies to explore the entire data set of 16,000 instead of sampling 500. However, the types of exploration and models used here could translate to the larger sample size.

Finally a limitation may be introduced as a result of the acknowledged variation in the administration of the YRBSS. States and local authorities are allowed to administer the survey as they see fit, which at times results in significant variation in sample selection and response.

6.8 Exercises

6.8.1 Conceptual Exercises

1. List the explanatory and response variable(s) for each research question.

    a. Are students with poor grades more likely to binge drink?
    b. What is the chance you are accepted into medical school given your GPA and MCAT scores?
    c. Is a single mom more likely to marry the baby's father if she has a boy?
    d. Are students participating in sports in college more or less likely to graduate?
    e. Is exposure to a particular chemical associated with a cancer diagnosis?

2. Interpret the odds ratios in the following abstract.

    Daycare Centers and Respiratory Health [Nafstad et al., 1999]

    - **Objective**. To estimate the effects of the type of daycare on respiratory health in preschool children.

    - **Methods**. A population-based, cross-sectional study of Oslo children born in 1992 was conducted at the end of 1996. A self-administered questionnaire inquired about daycare arrangements, environmental conditions, and family characteristics (n = 3853; response rate, 79%).

    - **Results**. In a logistic regression controlling for confounding, children in daycare centers had more often nightly cough (adjusted odds ratio, 1.89; 95% confidence interval 1.34-2.67), and blocked or runny nose without common cold (1.55; 1.07-1.61) during the past 12 months compared with children in home care.

TABLE 6.4: Data for Conceptual Exercise 4.

Turbine group	1	2	3	4
Humidity	Low	Low	High	High
n = number of turbine wheels	3	3	3	3
y = number of fissures	1	2	1	0

3. Construct a table and calculate the corresponding odds and odds ratios. Comment on the reported and calculated results in this *New York Times* article from Kolata [2009].

    • In November, the Centers for Disease Control and Prevention published a paper reporting that babies conceived with IVF, or with a technique in which sperm are injected directly into eggs, have a slightly increased risk of several birth defects, including a hole between the two chambers of the heart, a cleft lip or palate, an improperly developed esophagus and a malformed rectum. The study involved 9,584 babies with birth defects and 4,792 babies without. Among the mothers of babies without birth defects, 1.1% had used IVF or related methods, compared with 2.4% of mothers of babies with birth defects.

    • The findings are considered preliminary, and researchers say they believe IVF does not carry excessive risks. There is a 3% chance that any given baby will have a birth defect.

4. In a small pilot study, researchers compared two groups of 3 turbine wheels each under low humidity and two groups of 3 turbine wheels each under high-humidity conditions to determine if humidity is related to the number of fissures that occur. If Y = number of turbine wheels that develop fissures, then assume that $Y \sim$ Binomial($n = 3, p = p_L$) under low humidity, and $Y \sim$ Binomial($n = 3, p = p_H$) under high humidity, where $f(y; p) = \binom{n}{y} p^y (1-p)^{n-y}$. Write out the log-likelihood function $\log L(p_L, p_H)$, using the data in Table 6.4 and simplifying where possible.

6.8.2 Guided Exercises

1. **Soccer goals on target.** Data comes from an article in *Psychological Science* [Roskes et al., 2011]. The authors report on the success rate of penalty kicks that were on-target, so that either the keeper saved the shot or the shot scored, for FIFA World Cup shootouts between 1982 and 2010. They found that 18 out of 20 shots were

6.8 Exercises

scored when the goalkeeper's team was behind, 71 out of 90 shots were scored when the game was tied, and 55 out of 75 shots were scored with the goalkeeper's team ahead.

   a. Calculate the odds of a successful penalty kick for games in which the goalkeeper's team was behind, tied, or ahead. Then, construct empirical odds ratios for successful penalty kicks for (a) behind versus tied, and (b) tied versus ahead.
   b. Fit a model with the categorical predictor c("behind","tied","ahead") and interpret the exponentiated coefficients. How do they compare to the empirical odds ratios you calculated?

2. **Medical school admissions.** The data for Medical School Admissions is in MedGPA.csv, taken from undergraduates from a small liberal arts school over several years. We are interested in student attributes that are associated with higher acceptance rates.

   - Accept = accepted (A) into medical school or denied (D)
   - Acceptance = accepted (1) into medical school or denied (0)
   - Sex = male (M) or female (F)
   - BCPM = GPA in natural sciences and mathematics
   - GPA = overall GPA
   - VR = verbal reasoning subscale score of the MCAT
   - PS = physical sciences subscale score of the MCAT
   - WS = writing samples subscale score of the MCAT
   - BS = biological sciences subscale score of the MCAT
   - MCAT = MCAT total score
   - Apps = number of schools applied to

   Be sure to interpret model coefficients and associated tests of significance or confidence intervals when answering the following questions:

   a. Compare the relative effects of improving your MCAT score versus improving your GPA on your odds of being accepted to medical school.
   b. After controlling for MCAT and GPA, is the number of applications related to odds of getting into medical school?
   c. Is one MCAT subscale more important than the others?
   d. Is there any evidence that the effect of MCAT score or GPA differs for males and females?

3. **Moths.** An article in the *Journal of Animal Ecology* by Bishop [1972] investigated whether moths provide evidence of "survival of the fittest" with their camouflage traits. Researchers glued equal numbers of light and dark morph moths in lifelike positions on tree trunks at 7 locations from 0 to 51.2 km from Liverpool. They then

recorded the number of moths removed after 24 hours, presumably by predators. The hypothesis was that, since tree trunks near Liverpool were blackened by pollution, light morph moths would be more likely to be removed near Liverpool.

Data [Ramsey and Schafer, 2002] can be found in `moth.csv` and contains the variables below. In addition, R code at the end of the problem can be used to input the data and create additional useful variables.

- `MORPH` = light or dark
- `DISTANCE` = kilometers from Liverpool
- `PLACED` = number of moths of a specific morph glued to trees at that location
- `REMOVED` = number of moths of a specific morph removed after 24 hours

a. What are logits in this study?
b. Create an empirical logit plot of logits vs. distance by morph. What can we conclude from this plot?
c. Create a model with `DISTANCE` and `dark`. Interpret all the coefficients.
d. Create a model with `DISTANCE`, `dark`, and the interaction between both variables. Interpret all the coefficients.
e. Interpret a drop-in-deviance test and a Wald test to test the significance of the interaction term in (d).
f. Test the goodness-of-fit for the interaction model. What can we conclude about this model?
g. Is there evidence of overdispersion in the interaction model? What factors might lead to overdispersion in this case? Regardless of your answer, repeat (d) adjusting for overdispersion.
h. Compare confidence intervals for coefficients in your models from (g) and (d).
i. What happens if we expand the data set to contain one row per moth (968 rows)? Now we can run a logistic binary regression model. How does the logistic binary regression model compare to the binomial regression model? What are similarities and differences? Would there be any reason to run a logistic binomial regression rather than a logistic binary regression in a case like this? Some starter code can be found below the input code.

```
moth <- read_csv("data/moth.csv")
moth <- mutate(moth,
               notremoved = PLACED - REMOVED,
```

```
                    logit1 = log(REMOVED / notremoved),
                    prop1 = REMOVED / PLACED,
                    dark = ifelse(MORPH=="dark",1,0) )

mtemp1 = rep(moth$dark[1],moth$REMOVED[1])
dtemp1 = rep(moth$DISTANCE[1],moth$REMOVED[1])
rtemp1 = rep(1,moth$REMOVED[1])
mtemp1 = c(mtemp1,rep(moth$dark[1],
                        moth$PLACED[1]-moth$REMOVED[1]))
dtemp1 = c(dtemp1,rep(moth$DISTANCE[1],
                        moth$PLACED[1]-moth$REMOVED[1]))
rtemp1 = c(rtemp1,rep(0,moth$PLACED[1]-moth$REMOVED[1]))
for(i in 2:14)  {
  mtemp1 = c(mtemp1,rep(moth$dark[i],moth$REMOVED[i]))
  dtemp1 = c(dtemp1,rep(moth$DISTANCE[i],moth$REMOVED[i]))
  rtemp1 = c(rtemp1,rep(1,moth$REMOVED[i]))
  mtemp1 = c(mtemp1,rep(moth$dark[i],
                          moth$PLACED[i]-moth$REMOVED[i]))
  dtemp1 = c(dtemp1,rep(moth$DISTANCE[i],
                          moth$PLACED[i]-moth$REMOVED[i]))
  rtemp1 = c(rtemp1,rep(0,moth$PLACED[i]-moth$REMOVED[i]))  }
newdata = data.frame(removed=rtemp1,dark=mtemp1,dist=dtemp1)
newdata[1:25,]
cdplot(as.factor(rtemp1)~dtemp1)
```

4. **Birdkeeping and lung cancer: a retrospective observational study.** A 1972-1981 health survey in The Hague, Netherlands, discovered an association between keeping pet birds and increased risk of lung cancer. To investigate birdkeeping as a risk factor, researchers conducted a case-control study of patients in 1985 at four hospitals in The Hague. They identified 49 cases of lung cancer among patients who were registered with a general practice, who were age 65 or younger, and who had resided in the city since 1965. Each patient (case) with cancer was matched with two control subjects (without cancer) by age and sex. Further details can be found in Holst et al. [1988].

Age, sex, and smoking history are all known to be associated with lung cancer incidence. Thus, researchers wished to determine after age, sex, socioeconomic status, and smoking have been controlled for, is an additional risk associated with birdkeeping? The data [Ramsey and Schafer, 2002] is found in birdkeeping.csv, and the variables are listed below. In addition, R code at the end of the problem can be used to input the data and create additional useful variables.

- `female` = sex (1 = Female, 0 = Male)
- `age` = age, in years
- `highstatus` = socioeconomic status (1 = High, 0 = Low), determined by the occupation of the household's primary wage earner
- `yrsmoke` = years of smoking prior to diagnosis or examination
- `cigsday` = average rate of smoking, in cigarettes per day
- `bird` = indicator of birdkeeping (1 = Yes, 0 = No), determined by whether or not there were caged birds in the home for more than 6 consecutive months from 5 to 14 years before diagnosis (cases) or examination (controls)
- `cancer` = indicator of lung cancer diagnosis (1 = Cancer, 0 = No Cancer)

a. Perform an exploratory data analysis to see how each explanatory variable is related to the response (`cancer`). Summarize each relationship in one sentence.

- For quantitative explanatory variables (`age`, `yrsmoke`, `cigsday`), produce a cdplot, a boxplot, and summary statistics by cancer diagnosis.
- For categorical explanatory variables (`female` or `sex`, `highstatus` or `socioecon_status`, `bird` or `keep_bird`), produce a segmented bar chart and an appropriate table of proportions showing the relationship with cancer diagnosis.

b. In (a), you should have found no relationship between whether or not a patient develops lung cancer and either their age or sex. Why might this be? What implications will this have on your modeling?

c. Based on a two-way table with keeping birds and developing lung cancer from (a), find an unadjusted odds ratio comparing birdkeepers to non-birdkeepers and interpret this odds ratio in context. (Note: an *unadjusted* odds ratio is found by *not* controlling for any other variables.) Also, find an analogous relative risk and interpret it in context as well.

d. Are the elogits reasonably linear relating number of years smoked to the estimated log odds of developing lung cancer? Demonstrate with an appropriate plot.

e. Does there appear to be an interaction between number of years smoked and whether the subject keeps a bird? Demonstrate with an interaction plot and a coded scatterplot with empirical logits on the y-axis.

6.8 Exercises

Before answering the next questions, fit logistic regression models in R with `cancer` as the response and the following sets of explanatory variables:

- model1 = age, yrsmoke, cigsday, female, highstatus, bird
- model2 = yrsmoke, cigsday, highstatus, bird
- model4 = yrsmoke, bird
- model5 = the complete second order version of model4 (add squared terms and an interaction)
- model6 = yrsmoke, bird, yrsmoke:bird

f. Is there evidence that we can remove `age` and `female` from our model? Perform an appropriate test comparing model1 to model2; give a test statistic and p-value, and state a conclusion in context.

g. Is there evidence that the complete second order version of model4 improves its performance? Perform an appropriate test comparing model4 to model5; give a test statistic and p-value, and state a conclusion in context.

h. Carefully interpret each of the four model coefficients in model6 in context.

i. If you replaced `yrsmoke` everywhere it appears in model6 with a mean-centered version of `yrsmoke`, tell what would change among these elements: the 4 coefficients, the 4 p-values for coefficients, and the residual deviance.

j. model4 is a potential final model based on this set of explanatory variables. Find and carefully interpret 95% confidence intervals based on profile likelihoods for the coefficients of `yrsmoke` and `bird`.

k. How does the adjusted odds ratio for birdkeeping from model4 compare with the unadjusted odds ratio you found in (c)? Is birdkeeping associated with a significant increase in the odds of developing lung cancer, even after adjusting for other factors?

l. Use the categorical variable `years_factor` based on `yrsmoke` and replace `yrsmoke` in model4 with your new variable to create model4a. First, interpret the coefficient for `years_factorOver 25 years` in context. Then tell if you prefer model4 with years smoked as a numeric predictor or model4a with years smoked as a categorical predictor, and explain your reasoning.

m. Discuss the scope of inference in this study. Can we generalize our findings beyond the subjects in this study? Can we conclude that birdkeeping causes increased odds of developing lung cancer? Do you have other concerns with this study design or the analysis you carried out?

n. Read the article that appeared in the *British Medical Journal*. What similarities and differences do you see between their analyses and yours? What are a couple of things you learned from the article that weren't apparent in the short summary at the beginning of the assignment.

```
birds <- read_csv("data/birdkeeping.csv") %>%
  mutate(sex = ifelse(female == 1, "Female", "Male"),
         socioecon_status = ifelse(highstatus == 1,
                                   "High", "Low"),
         keep_bird = ifelse(bird == 1, "Keep Bird", "No Bird"),
         lung_cancer = ifelse(cancer == 1, "Cancer",
                              "No Cancer")) %>%
  mutate(years_factor = cut(yrsmoke,
                            breaks = c(-Inf, 0, 25, Inf),
                            labels = c("No smoking", "1-25 years",
                                       "Over 25 years")))
```

5. **2016 Election.** An award-winning[2] student project [Blakeman et al., 2018] examined driving forces behind Donald Trump's surprising victory in the 2016 Presidential Election, using data from nearly 40,000 voters collected as part of the 2016 Cooperative Congressional Election Survey (CCES). The student researchers investigated two theories: (1) Trump was seen as the candidate of change for voters experiencing economic hardship, and (2) Trump exploited voter fears about immigrants and minorities.

    The data set `electiondata.csv` has individual level data on voters in the 2016 Presidential Election, collected from the CCES and subsequently tidied. We will focus on the following variables:

    - Vote = 1 if Trump; 0 if another candidate
    - zfaminc = family income expressed as a z-score (number of standard deviations above or below the mean)
    - zmedinc = state median income expressed as a z-score
    - EconWorse = 1 if the voter believed the economy had gotten worse in the past 4 years; 0 otherwise
    - EducStatus = 1 if the voter had at least a bachelor's degree; 0 otherwise
    - republican = 1 if the voter identified as Republican; 0 otherwise

---

[2]https://www.causeweb.org/usproc/usclap/2018/spring/winners

- Noimmigrants = 1 if the voter supported at least 1 of 2 anti-immigrant policy statements; 0 if neither
- propforeign = proportion foreign born in the state
- evangelical = 1 if pew_bornagain is 2; otherwise 0

The questions below address Theory 1 (Economic Model). We want to see if there is significant evidence that voting for Trump was associated with family income level and/or with a belief that the economy became worse during the Obama Administration.

a. Create a plot showing the relationship between whether voters voted for Trump and their opinion about the status of the economy. What do you find?
b. Repeat (a) separately for Republicans and non-Republicans. Again describe what you find.
c. Create a plot with one observation per state showing the relationship between a state's median income and the log odds of a resident of that state voting for Trump. What can you conclude from this plot?

Answer (d)-(f) based on `model1a` below:

d. Interpret the coefficient for `zmedinc` in context.
e. Interpret the coefficient for `republican` in context.
f. Interpret the coefficient for `EconWorse:republican` in context. What does this allow us to conclude about Theory 1?

g. Repeat the above process for Theory 2 (Immigration Model). That is, produce meaningful exploratory plots, fit a model to the data with special emphasis on `Noimmigrants`, interpret meaningful coefficients, and state what can be concluded about Theory 2.
h. Is there any concern about the independence assumption in these models? We will return to this question in later chapters.

```
model1a <- glm(Vote01 ~ zfaminc + zmedinc + EconWorse +
    EducStatus + republican + EducStatus:republican +
    EconWorse:zfaminc + EconWorse:republican,
    family = binomial, data = electiondata)
summary(model1a)
```

6.8.3 Open-Ended Exercises

1. **2008 Presidential voting in Minnesota counties.** Data in `mn08.csv` contains results from the 2008 U.S. Presidential Election

by county in Minnesota, focusing on the two primary candidates (Democrat Barack Obama and Republican John McCain). You can consider the response to be either the percent of Obama votes in a county (binomial) or whether or not Obama had more votes than McCain (binary). Then build a model for your response using county-level predictors listed below. Interpret the results of your model.

- `County` = county name
- `Obama` = total votes for Obama
- `McCain` = total votes for McCain
- `pct_Obama` = percent of votes for Obama
- `pct_rural` = percent of county who live in a rural setting
- `medHHinc` = median household income
- `unemp_rate` = unemployment rate
- `pct_poverty` = percent living below the poverty line
- `medAge2007` = median age in 2007
- `medAge2000` = median age in 2000
- `Gini_Index` = measure of income disparity in a county
- `pct_native` = percent of native born residents

2. **Crime on campus.** The data set `c_data2.csv` contains statistics on violent crimes and property crimes for a sample of 81 U.S. colleges and universities. Characterize rates of violent crimes as a proportion of total crimes reported (i.e., `num_viol` / `total_crime`). Do they differ based on type of institution, size of institution, or region of the country?

- `Enrollment` = number of students enrolled
- `type` = university (U) or college (C)
- `num_viol` = number of violent crimes reported
- `num_prop` = number of property crimes reported
- `viol_rate_10000` = violent crime rate per 10,000 students enrolled
- `prop_rate_10000` = property crime rate per 10,000 students enrolled
- `total_crime` = total crimes reported (property and violent)
- `region` = region of the country

3. **NBA data.** Data in `NBA1718team.csv` [Kaggle, 2018b] looks at factors that are associated with a professional basketball team's winning percentage in the 2017-18 season. After thorough exploratory data analyses, create the best model you can to predict a team's winning percentage; be careful of collinearity between the covariates. Based on your EDA and modeling, describe the factors that seem to explain a team's success.

6.8 Exercises

- `win_pct` = Percentage of Wins,
- `FT_pct` = Average Free Throw Percentage per game,
- `TOV` = Average Turnovers per game,
- `FGA` = Average Field Goal Attempts per game,
- `FG` = Average Field Goals Made per game,
- `attempts_3P` = Average 3 Point Attempts per game,
- `avg_3P_pct` = Average 3 Point Percentage per game,
- `PTS` = Average Points per game,
- `OREB` = Average Offensive Rebounds per game,
- `DREB` = Average Defensive Rebounds per game,
- `REB` = Average Total Rebounds per game,
- `AST` = Average Assists per game,
- `STL` = Average Steals per game,
- `BLK` = Average Blocks per game,
- `PF` = Average Fouls per game,
- `attempts_2P` = Average 2 Point Attempts per game

4. **Trashball.** Great for a rainy day! A fun way to generate overdispersed binomial data. Each student crumbles an 8.5 by 11 inch sheet and tosses it from three prescribed distances ten times each. The response is the number of made baskets out of 10 tosses, keeping track of the distance. Have the class generate and collect potential covariates, and include them in your data set (e.g., years of basketball experience, using a tennis ball instead of a sheet of paper, height). Some sample analysis steps:

   a. Create scatterplots of logits vs. continuous predictors (distance, height, shot number, etc.) and boxplots of logit vs. categorical variables (sex, type of ball, etc.). Summarize important trends in one or two sentences.
   b. Create a graph with empirical logits vs. distance plotted separately by type of ball. What might you conclude from this plot?
   c. Find a binomial model using the variables that you collected. Give a brief discussion on your findings.

7
Correlated Data

7.1 Learning Objectives

After finishing this chapter, you should be able to:

- Given a data structure, recognize when there is potential for correlation.
- Identify observational units at varying levels.
- Provide reasons why correlated observations may cause problems when modeling if they are not addressed.
- Describe overdispersion and why it might occur.
- Understand how correlation in data can be taken into account using random effects models.
- Describe differences in the kind of inferences that can be made using fixed versus random effects models.
- Use software to fit a model taking correlation into account.

```
# Packages required for Chapter 7
library(gridExtra)
library(knitr)
library(kableExtra)
library(lme4)
library(ICC)
library(knitr)
library(tidyverse)
```

7.2 Introduction

Introductory statistics courses typically require responses which are approximately normal and independent of one another. We saw from the first chapters in this book that there are models for non-normal responses, so we have already

TABLE 7.1: Summary of simulations for Dams and Pups case study.

Scenario	Model	Model Name	β_0	SE β_0	t	p value	ϕ	Est prob	CI prob	Mean count	SD count	GOF p value
1a	Binomial	fit_1a_binom										
	Quasibinomial	fit_1a_quasi								X	X	X
1b	Binomial	fit_1b_binom										
	Quasibinomial	fit_1b_quasi								X	X	X

Scenario	Model	Model Name	β_1	SE β_1	t	p value	ϕ	Est odds ratio	CI odds ratio	Mean count Dose=1	SD count Dose=1	GOF p value
2a	Binomial	fit_2a_binom										
	Quasibinomial	fit_2a_quasi								X	X	X
2b	Binomial	fit_2b_binom										
	Quasibinomial	fit_2b_quasi								X	X	X

broadened the types of applications we can handle. In this chapter, we relax an additional assumption from introductory statistics, that of independence. Here we introduce methods that allow correlation to be taken into account. When modeling, correlation can be considered for normal and non-normal responses. Taken together, we can handle many more applications than the average introductory student.

In this chapter, we first focus on recognizing data structures that may imply correlation, introducing new terminology for discussing correlated data and its effects. Next, we consider potential problems correlated outcomes may cause and why we need to take correlation into account when modeling. Models which take correlation into account are then described and fit with the help of R.

This chapter will feel different than other chapters in this book. In order to understand correlated data and contrast it with independent observations, we will be exploring several simulations and associated thought questions. In that way, you should develop a feel for correlated data which will carry through into the remaining chapters of this book. In addition, as you proceed through the simulations, fill out Table 7.1; this will help you see patterns as you compare analyses which account for correlation with those that do not.

7.3 Recognizing Correlation

Correlated data is encountered in nearly every field. In education, student scores from a particular teacher are typically more similar than scores of other students who have had a different teacher. During a study measuring depression indices weekly over the course of a month, we usually find that four measures for the same patient tend to be more similar than depression indices from other patients. In political polling, opinions from members of

7.5 Case Study: Dams and Pups   195

the same household are usually more similar than opinions of members from other randomly selected households. The structure of these data sets suggest inherent patterns of similarities or correlation among outcomes. This kind of correlation specifically concerns correlation of observations *within the same teacher or patient or household* and is referred to as **intraclass correlation**.

Correlated data often takes on a multilevel structure. That is, population elements are grouped into aggregates, and we often have information on both the individual elements and the aggregated groups. For instance, students are grouped by teacher, weekly depression measures are grouped by patient, and survey respondents are grouped by household. In these cases, we refer to **levels** of measurement and observational units at each level. For example, students might represent **level-one observational units**, while teachers represent **level-two observational units**, where **level one** is the most basic level of observation, and level-one observations are aggregated to form **level-two** observations. Then, if we are modeling a response such as test score, we may want to examine the effects of student characteristics such as sex and ethnicity, and teacher characteristics such as years of experience. Student characteristics would be considered **level-one covariates**, while teacher characteristics would be **level-two covariates**.

7.4 Case Study: Dams and Pups

A **teratogen** is a substance or exposure that can result in harm to a developing fetus. An experiment can be conducted to determine whether increasing levels of a potential teratogen results in increasing probability of defects in rat pups. We will simulate the results of an experiment in which 24 dams (mother rats) are randomized to four groups: 3 groups of 6 dams each are administered a potential teratogen in high (3 mg), medium (2 mg), or low (1 mg) doses, and the fourth group serves as the control. Each of the 24 dams produces 10 rat pups, and each pup is examined for the presence of an anomaly or deformity.

7.5 Sources of Variability

Before we analyze data from our simulated experiment, let's step back and look at the big picture. Statistics is all about analyzing and explaining variability, so let's consider what sources of variability we have in the dams and pups example. There are several reasons why the counts of the number of defective pups might differ from dam to dam, and it is helpful to explicitly identify these reasons in order to determine how the dose levels affect the pups while also accommodating correlation.

Dose effect The dams and pups experiment is being carried out to determine whether different dose levels affect the development of defects differently. Of particular interest is determining whether a **dose-response** effect is present. A dose-response effect is evident when dams receiving higher dose levels produce higher proportions of pups with defects. Knowing defect rates at specific dose levels is typically of interest within this experiment and beyond. Publishing the defect rates for each dose level in a journal paper, for example, would be of interest to other teratologists. For that reason, we refer to dose level effects as **fixed effects**.

Dams (litter) effect In many settings like this, there is a litter effect as well. For example, some dams may exhibit a propensity to produce pups with defects, while others rarely produce litters with defective pups. That is, observations on pups within the same litter are likely to be similar or correlated. Unlike the dose effect, teratologists reading experiment results are not interested in the estimated probability of defect for each dam in the study, and we would not report these estimated probabilities in a paper. However, there may be interest in the *variability* in litter-specific defect probabilities; accounting for dam-to-dam variability reduces the amount of unexplained variability and leads to more precise estimates of fixed effects like dose. Often this kind of effect is modeled using the idea that randomly selected dams produce **random effects**. This provides one way in which to model correlated data, in this case the correlation between pups from the same dam. We elaborate on this idea throughout the remainder of the text.

Pup-to-pup variability The within litter pup-to-pup differences reflect random, unexplained variation in the model.

7.6 Scenario 1: No Covariates

In Scenario 1, we will ignore dose and assume that dose has no effect on the probability of a deformed pup. You can follow along with the simulation (and modify it as desired) in the *Rmd* file for this chapter (note that some lines of code are run but not printed in the chapter text), and fill out Table 7.1.

First, we will consider Scenario 1a where each dam's probability of producing a deformed pup is $p = 0.5$; thus, each dam's log odds are 0. We then will compare this to Scenario 1b, where dams' probabilities of producing a deformed

7.6 Scenario 1: No Covariates

pup follow a beta distribution where $\alpha = \beta = 0.5$ (which has expected value 0.5). So, in Scenario 1b, each dam has a different probability of producing a deformed pup, but the probabilities average out to 0.5, whereas in Scenario 1a each dam has an identical probability of producing a deformed pup (0.5). Figure 7.1 illustrates these two scenarios; every dam in Scenario 1a has a probability of 0.5 (the thick vertical line), while probabilities in Scenario 1b are selected from the black dashed distribution, so that probabilities near 0 and 1 are more likely than probabilities near 0.5. The histogram shows 24 randomly generated probabilities under one run of Scenario 1b.

The R code below produces a simulated number of deformed pups for each of the 24 dams under Scenario 1a and Scenario 1b, and Figure 7.2 displays distributions of counts of deformed pups per dam for the two simulations. In Scenario 1a, the mean number of deformed pups is 5.17 with standard deviation 1.49. In Scenario 1b, the mean number of deformed pups is 5.67 with standard deviation 4.10.

```
pi_1a <- rep(0.5, 24)
count_1a <- rbinom(24, 10, pi_1a)

pi_1b <- rbeta(24,.5,.5)
count_1b <- rbinom(24, 10, pi_1b)
```

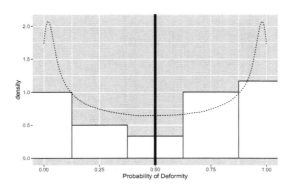

FIGURE 7.1: Dam probabilities in Scenario 1.

Thought Questions

1. Will the counts of deformed pups for dams in Scenario 1a behave like a binomial distribution with $n = 10$ and $p = 0.5$ (that is, like counting heads in 10 flips of a fair coin)? Why or why not?

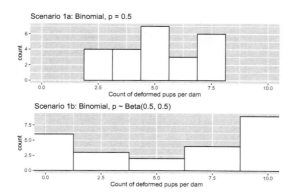

FIGURE 7.2: Counts of deformed pups per dam under Scenarios 1a and 1b.

2. Will the counts of deformed pups for dams in Scenario 1b behave like a binomial distribution with $n = 10$ and $p = 0.5$ (that is, like counting heads in 10 flips of a fair coin)? If not, extend the coin flipping analogy to Scenario 1b.

3. Is Scenario 1b realistic? Why might some dams have higher probabilities than others?

If we were to model the number of deformed pups per dam in Scenario 1a, we could ignore the potential of a dam effect (since all dams behave the same) and proceed with regular binomial regression as in Chapter 6. Since we have no predictors, we would start with the model:

$$\log\left(\frac{\hat{p}}{1-\hat{p}}\right) = \hat{\beta}_0, \text{ where } \hat{\beta}_0 = 0.067$$

which produces an estimated odds of deformity $\hat{p}/(1-\hat{p}) = e^{0.067} = 1.069$ and estimated probability $\hat{p} = 0.517$. Creating 95% confidence intervals using a profile likelihood approach, we get:

$$95\% \text{ CI for } p/(1-p) = (0.830, 1.378)$$
$$95\% \text{ CI for } p = (0.454, 0.579)$$

However, we can account for potential overdispersion with a **quasibinomial model**, just as we did in Section 6.5.8, in case the observed variance is larger than the variance under a binomial model. Quasibinomial regression yields the same estimate for β_0 as the binomial regression model ($\hat{\beta}_0 = 0.067$), but we now have overdispersion paramater $\hat{\phi} = 0.894$. This gives us the following 95% profile likelihood-based confidence intervals:

$$95\% \text{ CI for } p/(1-p) = (0.841, 1.359)$$
$$95\% \text{ CI for } p = (0.457, 0.576)$$

Turning to Scenario 1b, where each dam has a unique probability of producing a pup with a deformity based on a beta distribution, we can fit binomial and quasibinomial models as well.

A binomial model gives regression equation

$$\log\left(\frac{\hat{p}}{1-\hat{p}}\right) = 0.268,$$

with associated profile likelihood 95% confidence intervals:

$$95\% \text{ CI for } p/(1-p) = (1.014, 1.691)$$
$$95\% \text{ CI for } p = (0.504, 0.628)$$

We could compare this to a quasibinomial model. With overdispersion paramater $\hat{\phi} = 6.858$, we now have profile likelihood-based intervals:

$$95\% \text{ CI for } p/(1-p) = (0.673, 2.594)$$
$$95\% \text{ CI for } p = (0.402, 0.722)$$

Thought Questions

4. Describe how the quasibinomial analysis of Scenario 1b differs from the binomial analysis of the same simulated data. Refer to Table 7.1 when answering this question; you will need to run the R code in the R markdown file for this chapter to completely fill out the table. Do confidence intervals contain the true model parameters?

5. Why are differences between quasibinomial and binomial models of Scenario 1a less noticeable than the differences in Scenario 1b?

7.7 Scenario 2: Dose Effect

In Scenario 1, each dam's probability of producing a deformed pup was independent of their dosage of the teratogen. In Scenario 2 we allow for a dose effect. To recall, in this hypothetical experiment we have 24 total dams evenly split into 4 groups receiving either no dose (coded as dose = 0 mg), a low dose (dose = 1), a medium dose (dose = 2), or a high dose (dose = 3) of the teratogen.

We will suppose that true probability that a dam's pup has a deformity is related to the dose the dam received through this model:

$$\log\left(\frac{p}{1-p}\right) = -2 + 1.33 \text{ dose}$$

That is, we assume that the log odds of a deformity is linearly related to dose through the equation above, and the odds of a deformity are 3.79 times greater ($e^{1.33}$) for each 1-mg increase in dose.

In Scenario 2a, a dam who received a dose of x would have probability

$$p = P(\text{deformity} \mid \text{dose} = x) = e^{-2+1.33x}/(1 + e^{-2+1.33x}).$$

Thus, dams who received doses of 0, 1, 2, and 3 mg would have probabilities 0.12, 0.34, 0.66, and 0.88, respectively, under Scenario 2a.

In Scenario 2b, each dam who received a dose of x has probability of deformity randomly chosen from a beta distribution where $\alpha = 2p/(1-p)$ and $\beta = 2$. These beta distribution parameters ensure that, on average, dams with a dose x in Scenario 2b have the same probability of a deformed pup as dams with dose x in Scenario 2a. For example, dams receiving the 1-mg dosage under Scenario 2b would have probabilities following a beta distribution with $\alpha = 2(0.34)/(1-0.34) = 1.03$ and $\beta = 2$, which has mean $\frac{\alpha}{\alpha+\beta} = 0.34$. The big difference is that *all* dams receiving the 1-mg dosage in Scenario 2a have probability 0.34 of a deformed pup, whereas dams receiving the 1-mg dosage in Scenario 2b each have a unique probability of a deformed pup, but those probabilities average out to 0.34.

Figure 7.3 displays histograms for each dosage group of each dam's probability of producing deformed pups under Scenario 2b as well as theoretical distributions of probabilities. A vertical line is displayed at each hypothetical distribution's mean; the vertical line represents the fixed probability of a deformed pup for all dams under Scenario 2a.

```
set.seed(1)

dose <- c(rep(0,6),rep(1,6),rep(2,6),rep(3,6))

pi_2a <- exp(-2+4/3*dose)/(1+exp(-2+4/3*dose))
count_2a <- rbinom(24, 10, pi_2a)

b <- 2
a <- b*pi_2a / (1-pi_2a)
pi_2b <- rbeta(24, a, b)
count_2b <- rbinom(24, 10, pi_2b)
```

7.7 Scenario 2: Dose Effect

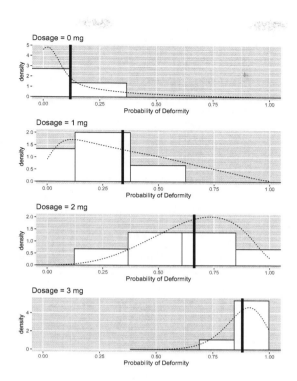

FIGURE 7.3: Observed (histograms) and theoretical (density curves) distributions of dams' probabilities of producing deformed pups by dose group in Scenario 2b. The thick vertical lines represent the fixed probabilities of a deformed pup by dose group in Scenario 2a.

TABLE 7.2: Summary statistics of Scenario 2 by dose.

Dosage	Scenario 2a				Scenario 2b			
	Mean p	SD p	Mean Count	SD Count	Mean p	SD p	Mean Count	SD Count
0	0.119	0	1.333	1.366	0.061	0.069	0.500	0.837
1	0.339	0	3.167	1.835	0.239	0.208	3.500	2.881
2	0.661	0	5.833	1.472	0.615	0.195	5.833	1.941
3	0.881	0	8.833	1.169	0.872	0.079	8.833	1.169

Thought Questions

6. Compare and contrast the probabilities associated with the 24 dams under Scenarios 1a and 1b to the probabilities under Scenarios 2a and 2b.

7. In Scenario 2a, dams produced 4.79 deformed pups on average, with standard deviation 3.20. Scenario 2b saw an average of 4.67 with standard deviation 3.58. Explain why comparisons by dose are more meaningful than these overall comparisons. You might refer to the results in Table 7.2.

8. In Table 7.1, predict what you'll see in the column headed "CI_odds_ratio". Among the 4 entries: What can you say about the center and the width of the confidence intervals? Which will be similar and why? Which will be different and how?

We first model Scenario 2a without adjusting for potential overdispersion. Binomial regression gives us the model:

$$\log\left(\frac{\hat{p}}{1-\hat{p}}\right) = -2.02 + 1.26 \text{ dose} \tag{7.1}$$

Equation (7.1) has associated odds ratio corresponding to a 1-mg dose increase of $e^{\beta_1} = 3.54$ with 95% profile likelihood confidence interval $(2.61, 4.96)$. We can be 95% confident that odds of deformity are between 2.61 and 4.96 times higher for each 1-mg increase in dose.

Alternatively, we can use a quasibinomial model to account for any overdispersion in the binomial model. With $\widehat{\phi} = 1.27$, we have the 95% profile likelihood confidence interval for e_1^{β}: $(2.51, 5.19)$.

Turning to Scenario 2b, where probabilities were different between dams receiving the same dosage, we have the binomial model

$$\log\left(\frac{\hat{p}}{1-\hat{p}}\right) = -2.41 + 1.46 \text{ dose}$$

Generating a 95% confidence interval for e^{β_1}, we have: $(3.09, 6.27)$.

If we use a quasibinomial model, we find overdispersion paramater $\widehat{\phi} = 1.93$, yielding 95% confidence interval for e^{β_1}: $(2.74, 7.35)$.

Thought Questions

9. Describe how the quasibinomial analysis of Scenario 2b differs from the binomial analysis of the same simulated data. Refer to Table 7.1 when answering this question; you will need to run the R code in

the R markdown file for this chapter to completely fill out the table. Do confidence intervals contain the true model parameters?

10. Why are differences between quasibinomial and binomial models of Scenario 2a less noticeable than the differences in Scenario 2b?

11. Why does Scenario 2b contain correlated data that we must account for, while Scenario 2a does not?

7.8 Case Study: Tree Growth

A student research team at St. Olaf College contributed to the efforts of biologist Kathy Shea to investigate a rich data set concerning forestation in the surrounding land [Eisinger et al., 2011]. Here is a paragraph from the introduction to their project report:

> Much of south-central Minnesota was comprised of maple-basswood forest prior to agricultural development and settlement. Currently, the forests that remain are highly fragmented and disturbed by human activities. Original land surveys describe the forest, also known as the Big Woods ecosystem, as dominated by elm, maple, oak, and basswood. In order to recover the loss of ecosystem services that forests provide, many new forested areas have been established through restoration.

Tubes were placed on trees in some locations or *transects* but not in others. One research question is whether tree growth in the first year is affected by the presence of tubes. This analysis has a structure similar to the dams and pups; the two study designs are depicted in Figure 7.4.

Some transects were assigned to have tubes on all of their trees, and other transects had no tubes on all of their trees, just as every dam assigned to a certain group received the same dose. Within a transect, each tree's first year of growth was measured, much like the presence or absence of a defect was noted for every pup within a dam. Although the response in the tree tube study is continuous (and somewhat normally distributed) rather than binary as in the dams and pups study, we can use methods to account for correlation of trees within a transect, just as we accounted for correlation of pups within dams.

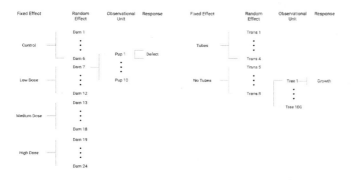

FIGURE 7.4: Data structures in the Dams and Pups (left) and Tree Growth (right) case studies.

7.8.1 Format of the Data Set

We will consider a subset of the full data set in treetube.csv for illustration purposes here: the 382 trees with heights recorded in both 1990 and 1991. Thus, we will consider the following variables:

- id = a unique identifier for each tree
- transect = a unique identifier for each transect containing several trees
- species = tree species
- tubes = an indicator variable for the presence or absence of tubes for a given transect
- height91 = first year height for each tree in meters
- height90 = baseline height for each tree in meters
- growth_yr1 = height91 - height90, in meters

A sample of 10 observations are displayed in Table 7.3.

This portion of the data indicates that the four trees in transect 18 have tubes, while the other 6 trees listed do not. The concern with this kind of data structure is that trees from the same transect may be more similar or correlated with one another, in contrast to trees from other transects. This could be true for a variety of reasons: some transects may receive less sun than others, or irrigation of the soil may differ from transect to transect. These unmeasured but possibly influential factors may imply a correlation among trees within transects. In that case, we would not have independent pieces of information, so that the number of trees within a transect would overstate the amount of independent information. To prepare for an analysis of this potentially correlated data, we examine the sources of variability in first-year tree growth.

7.8 Case Study: Tree Growth   205

TABLE 7.3: A sample of 10 trees and their growth from 1990 to 1991.

id	transect	species	tubes	height91	height90	growth_yr1
398	14	Black Walnut	0	0.332	0.200	0.132
402	14	Black Walnut	0	0.354	0.330	0.024
450	18	Black Walnut	1	0.214	0.174	0.040
451	18	Black Walnut	1	0.342	0.289	0.053
453	18	Black Walnut	1	0.395	0.205	0.190
458	18	Black Walnut	1	0.420	0.290	0.130
560	22	Black Walnut	0	0.400	0.295	0.105
564	22	Black Walnut	0	0.549	0.390	0.159
569	22	Black Walnut	0	0.340	0.270	0.070
571	22	Black Walnut	0	0.394	0.271	0.123

7.8.2 Sources of Variability

First year tree growth may vary because of:

Tube effects A purpose of this analysis is to determine whether tubes affect first-year tube growth. Differences in mean growth based on the presence or absence of tubes would be of interest to researchers, and they would be included in a publication for this analysis. For this reason, tube effects are referred to as **fixed effects**. This is analogous to the dose effect in the dams and pups example.

Transect effects For some of the factors previously mentioned such as sun exposure or water availability, first-year growth may vary by transect. Knowing which specific transects produce greater growth is not of interest and would not appear in a publication of this study. These **random effects** are analogous to dam effects which were not of inherent interest, but which we nevertheless wished to account for.

Tree-to-tree variability within transects There is inherent variability in tree growth even when they are subject to the same transect and treatment effects. This variability remains unexplained in our model, although we will attempt to explain some of it with covariates such as species.

Data sets with this kind of structure are often referred to as **multilevel data**, and the remaining chapters delve into models for multilevel data in gory detail. With a continuous response variable, we will actually add random effects for transects to a more traditional linear least squares regression model rather than estimate an overdispersion parameter as with a binary response. Either way, if observations are really correlated, proper accounting will lead to larger standard errors for model coefficients and larger (but more appropriate) p-values for testing the significance of those coefficients.

7.8.3 Analysis Preview: Accounting for Correlation

Attempting to model the effects of tubes on tree growth, we could use LLSR which yields the model:

$$\hat{\text{Growth}} = 0.106 - 0.040 \text{ Tube}$$

```
tube_linear <- lm(growth_yr1 ~ tubes, data = treetubes_yr1)
```

```
##               Estimate Std. Error t value  Pr(>|t|)
## (Intercept)   0.10585   0.005665  18.685   9.931e-56
## tubes        -0.04013   0.041848  -0.959   3.382e-01

##  R squared =  0.002414
##  Residual standard error =  0.1097
```

However, the LLSR model assumes that all observations are independent, including trees in the same transect. One way to account for potential correlation is to estimate an additional parameter for the transect-to-transect variance. In other words, we are allowing for a random effect that each transect contributes to the overall variability. The multilevel model below does just that.

```
tube_multi1 <- lmer(growth_yr1 ~ tubes + (1|transect),
                    data = treetubes_yr1)
```

```
##               Estimate Std. Error t value
## (Intercept)   0.10636   0.01329   8.005
## tubes        -0.04065   0.05165  -0.787

##  Groups    Name         Variance Std.Dev.
##  transect  (Intercept)  0.00084  0.029
##  Residual               0.01155  0.107
```

As we saw in the case of the binary outcomes, the standard error for the coefficients is larger when we take correlation into account. The t-statistic for tubes is smaller, reducing our enthusiasm for the tubes effect. This conservative approach occurs because the observations within a transect are correlated and therefore not independent as assumed in the naive model. We will study these models in depth in the remaining chapters.

7.9 Summary

The most important idea from this chapter is that structures of data sets may imply that outcomes are correlated. Correlated outcomes provide less information than independent outcomes, resulting in effective sample sizes that are less than the total number of observations. Neglecting to take into account correlation may lead to underestimating standard errors of coefficients, overstating significance and precision. Correlation is likely and should be accounted for if basic observational units (e.g., pups, trees) are aggregated in ways that would lead us to expect units within groups to be similar.

We have mentioned two ways to account for correlation: incorporate a dispersion parameter or include random effects. In the following chapters, we will primarily focus on models with random effects. In fact, there are even more ways to account for correlation, including inflating the variance using Huber-White estimators (aka Sandwich estimators), and producing corrected variances using bootstrapping. These are beyond the scope of this text.

7.10 Exercises

7.10.1 Conceptual Exercises

1. **Examples with correlated data.** For each of the following studies:
    - Identify the most basic observational units
    - Identify the grouping units (could be multiple levels of grouping)
    - State the response(s) measured and variable type (normal, binary, Poisson, etc.)
    - Write a sentence describing the within-group correlation.
    - Identify fixed and random effects

    a. *Nurse stress study.* Four wards were randomly selected at each of 25 hospitals and randomly assigned to offer a stress reduction program for nurses on the ward or to serve as a control. At the conclusion of the study period, a random sample of 10 nurses from each ward completed a test to measure job-related stress. Factors assumed to be related include nurse experience, age, hospital size and type of ward.

b. *Epilepsy study.* Researchers conducted a randomized controlled study where patients were randomly assigned to either an anti-epileptic drug or a placebo. For each patient, the number of seizures at baseline was measured over a 2-week period. For four consecutive visits the number of seizures were determined over the past 2-week period. Patient age and sex along with visit number were recorded.

c. *Cockroaches!* For a study of cockroach infestation, traps were set up in the kitchen, bathroom, and bedroom in a random sample of 100 New York City apartments. The goal is to estimate cockroach infestation levels given tenant income and age of the building.

d. *Prairie restoration.* Researchers at a small Midwestern college decided to experimentally explore the underlying causes of variation in soil reconstruction projects in order to make future projects more effective. Introductory ecology classes were organized to collect weekly data on plants in pots containing soil samples. Data will be examined to compare:

- germination and growth of two species of prairie plants—leadplants (*Amorpha canescens*) and coneflowers (*Ratibida pinnata*).
- soil from a cultivated (agricultural) field, a natural prairie, and a restored (reconstructed) prairie.
- the effect of sterilization, since half of the sampled soil was sterilized to determine if rhizosphere differences were responsible for the observed variation.

e. *Radon in Minnesota.* Radon is a carcinogen – a naturally occurring radioactive gas whose decay products are also radioactive – known to cause lung cancer in high concentrations. The EPA sampled more than 80,000 homes across the U.S. Each house came from a randomly selected county and measurements were made on each level of each home. Uranium measurements at the county level were included to improve the radon estimates.

f. *Teen alcohol use.* Curran et al. [1997] collected data on 82 adolescents at three time points starting at age 14 to assess factors that affect teen drinking behavior. Key variables in the data set `alcohol.csv` (source: Singer and Willett [2003]) are as follows:

- `id` = numerical identifier for subject
- `age` = 14, 15, or 16
- `coa` = 1 if the teen is a child of an alcoholic parent; 0 otherwise
- `male` = 1 if male; 0 if female
- `peer` = a measure of peer alcohol use, taken when each subject was 14. This is the square root of the sum of two 6-point

7.10 Exercises

items about the proportion of friends who drink occasionally or regularly.
- alcuse = the primary response. Four items—(a) drank beer or wine, (b) drank hard liquor, (c) 5 or more drinks in a row, and (d) got drunk—were each scored on an 8-point scale, from 0="not at all" to 7="every day". Then alcuse is the square root of the sum of these four items.

Primary research questions included:

- do trajectories of alcohol use differ by parental alcoholism?
- do trajectories of alcohol use differ by peer alcohol use?

2. **More dams and pups** Describe how to generalize the pup and dam example by allowing for different size litters.

7.10.2 Guided Exercises

1. **Exploring Beta distributions.** In the Dams and Pups Case Study, we use the beta distribution to randomly select the probability that a dam produces a defective pup. While the beta distribution is described in Chapter 3, it can be valuable to play with the parameters α and β to see what distribution ranges and shapes are possible. Some basic R code for plotting a beta density curve can be found at the end of this problem.

    a. What values do beta random variables take on?
    b. What do these values represent for the dams and pups simulation?
    c. Do the possible values depend on α or β?
    d. What is a feature of the beta density when $\alpha = \beta$?
    e. What happens to the density when $\alpha \neq \beta$?
    f. How does the magnitude of α or β affect the density?
    g. How does the difference between α and β affect the density?
    h. If you wanted to simulate dams with mostly low probabilities of defects and a few with very high probabilities, how would you do it?
    i. If you wanted to simulate dams with mostly high probabilities of defects and a few with very low probabilities, how would you do it?
    j. If you wanted to simulate a population of dams where half of the probabilities of defects are very high and half are very low, how would you do it?
    k. How might you decide on values for α and β if you have run

a preliminary experiment and gathered data on the number of dams with deformed pups?

```
# inputs for the beta distribution must be between 0 and 1
p <- seq(0,1,by=.05)

# To plot a beta density use dbeta; here I selected a=5, b=1
density <- dbeta(p,5,1)
plot(p, density, type = "l")
```

2. **Dams and pups (continued).** Modify the dams and pups simulation in the following ways. In each case, produce plots and describe the results of your modified simulation.
    a. Pick a different beta distribution for Scenario 1b.
    b. Center the beta distributions in Scenarios 1a and 1b somewhere other than 0.5.
    c. Repeat Scenario 2a with 3 doses and an underlying logistic model of your own choosing. Then create beta distributions as in Scenario 2b to match your 3 doses.

7.10.3 Note on Correlated Binary Outcomes

The correlated binomial counts simulated in the Dams and Pups Case Study are in fact beta-binomial random variables like those simulated in the Guided Exercises from Chapter 3. In fact, we could use the form of a beta-binomial pdf to model overdispersed binomial variables. Unlike the more generic form of accounting for correlation using dispersion parameter estimates, beta-binomial models are more specific and highly parameterized. This approach involves more assumptions but may also yield more information than the quasi-likelihood approach. If the beta-binomial model is incorrect, however, our results may be misleading. That said, the beta-binomial structure is quite flexible and conforms to many situations.

8
Introduction to Multilevel Models

8.1 Learning Objectives

After finishing this chapter, you should be able to:
- Recognize when response variables and covariates have been collected at multiple (nested) levels.
- Apply exploratory data analysis techniques to multilevel data.
- Write out a multilevel statistical model, including assumptions about variance components, in both by-level and composite forms.
- Interpret model parameters (including fixed effects and variance components) from a multilevel model, including cases in which covariates are continuous, categorical, or centered.
- Understand the taxonomy of models, including why we start with an unconditional means model.
- Select a final model, using criteria such as AIC, BIC, and deviance.

```
# Packages required for Chapter 8
library(MASS)
library(gridExtra)
library(mnormt)
library(lme4)
library(knitr)
library(kableExtra)
library(tidyverse)
```

8.2 Case Study: Music Performance Anxiety

Stage fright can be a serious problem for performers, and understanding the personality underpinnings of performance anxiety is an important step in determining how to minimize its impact. Sadler and Miller [2010] studied the emotional state of musicians before performances and factors which may affect their emotional state. Data was collected by having 37 undergraduate music majors from a competitive undergraduate music program fill out diaries prior to performances over the course of an academic year. In particular, study participants completed a Positive Affect Negative Affect Schedule (PANAS) before each performance. The PANAS instrument provided two key outcome measures: negative affect (a state measure of anxiety) and positive affect (a state measure of happiness). We will focus on negative affect as our primary response measuring performance anxiety.

Factors which were examined for their potential relationships with performance anxiety included: performance type (solo, large ensemble, or small ensemble); audience (instructor, public, students, or juried); if the piece was played from memory; age; gender; instrument (voice, orchestral, or keyboard); and, years studying the instrument. In addition, the personalities of study participants were assessed at baseline through the Multidimensional Personality Questionnaire (MPQ). The MPQ provided scores for one lower-order factor (absorption) and three higher-order factors: positive emotionality (PEM—a composite of well-being, social potency, achievement, and social closeness); negative emotionality (NEM—a composite of stress reaction, alienation, and aggression); and, constraint (a composite of control, harm avoidance, and traditionalism).

Primary scientific hypotheses of the researchers included:
- Lower music performance anxiety will be associated with lower levels of a subject's negative emotionality.
- Lower music performance anxiety will be associated with lower levels of a subject's stress reaction.
- Lower music performance anxiety will be associated with greater number of years of study.

8.3 Initial Exploratory Analyses

TABLE 8.1: A snapshot of selected variables from the first three and the last three observations in the Music Performance Anxiety case study.

Obs	id	diary	perf_type	memory	na	gender	instrument	mpqab	mpqpem	mpqnem
1	1	1	Solo	Unspecified	11	Female	voice	16	52	16
2	1	2	Large Ensemble	Memory	19	Female	voice	16	52	16
3	1	3	Large Ensemble	Memory	14	Female	voice	16	52	16
495	43	2	Solo	Score	13	Female	voice	31	64	17
496	43	3	Small Ensemble	Memory	19	Female	voice	31	64	17
497	43	4	Solo	Score	11	Female	voice	31	64	17

8.3 Initial Exploratory Analyses

8.3.1 Data Organization

Our examination of the data from Sadler and Miller [2010] in `musicdata.csv` will focus on the following key variables:

- `id` = unique musician identification number
- `diary` = cumulative total of diaries filled out by musician
- `perf_type` = type of performance (Solo, Large Ensemble, or Small Ensemble)
- `audience` = who attended (Instructor, Public, Students, or Juried)
- `memory` = performed from Memory, using Score, or Unspecified
- `na` = negative affect score from PANAS
- `gender` = musician gender
- `instrument` = Voice, Orchestral, or Piano
- `mpqab` = absorption subscale from MPQ
- `mpqpem` = positive emotionality (PEM) composite scale from MPQ
- `mpqnem` = negative emotionality (NEM) composite scale from MPQ

Sample rows containing selected variables from our data set are illustrated in Table 8.1; note that each subject (id) has one row for each unique diary entry.

As with any statistical analysis, our first task is to explore the data, examining distributions of individual responses and predictors using graphical and numerical summaries, and beginning to discover relationships between variables. With multilevel models, exploratory analyses must eventually account for the level at which each variable is measured. In a two-level study such as this one, **Level One** will refer to variables measured at the most frequently occurring observational unit, while **Level Two** will refer to variables measured on larger observational units. For example, in our study on music performance anxiety, many variables are measured at every performance. These "Level One" variables include:

- negative affect (our response variable)
- performance characteristics (type, audience, if music was performed from memory)
- number of previous performances with a diary entry

However, other variables measure characteristics of study participants that remain constant over all performances for a particular musician; these are considered "Level Two" variables and include:

- demographics (age and gender of musician)
- instrument used and number of previous years spent studying that instrument
- baseline personality assessment (MPQ measures of positive emotionality, negative emotionality, constraint, stress reaction, and absorption)

8.3.2 Exploratory Analyses: Univariate Summaries

Because of this data structure—the assessment of some variables on a performance-by-performance basis and others on a subject-by-subject basis—we cannot treat our data set as consisting of 497 independent observations. Although negative affect measures from different subjects can reasonably be assumed to be independent (unless, perhaps, the subjects frequently perform in the same ensemble group), negative affect measures from different performances by the same subject are not likely to be independent. For example, some subjects tend to have relatively high performance anxiety across all performances, so that knowing their score for Performance 3 was 20 makes it more likely that their score for Performance 5 is somewhere near 20 as well. Thus, we must carefully consider our exploratory data analysis, recognizing that certain plots and summary statistics may be useful but imperfect in light of the correlated observations.

First, we will examine each response variable and potential covariate individually. Continuous variables can be summarized using histograms and summaries of center and spread; categorical variables can be summarized with tables and possibly bar charts. When examining Level One covariates and responses, we will begin by considering all 497 observations, essentially treating each performance by each subject as independent even though we expect observations from the same musician to be correlated. Although these plots will contain dependent points, since each musician provides data for up to 15 performances, general patterns exhibited in these plots tend to be real. Alternatively, we can calculate mean scores across all performances for each of the 37 musicians so that we can more easily consider each plotted point to be independent. The disadvantage of this approach would be lost information which, in a study such as this with a relatively small number of musicians each being observed over many performances, could be considerable. In addition, if the sample sizes varied greatly by subject, a mean based on 1 observation would be given

8.3 Initial Exploratory Analyses

equal weight to a mean based on 15 observations. Nevertheless, both types of exploratory plots typically illustrate similar relationships.

In Figure 8.1 we see histograms for the primary response (negative affect); plot (a) shows all 497 (dependent) observations, while plot (b) shows the mean negative affect for each of the 37 musicians across all their performances. Through plot (a), we see that performance anxiety (negative affect) across all performances follows a right-skewed distribution with a lower bound of 10 (achieved when all 10 questions are answered with a 1). Plot (b) shows that mean negative affect is also right-skewed (although not as smoothly decreasing in frequency), with range 12 to 23.

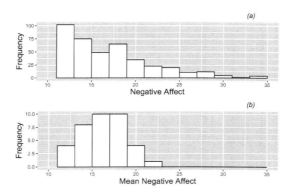

FIGURE 8.1: Histogram of the continuous Level One response (negative effect). Plot (a) contains all 497 performances across the 37 musicians, while plot (b) contains one observation per musician (the mean negative affect across all performances).

We can also summarize categorical Level One covariates across all (possibly correlated) observations to get a rough relative comparison of trends. A total of 56.1% of the 497 performances in our data set were solos, while 27.3% were large ensembles and 16.5% were small ensembles. The most common audience type was a public performance (41.0%), followed by instructors (30.0%), students (20.1%), and finally juried recitals (8.9%). In 30.0% of performances, the musician played by memory, while 55.1% used the score and 14.9% of performances were unspecified.

To generate an initial examination of Level Two covariates, we consider a data set with just one observation per subject, since Level Two variables are constant over all performances from the same subject. Then, we can proceed as we did with Level One covariates—using histograms to illustrate the distributions of continuous covariates (see Figure 8.2) and tables to summarize categorical covariates. For example, we learn that the majority of subjects have positive emotionality scores between 50 and 60, but that several subjects fall into a lengthy lower tail with scores between 20 and 50. A summary of categorical Level Two covariates reveals that among the 37 subjects (26 female and 11 male), 17 play an orchestral instrument, 15 are vocal performers, and 5 play a keyboard instrument.

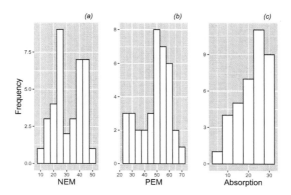

FIGURE 8.2: Histograms of the 3 continuous Level Two covariates (negative emotionality (NEM), positive emotionality (PEM), and absorption). Each plot contains one observation per musician.

8.3.3 Exploratory Analyses: Bivariate Summaries

The next step in an initial exploratory analysis is the examination of numerical and graphical summaries of relationships between model covariates and responses. In examining these bivariate relationships, we hope to learn: (1) if there is a general trend suggesting that as the covariate increases the response either increases or decreases, (2) if subjects at certain levels of the covariate tend to have similar mean responses (low variability), and (3) if the variation in the response differs at different levels of the covariate (unequal variability).

As with individual variables, we will begin by treating all 497 performances recorded as independent observations, even though blocks of 15 or so performances were performed by the same musician. For categorical Level One covariates, we can generate boxplots against negative affect as in Figure 8.3, plots (a) and (b). From these boxplots, we see that lower levels of performance anxiety seem to be associated with playing in large ensembles and playing in front of an instructor. For our lone continuous Level One covariate (number of previous performances), we can generate a scatterplot against negative affect as in plot (c) from Figure 8.3, adding a fitted line to illustrate general trends upward or downward. From this scatterplot, we see that negative affect seems to decrease slightly as a subject has more experience.

8.3 Initial Exploratory Analyses

To avoid the issue of dependent observations in our three plots from Figure 8.3, we could generate separate plots for each subject and examine trends within and across subjects. These "lattice plots" are illustrated in Figures 8.4, 8.5, and 8.6; we discuss such plots more thoroughly in Chapter 9. While general trends are difficult to discern from these lattice plots, we can see the variety in subjects in sample size distributions and overall level of performance anxiety. In particular, in Figure 8.6, we notice that linear fits for many subjects illustrate the same slight downward trend displayed in the overall scatterplot in Figure 8.3, although some subjects experience increasing anxiety and others exhibit non-linear trends. Having an idea of the range of individual trends will be important when we begin to draw overall conclusions from this study.

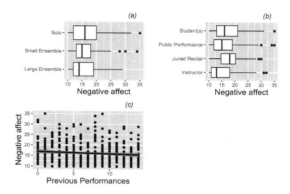

FIGURE 8.3: Boxplots of two categorical Level One covariates (performance type (a) and audience type (b)) vs. model response, and scatterplot of one continuous Level One covariate (number of previous diary entries (c)) vs. model response (negative affect). Each plot contains one observation for each of the 497 performances.

In Figure 8.7, we use boxplots to examine the relationship between our primary categorical Level Two covariate (instrument) and our continuous model response. Plot (a) uses all 497 performances, while plot (b) uses one observation per subject (the mean performance anxiety across all performances) regardless of how many performances that subject had. Naturally, plot (b) has a more condensed range of values, but both plots seem to support the notion that performance anxiety is slightly lower for vocalists and maybe a bit higher for keyboardists.

In Figure 8.8, we use scatterplots to examine the relationships between continuous Level Two covariates and our model response. Performance anxiety appears to vary little with a subject's positive emotionality, but there is some evidence to suggest that performance anxiety increases with increasing negative emotionality and absorption level. Plots based on mean negative affect, with one observation per subject, support conclusions based on plots with all

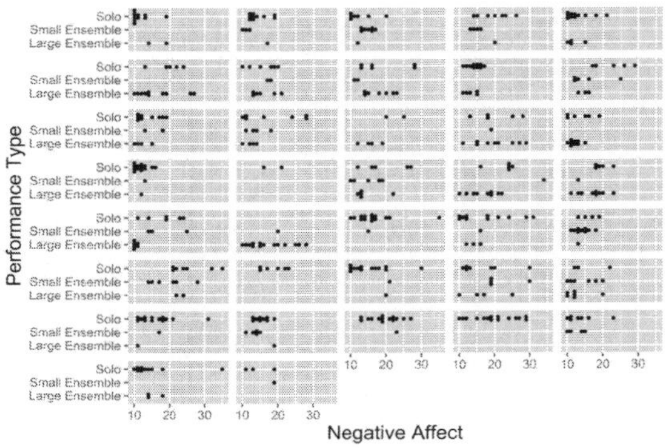

FIGURE 8.4: Lattice plot of performance type vs. negative affect, with separate dotplots by subject.

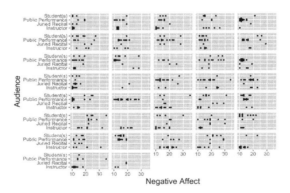

FIGURE 8.5: Lattice plot of audience type vs. negative affect, with separate dotplots by subject.

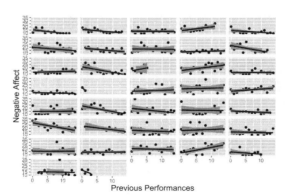

FIGURE 8.6: Lattice plot of previous performances vs. negative affect, with separate scatterplots with fitted lines by subject.

8.3 Initial Exploratory Analyses

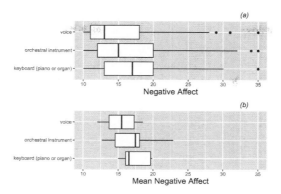

FIGURE 8.7: Boxplots of the categorical Level Two covariate (instrument) vs. model response (negative affect). Plot (a) is based on all 497 observations from all 37 subjects, while plot (b) uses only one observation per subject.

observations from all subjects; indeed the overall relationships are in the same direction and of the same magnitude.

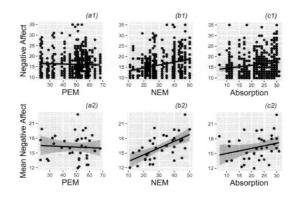

FIGURE 8.8: Scatterplots of continuous Level Two covariates (positive emotionality (PEM), negative emotionality (NEM), and absorption) vs. model response (negative affect). The top plots (a1, b1, c1) are based on all 497 observations from all 37 subjects, while the bottom plots (a2, b2, c2) use only one observation per subject.

Of course, any graphical analysis is exploratory, and any notable trends at this stage should be checked through formal modeling. At this point, a statistician begins to ask familiar questions such as:

- which characteristics of individual performances are most associated with performance anxiety?

- which characteristics of study participants are most associated with performance anxiety?
- are any of these associations statistically significant?
- does the significance remain after controlling for other covariates?
- how do we account for the lack of independence in performances by the same musician?

As you might expect, answers to these questions will arise from proper consideration of variability and properly identified statistical models.

8.4 Two-Level Modeling: Preliminary Considerations

8.4.1 Ignoring the Two-Level Structure (not recommended)

Armed with any statistical software package, it would be relatively simple to take our complete data set of 497 observations and run a multiple linear least squares regression model seeking to explain variability in negative affect with a number of performance-level or musician-level covariates. As an example, output from a model with two binary covariates (Does the subject play an orchestral instrument? Was the performance a large ensemble?) is presented below. Do you see any problems with this approach?

```
# Linear least square regression model with LINE conditions
modelc0 <- lm(na ~ orch + large + orch:large, data = music)
```

```
##              Estimate Std. Error  t value    Pr(>|t|)
## (Intercept)  15.7212     0.3591  43.7785   5.548e-172
## orch          1.7887     0.5516   3.2426    1.265e-03
## large        -0.2767     0.7910  -0.3498    7.266e-01
## orch:large   -1.7087     1.0621  -1.6088    1.083e-01

##   R squared = 0.02782
##   Residual standard error = 5.179
```

Other than somewhat skewed residuals, residual plots (not shown) do not indicate any major problems with the LLSR model. However, another key assumption in these models is the independence of all observations. While we might reasonably conclude that responses from different study participants are

8.4 Two-Level Modeling: Preliminary Considerations

independent (although possibly not if they are members of the same ensemble group), it is not likely that the 15 or so observations taken over multiple performances from a single subject are similarly independent. If a subject begins with a relatively high level of anxiety (compared to other subjects) before their first performance, chances are good that they will have relatively high anxiety levels before subsequent performances. Thus, multiple linear least squares regression using all 497 observations is not advisable for this study (or multilevel data sets in general).

8.4.2 A Two-Stage Modeling Approach (better but imperfect)

If we assume that the 37 study participants can reasonably be considered to be independent, we could use traditional linear least squares regression techniques to analyze data from this study if we could condense each subject's set of responses to a single meaningful outcome. Candidates for this meaningful outcome include a subject's last performance anxiety measurement, average performance anxiety, minimum anxiety level, etc. For example, in clinical trials, data is often collected over many weekly or monthly visits for each patient, except that many patients will drop out early for many reasons (e.g., lack of efficacy, side effects, personal reasons). In these cases, treatments are frequently compared using "last-value-carried-forward" methods—the final visit of each patient is used as the primary outcome measure, regardless of how long they remained in the study. However, "last-value-carried-forward" and other summary measures feel inadequate, since we end up ignoring much of the information contained in the multiple measures for each individual. A more powerful solution is to model performance anxiety at multiple levels.

We will begin by considering all performances by a single individual. For instance, consider the 15 performances for which Musician #22 recorded a diary, illustrated in Table 8.2.

Does this musician tend to have higher anxiety levels when he is playing in a large ensemble or playing in front of fellow students? Which factor is the biggest determinant of anxiety for a performance by Musician #22? We can address these questions through multiple LLSR applied to only Musician #22's data, using appropriate indicator variables for factors of interest.

Let Y_{22j} be the performance anxiety score of Musician #22 before performance j. Consider the observed performances for Musician #22 to be a random sample of all conceivable performances by that subject. If we are initially interested in examining the effect of playing in a large ensemble, we can model the performance anxiety for Musician #22 according to the model:

$$Y_{22j} = a_{22} + b_{22}\text{LargeEns}_{22j} + \epsilon_{22j} \text{ where } \epsilon_{22j} \sim N(0, \sigma^2) \text{ and} \qquad (8.1)$$

TABLE 8.2: Data from the 15 performances of Musician #22.

	id	diary	perform_type	audience	na	instrument
240	22	1	Solo	Instructor	24	orchestral instrument
241	22	2	Large Ensemble	Public Performance	21	orchestral instrument
242	22	3	Large Ensemble	Public Performance	14	orchestral instrument
243	22	4	Large Ensemble	Public Performance	15	orchestral instrument
244	22	5	Large Ensemble	Public Performance	10	orchestral instrument
245	22	6	Solo	Instructor	24	orchestral instrument
246	22	7	Solo	Student(s)	24	orchestral instrument
247	22	8	Solo	Instructor	16	orchestral instrument
248	22	9	Small Ensemble	Public Performance	34	orchestral instrument
249	22	10	Large Ensemble	Public Performance	22	orchestral instrument
250	22	11	Large Ensemble	Public Performance	19	orchestral instrument
251	22	12	Large Ensemble	Public Performance	18	orchestral instrument
252	22	13	Large Ensemble	Public Performance	12	orchestral instrument
253	22	14	Large Ensemble	Public Performance	19	orchestral instrument
254	22	15	Solo	Instructor	25	orchestral instrument

$$\text{LargeEns}_j = \begin{cases} 1 & \text{if perf-type = Large Ensemble} \\ 0 & \text{if perf-type = Solo or Small Ensemble} \end{cases}$$

The parameters in this model (a_{22}, b_{22}, and σ^2) can be estimated through least squares methods. a_{22} represents the true intercept for Musician #22—the expected anxiety score for Musician #22 when performance type is a Solo or Small Ensemble (LargeEns = 0), or the true average anxiety for Musician #22 over all Solo or Small Ensemble performances he may conceivably give. b_{22} represents the true slope for Musician #22—the expected increase in performance anxiety for Musician #22 when performing as part of a Large Ensemble rather than in a Small Ensemble or as a Solo, or the true average difference in anxiety scores for Musician #22 between Large Ensemble performances and other types. Finally, the ϵ_{22j} terms represent the deviations of Musician #22's actual performance anxiety scores from the expected scores under this model—the part of Musician #22's anxiety before performance j that is not explained by performance type. The variability in these deviations from the regression model is denoted by σ^2.

For Subject 22, we estimate $\hat{a}_{22} = 24.5$, $\hat{b}_{22} = -7.8$, and $\hat{\sigma} = 4.8$. Thus, according to our simple linear regression model, Subject 22 had an estimated anxiety score of 24.5 before Solo and Small Ensemble performances, and 16.7 (7.8 points lower) before Large Ensemble performances. With an R^2 of 0.425, the regression model explains a moderate amount (42.5%) of the performance-to-performance variability in anxiety scores for Subject 22, and the trend toward lower scores for large ensemble performances is statistically significant at the 0.05 level (t(13)=-3.10, p=.009).

8.4 Two-Level Modeling: Preliminary Considerations

```
regr.id22 = lm(na ~ large, data = id22)
```

```
##               Estimate Std. Error t value  Pr(>|t|)
## (Intercept)    24.500       1.96  12.503  1.275e-08
## large          -7.833       2.53  -3.097  8.504e-03

##   R squared =  0.4245
##   Residual standard error =  4.8
```

We could continue trying to build a better model for Subject 22, adding indicators for audience and memory, and even adding a continuous variable representing the number of previous performances where a diary was kept. As our model R-squared value increased, we would be explaining a larger proportion of Subject 22's performance-to-performance variability in anxiety. It would not, however, improve our model to incorporate predictors for age, gender, or even negative emotionality based on the MPQ—why is that?

For the present time, we will model Subject 22's anxiety scores for his 15 performances using the model given by Equation (8.1), with a lone indicator variable for performing in a Large Ensemble. We can then proceed to fit the LLSR model in Equation (8.1) to examine the effect of performing in a Large Ensemble for each of the 37 subjects in this study. These are called **Level One models**. As displayed in Figure 8.9, there is considerable variability in the fitted intercepts and slopes among the 37 subjects. Mean performance anxiety scores for Solos and Small Ensembles range from 11.6 to 24.5, with a median score of 16.7, while mean differences in performance anxiety scores for Large Ensembles compared to Solos and Small Ensembles range from -7.9 to 5.0, with a median difference of -1.7. Can these differences among individual musicians be explained by (performance-invariant) characteristics associated with each individual, such as gender, age, instrument, years studied, or baseline levels of personality measures? Questions like these can be addressed through further statistical modeling.

As an illustration, we can consider whether or not there are significant relationships between individual regression parameters (intercepts and slopes) and instrument played. From a modeling perspective, we would build a system of two **Level Two models** to predict the fitted intercept (a_i) and fitted slopes (b_i) for Subject i:

$$a_i = \alpha_0 + \alpha_1 \text{Orch}_i + u_i \qquad (8.2)$$
$$b_i = \beta_0 + \beta_1 \text{Orch}_i + v_i \qquad (8.3)$$

where $\text{Orch}_i = 1$ if Subject i plays an orchestral instrument and $\text{Orch}_i = 0$ if Subject i plays keyboard or is a vocalist. Note that, at Level Two, our response variables are not observed measurements such as performance anxiety scores,

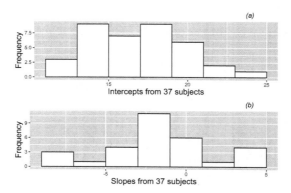

FIGURE 8.9: Histograms of intercepts and slopes from fitting simple regression models by subject, where each model contained a single binary predictor indicating if a performance was part of a large ensemble.

but rather the fitted regression coefficients from the Level One models fit to each subject. (Well, in our theoretical model, the responses are actually the true intercepts and slopes from Level One models for each subject, but in reality, we have to use our estimated slopes and intercepts.)

Exploratory data analysis (see boxplots by instrument in Figure 8.10) suggests that subjects playing orchestral instruments have higher intercepts than vocalists or keyboardists, and that orchestral instruments are associated with slightly lower (more negative) slopes, although with less variability that the slopes of vocalists and keyboardists. These trends are borne out in regression modeling. If we fit Equations (8.2) and (8.3) using fitted intercepts and slopes as our response variables, we obtain the following estimated parameters: $\hat{\alpha}_0 = 16.3$, $\hat{\alpha}_1 = 1.4$, $\hat{\beta}_0 = -0.8$, and $\hat{\beta}_1 = -1.4$. Thus, the intercept (a_i) and slope (b_i) for Subject i can be modeled as:

$$\hat{a}_i = 16.3 + 1.4 \text{Orch}_i + u_i \tag{8.4}$$
$$\hat{b}_i = -0.8 - 1.4 \text{Orch}_i + v_i$$

where a_i is the true mean negative affect when Subject i is playing solos or small ensembles, and b_i is the true mean difference in negative affect for Subject i between large ensembles and other performance types. Based on these models, average performance anxiety before solos and small ensembles is 16.3 for vocalists and keyboardists, but 17.7 (1.4 points higher) for orchestral instrumentalists. Before playing in large ensembles, vocalists and instrumentalists have performance anxiety (15.5) which is 0.8 points lower, on average, than before solos and small ensembles, while subjects playing orchestral instruments experience an average difference of 2.2 points, producing an average

8.5 Two-Level Modeling: A Unified Approach

performance anxiety of 15.5 before playing in large ensembles just like subjects playing other instruments. However, the difference between orchestral instruments and others does not appear to be statistically significant for either intercepts (t=1.424, p=0.163) or slopes (t=-1.168, p=0.253).

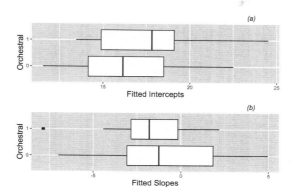

FIGURE 8.10: Boxplots of fitted intercepts, plot (a), and slopes, plot (b), by orchestral instrument (1) vs. keyboard or vocalist (0).

This two-stage modeling process does have some drawbacks. Among other things, (1) it weights every subject the same regardless of the number of diary entries we have, (2) it responds to missing individual slopes (from 7 subjects who never performed in a large ensemble) by simply dropping those subjects, and (3) it does not share strength effectively across individuals. These issues can be better handled through a unified multilevel modeling framework which we will develop in the next section.

8.5 Two-Level Modeling: A Unified Approach

8.5.1 Our Framework

For the unified approach, we will still envision two levels of models as in Section 8.4.2, but we will use likelihood-based methods for parameter estimation rather than ordinary least squares to address the drawbacks associated with the two-stage approach. To illustrate the unified approach, we will first generalize the models presented in Section 8.4.2. Let Y_{ij} be the performance anxiety score of the ith subject before performance j. If we are initially interested in examining the effects of playing in a large ensemble and playing an orchestral instrument, then we can model the performance anxiety for Subject i in performance j with the following system of equations:

- Level One:
$$Y_{ij} = a_i + b_i \text{LargeEns}_{ij} + \epsilon_{ij}$$
- Level Two:
$$a_i = \alpha_0 + \alpha_1 \text{Orch}_i + u_i$$
$$b_i = \beta_0 + \beta_1 \text{Orch}_i + v_i,$$

In this system, there are 4 key **fixed effects** to estimate: α_0, α_1, β_0 and β_1. Fixed effects are the fixed but unknown population effects associated with certain covariates. The intercepts and slopes for each subject from Level One, a_i and b_i, don't need to be formally estimated as we did in Section 8.4.2; they serve to conceptually connect Level One with Level Two. In fact, by substituting the two Level Two equations into the Level One equation, we can view this two-level system of models as a single **Composite Model** without a_i and b_i:

$$Y_{ij} = [\alpha_0 + \alpha_1 \text{Orch}_i + \beta_0 \text{LargeEns}_{ij} + \beta_1 \text{Orch}_i \text{LargeEns}_{ij}]$$
$$+ [u_i + v_i \text{LargeEns}_{ij} + \epsilon_{ij}]$$

From this point forward, when building multilevel models, we will use Greek letters (such as α_0) to denote final fixed effects model parameters to be estimated empirically, and Roman letters (such as a_0) to denote preliminary fixed effects parameters at lower levels. Variance components that will be estimated empirically will be denoted with σ or ρ, while terms such as ϵ and u_i represent error terms. In our framework, we can estimate final parameters directly without first estimating preliminary parameters, which can be seen with the Composite Model formulation (although we can obtain estimates of preliminary parameters in those occasional cases when they are of interest to us). Note that when we model a slope term like b_i from Level One using Level Two covariates like Orch_i, the resulting Composite Model contains a **cross-level interaction term**, denoting that the effect of LargeEns_{ij} depends on the instrument played.

Furthermore, with a binary predictor at Level Two such as instrument, we can write out what our Level Two model looks like for those who play keyboard or are vocalists ($\text{Orch}_i = 0$) and those who play orchestral instruments ($\text{Orch}_i = 1$):

- Keyboardists and Vocalists ($\text{Orch}_i = 0$)

$$a_i = \alpha_0 + u_i$$
$$b_i = \beta_0 + v_i$$

- Orchestral Instrumentalists ($\text{Orch}_i = 1$)

$$a_i = (\alpha_0 + \alpha_1) + u_i$$
$$b_i = (\beta_0 + \beta_1) + v_i$$

8.5 Two-Level Modeling: A Unified Approach

Writing the Level Two model in this manner helps us interpret the model parameters from our two-level model. In this case, even the Level One covariate is binary, so that we can write out expressions for mean performance anxiety based on our model for four different combinations of instrument played and performance type:

- Keyboardists or vocalists playing solos or small ensembles: α_0
- Keyboardists or vocalists playing large ensembles: $\alpha_0 + \beta_0$
- Orchestral instrumentalists playing solos or small ensembles: $\alpha_0 + \alpha_1$
- Orchestral instrumentalists playing large ensembles: $\alpha_0 + \alpha_1 + \beta_0 + \beta_1$

8.5.2 Random vs. Fixed Effects

Before we can use likelihood-based methods to estimate our model parameters, we still must define the distributions of our error terms. The error terms ϵ_{ij}, u_i, and v_i represent random effects in our model. In multilevel models, it is important to distinguish between fixed and random effects. Typically, **fixed effects** describe levels of a factor that we are specifically interested in drawing inferences about, and which would not change in replications of the study. For example, in our music performance anxiety case study, the levels of performance type will most likely remain as solos, small ensembles, and large ensembles even in replications of the study, and we wish to draw specific conclusions about differences between these three types of performances. Thus, performance type would be considered a fixed effect. On the other hand, **random effects** describe levels of a factor which can be thought of as a sample from a larger population of factor levels; we are not typically interested in drawing conclusions about specific levels of a random effect, although we are interested in accounting for the influence of the random effect in our model. For example, in our case study, the different musicians included can be thought of as a random sample from a population of performing musicians. Although our goal is not to make specific conclusions about differences between any two musicians, we do want to account for inherent differences among musicians in our model, and by doing so, we will be able to draw more precise conclusions about our fixed effects of interest. Thus, musician would be considered a random effect.

8.5.3 Distribution of Errors: Multivariate Normal

As part of our multilevel model, we must provide probability distributions to describe the behavior of random effects. Typically, we assume that random effects follow a normal distribution with mean 0 and a variance parameter which must be estimated from the data. For example, at Level One, we will assume that the errors associated with each performance of a particular musician can be described as: $\epsilon_{ij} \sim N(0, \sigma^2)$. At Level Two, we have one error term (u_i) associated with subject-to-subject differences in intercepts, and one error term (v_i) associated with subject-to-subject differences in slopes. That is, u_i represents the deviation of Subject i from the mean performance anxiety before solos and small ensembles after accounting for their instrument, and v_i represents the deviation of Subject i from the mean difference in performance anxiety between large ensembles and other performance types after accounting for their instrument.

In modeling the random behavior of u_i and v_i, we must also account for the possibility that random effects at the same level might be correlated. Subjects with higher baseline performance anxiety have a greater capacity for showing decreased anxiety in large ensembles as compared to solos and small ensembles, so we might expect that subjects with larger intercepts (performance anxiety before solos and small ensembles) would have smaller slopes (indicating greater decreases in anxiety before large ensembles). In fact, our fitted Level One intercepts and slopes in this example actually show evidence of a fairly strong negative correlation ($r = -0.525$, see Figure 8.11).

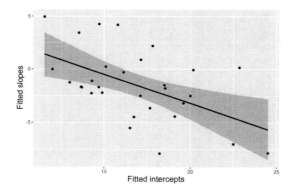

FIGURE 8.11: Scatterplot with fitted regression line for estimated intercepts and slopes (one point per subject).

To allow for this correlation, the error terms at Level Two can be assumed to follow a **multivariate normal distribution** in our unified multilevel model. Mathematically, we can express this as:

$$\left[\begin{array}{c} u_i \\ v_i \end{array} \right] \sim N \left(\left[\begin{array}{c} 0 \\ 0 \end{array} \right], \left[\begin{array}{cc} \sigma_u^2 & \rho_{uv}\sigma_u\sigma_v \\ \rho_{uv}\sigma_u\sigma_v & \sigma_v^2 \end{array} \right] \right)$$

where σ_u^2 is the variance of the u_i terms, σ_v^2 is the variance of the v_i terms,

8.5 Two-Level Modeling: A Unified Approach

and $\sigma_{uv} = \rho_{uv}\sigma_u\sigma_v$ is the covariance between the u_i and the v_i terms (i.e., how those two terms vary together).

Note that the correlation ρ_{uv} between the error terms is simply the covariance $\sigma_{uv} = \rho_{uv}\sigma_u\sigma_v$ converted to a $[-1, 1]$ scale through the relationship:

$$\rho_{uv} = \frac{\sigma_{uv}}{\sigma_u\sigma_v}$$

With this expression, we are allowing each error term to have its own variance (around a mean of 0) and each pair of error terms to have its own covariance (or correlation). Thus, if there are n equations at Level Two, we can have n variance terms and $n(n-1)/2$ covariance terms for a total of $n + n(n-1)/2$ variance components. These variance components are organized in matrix form, with variance terms along the diagonal and covariance terms in the off-diagonal. In our small example, we have $n = 2$ equations at Level Two, so we have 3 variance components to estimate—2 variance terms (σ_u^2 and σ_v^2) and 1 correlation (ρ_{uv}).

The multivariate normal distribution with $n = 2$ is illustrated in Figure 8.12 for two cases: (a) the error terms are uncorrelated ($\sigma_{uv} = \rho_{uv} = 0$), and (b) the error terms are positively correlated ($\sigma_{uv} > 0$ and $\rho_{uv} > 0$). In general, if the errors in intercepts (u_i) are placed on the x-axis and the errors in slopes (v_i) are placed on the y-axis, then σ_u^2 measures spread in the x-direction and σ_v^2 measures spread in the y-direction, while σ_{uv} measures tilt. Positive tilt ($\sigma_{uv} > 0$) indicates a tendency for errors from the same subject to both be positive or both be negative, while negative tilt ($\sigma_{uv} < 0$) indicates a tendency for one error from a subject to be positive and the other to be negative. In Figure 8.12, $\sigma_u^2 = 4$ and $\sigma_v^2 = 1$, so both contour plots show a greater range of errors in the x-direction than the y-direction. Ellipses near the center of the contour plot indicate pairs of u_i and v_i that are more likely. In Figure 8.12 (a) $\sigma_{uv} = \rho_{uv} = 0$, so the axes of the contour plot correspond to the x- and y-axes, but in Figure 8.12 (b) $\sigma_{uv} = 1.5$, so the contour plot tilts up, reflecting a tendency for high values of u_i to be associated with high values of v_i.

8.5.4 Technical Issues when Testing Parameters (optional)

Now, our relatively simple two-level model has 8 parameters that need to be estimated: 4 fixed effects (α_0, α_1, β_0, and β_1), and 4 variance components (σ^2, σ_u^2, σ_v^2, and σ_{uv}). Note that we use the term **variance components** to signify model parameters that describe the behavior of random effects. We can use statistical software, such as the lmer() function from the lme4 package in R, to obtain parameter estimates using our 497 observations. The most common methods for estimating model parameters—both fixed effects and variance components—are maximum likelihood (ML) and **restricted maximum likelihood (REML)**. The method of ML was introduced in

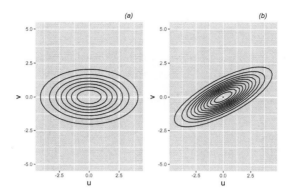

FIGURE 8.12: Contour plots illustrating a multivariate normal density with (a) no correlation between error terms, and (b) positive correlation between error terms.

Chapter 2, where parameter estimates are chosen to maximize the value of the likelihood function based on observed data. REML is conditional on the fixed effects, so that the part of the data used for estimating variance components is separated from that used for estimating fixed effects. Thus REML, by accounting for the loss in degrees of freedom from estimating the fixed effects, provides an unbiased estimate of variance components, while ML estimators for variance components are biased under assumptions of normality, since they use estimated fixed effects rather than the true values. REML is preferable when the number of parameters is large or the primary interest is obtaining estimates of model parameters, either fixed effects or variance components associated with random effects. ML should be used if nested fixed effects models are being compared using a likelihood ratio test, although REML is fine for nested models of random effects (with the same fixed effects model). In this text, we will typically report REML estimates unless we are specifically comparing nested models with the same random effects. In most case studies and most models we consider, there is very little difference between ML and REML parameter estimates. Additional details are beyond the scope of this book [Singer and Willett, 2003].

Note that the multilevel output shown beginning in the next section contains no p-values for performing hypothesis tests. This is primarily because the exact distribution of the test statistics under the null hypothesis (no fixed effect) is unknown, primarily because the exact degrees of freedom is not known [Bates et al., 2015]. Finding good approximate distributions for test statistics (and thus good approximate p-values) in multilevel models is an area of active research. In most cases, we can simply conclude that t-values (ratios of parameter estimates to estimated standard errors) with absolute value above 2 indicate significant

8.5 Two-Level Modeling: A Unified Approach

evidence that a particular model parameter is different than 0. Certain software packages will report p-values corresponding to hypothesis tests for parameters of fixed effects; these packages are typically using conservative assumptions, large-sample results, or approximate degrees of freedom for a t-distribution. In Section 1.6.5, we introduced the bootstrap as a non-parametric, computational approach for producing confidence intervals for model parameters. In addition, in Section 9.6.4, we will introduce a method called the parametric bootstrap which is being used more frequently by researchers to better approximate the distribution of the likelihood test statistic and produce more accurate p-values by simulating data under the null hypothesis [Efron, 2012].

8.5.5 An Initial Model with Parameter Interpretations

The output below contains likelihood-based estimates of our 8 parameters from a two-level model applied to the music performance anxiety data:

```
     Linear mixed model fit by REML ['lmerMod']
A)   Formula: na ~ orch + large + orch:large + (large | id)
        Data: music
B)   REML criterion at convergence: 2987

B2)       AIC       BIC    logLik  deviance  df.resid
         3007      3041     -1496      2991       489

     Random effects:
     Groups   Name          Variance  Std.Dev.  Corr
C)   id       (Intercept)    5.655    2.378
D)            large          0.452    0.672    -0.63
E)   Residual                21.807   4.670
F)   Number of obs: 497, groups:  id, 37

     Fixed effects:
                 Estimate  Std. Error  t value
G)   (Intercept)  15.930     0.641      24.83
H)   orch          1.693     0.945       1.79
I)   large        -0.911     0.845      -1.08
J)   orch:large   -1.424     1.099      -1.30
```

This output (except for the capital letters along the left column) was specifically generated by the lmer() function in R; multilevel modeling results from other packages will contain similar elements. Because we will use lmer() output to summarize analyses of case studies in this and following sections, we will spend a little time now orienting ourselves to the most important features in this output.

- A: How our multilevel model is written in R, based on the composite model formulation. For more details, see Section 8.12.
- B: Measures of model performance. Since this model was fit using REML, this line only contains the REML criterion.
- B2: If the model is fit with ML instead of REML, the measures of performance will contain AIC, BIC, deviance, and the log-likelihood.
- C: Estimated variance components ($\hat{\sigma}_u^2$ and $\hat{\sigma}_u$) associated with the intercept equation in Level Two.
- D: Estimated variance components ($\hat{\sigma}_v^2$ and $\hat{\sigma}_v$) associated with the large ensemble effect equation in Level Two, along with the estimated correlation ($\hat{\rho}_{uv}$) between the two Level Two error terms.
- E: Estimated variance components ($\hat{\sigma}^2$ and $\hat{\sigma}$) associated with the Level One equation.
- F: Total number of performances where data was collected (Level One observations = 497) and total number of subjects (Level Two observations = 37).
- G: Estimated fixed effect ($\hat{\alpha}_0$) for the intercept term, along with its standard error and t-value (which is the ratio of the estimated coefficient to its standard error). As described in Section 8.5.4, no p-value testing the significance of the coefficient is provided because the exact null distribution of the t-value is unknown.
- H: Estimated fixed effect ($\hat{\alpha}_1$) for the orchestral instrument effect, along with its standard error and t-value.
- I: Estimated fixed effect ($\hat{\beta}_0$) for the large ensemble effect, along with its standard error and t-value.
- J: Estimated fixed effect ($\hat{\beta}_1$) for the interaction between orchestral instruments and large ensembles, along with its standard error and t-value.

Assuming the 37 musicians in this study are representative of a larger population of musicians, parameter interpretations for our 8 model parameters are given below:

- Fixed effects:
    - $\hat{\alpha}_0 = 15.9$. The estimated mean performance anxiety for solos and small ensembles (Large=0) for keyboard players and vocalists (Orch=0) is 15.9.
    - $\hat{\alpha}_1 = 1.7$. Orchestral instrumentalists have an estimated mean performance anxiety for solos and small ensembles which is 1.7 points higher than keyboard players and vocalists.
    - $\hat{\beta}_0 = -0.9$. Keyboard players and vocalists have an estimated mean decrease in performance anxiety of 0.9 points when playing in large ensembles instead of solos or small ensembles.
    - $\hat{\beta}_1 = -1.4$. Orchestral instrumentalists have an estimated mean decrease in performance anxiety of 2.3 points when playing in large ensembles

8.6 Two-Level Modeling: A Unified Approach

TABLE 8.3: Comparison of estimated coefficients and standard errors from the approaches mentioned in this section.

Variable	Independence	TwoStage	LVCF	Multilevel
Intercept	15.72(0.36)	16.28(0.67)	15.20(1.25)	15.93(0.64)
Orch	1.79(0.55)	1.41(0.99)	1.45(1.84)	1.69(0.95)
Large	-0.28(0.79)	-0.77(0.85)	-	-0.91(0.85)
Orch*Large	-1.71(1.06)	-1.41(1.20)	-	-1.42(1.10)

instead of solos or small ensembles, 1.4 points greater than the mean decrease among keyboard players and vocalists.
- Variance components
  - $\hat{\sigma}_u = 2.4$. The estimated standard deviation of performance anxiety levels for solos and small ensembles is 2.4 points, after controlling for instrument played.
  - $\hat{\sigma}_v = 0.7$. The estimated standard deviation of differences in performance anxiety levels between large ensembles and other performance types is 0.7 points, after controlling for instrument played.
  - $\hat{\rho}_{uv} = -0.64$. The estimated correlation between performance anxiety scores for solos and small ensembles and increases in performance anxiety for large ensembles is -0.64, after controlling for instrument played. Those subjects with higher performance anxiety scores for solos and small ensembles tend to have greater decreases in performance anxiety for large ensemble performances.
  - $\hat{\sigma} = 4.7$. The estimated standard deviation in residuals for the individual regression models is 4.7 points.

Table 8.3 shows a side-by-side comparison of estimated coefficients from the approaches described to this point. Underlying assumptions, especially regarding the error and correlation structure, differ, and differences in estimated effects are potentially meaningful. Note that some standard errors are greatly *underestimated* under independence, and that no Level One covariates (such as performance type) can be analyzed under a method such as last-visit-carried-forward which uses one observation per subject. Moving forward, we will employ the unified multilevel approach to maximize the information being used to estimate model parameters and to remain faithful to the structure of the data.

Two-level modeling as done with the music performance anxiety data usually involves fitting a number of models. Subsequent sections will describe a process of starting with the simplest two-level models and building toward a final model which addresses the research questions of interest.

8.6 Building a Multilevel Model

8.6.1 Model Building Strategy

Initially, it is advisable to first fit some simple, preliminary models, in part to establish a baseline for evaluating larger models. Then, we can build toward a final model for description and inference by attempting to add important covariates, centering certain variables, and checking model assumptions. In this study, we are particularly interested in Level Two covariates—those subject-specific variables that provide insight into why individuals react differently in anxiety-inducing situations. To get more precise estimates of the effect of Level Two covariates, we also want to control for Level One covariates that describe differences in individual performances.

Our strategy for building multilevel models will begin with extensive exploratory data analysis at each level. Then, after examining models with no predictors to assess variability at each level, we will first focus on creating a Level One model, starting simple and adding terms as necessary. Next, we will move to Level Two models, again starting simple and adding terms as necessary, beginning with the equation for the intercept term. Finally, we will examine the random effects and variance components, beginning with a full set of error terms and then removing covariance terms and variance terms where advisable (for instance, when parameter estimates are failing to converge or producing impossible or unlikely values). This strategy follows closely with that described by Raudenbush and Bryk [2002] and used by Singer and Willett [2003]. Singer and Willett further find that the modeled error structure rarely matters in practical contexts. Other model building approaches are certainly possible. Diggle et al. [2002], for example, begins with a saturated fixed effects model, determines variance components based on that, and then simplifies the fixed part of the model after fixing the random part.

8.6.2 An Initial Model: Random Intercepts

The first model fit in almost any multilevel context should be the **unconditional means model**, also called a **random intercepts model**. In this model, there are no predictors at either level; rather, the purpose of the unconditional means model is to assess the amount of variation at each level—to compare variability within subject to variability between subjects. Expanded models will then attempt to explain sources of between and within subject variability.

The unconditional means (random intercepts) model, which we will denote as Model A, can be specified either using formulations at both levels:

8.6 Building a Multilevel Model

- Level One:
$$Y_{ij} = a_i + \epsilon_{ij} \text{ where } \epsilon_{ij} \sim N(0, \sigma^2)$$

- Level Two:
$$a_i = \alpha_0 + u_i \text{ where } u_i \sim N(0, \sigma_u^2)$$

or as a composite model:
$$Y_{ij} = \alpha_0 + u_i + \epsilon_{ij}$$

In this model, the performance anxiety scores of subject i are not a function of performance type or any other Level One covariate, so that a_i is the true mean response of all observations for subject i. On the other hand, α_0 is the grand mean – the true mean of all observations across the entire population. Our primary interest in the unconditional means model is the variance components – σ^2 is the within-person variability, while σ_u^2 is the between-person variability. The name **random intercepts model** then arises from the Level Two equation for a_i: each subject's intercept is assumed to be a random value from a normal distribution centered at α_0 with variance σ_u^2.

Using the composite model specification, the unconditional means model can be fit to the music performance anxiety data using statistical software:

```
#Model A (Unconditional means model)
model.a <- lmer(na ~ 1 + (1 | id), REML = T, data = music)
```

```
##  Groups    Name         Variance  Std.Dev.
##  id        (Intercept)   4.95     2.22
##  Residual               22.46     4.74

##  Number of Level Two groups = 37

##               Estimate  Std. Error  t value
##  (Intercept)    16.24     0.4279    37.94
```

From this output, we obtain estimates of our three model parameters:

- $\hat{\alpha}_0 = 16.2 =$ the estimated mean performance anxiety score across all performances and all subjects.
- $\hat{\sigma}^2 = 22.5 =$ the estimated variance in within-person deviations.
- $\hat{\sigma}_u^2 = 5.0 =$ the estimated variance in between-person deviations.

The relative levels of between- and within-person variabilities can be compared through the **intraclass correlation coefficient**:

$$\hat{\rho} = \frac{\text{Between-person variability}}{\text{Total variability}} = \frac{\hat{\sigma}_u^2}{\hat{\sigma}_u^2 + \hat{\sigma}^2} = \frac{5.0}{5.0 + 22.5} = .182.$$

Thus, 18.2% of the total variability in performance anxiety scores are attributable to differences among subjects. In this particular model, we can also say that the average correlation for any pair of responses from the same individual is a moderately low .182. As ρ approaches 0, responses from an individual are essentially independent and accounting for the multilevel structure of the data becomes less crucial. However, as ρ approaches 1, repeated observations from the same individual essentially provide no additional information and accounting for the multilevel structure becomes very important. With ρ near 0, the **effective sample size** (the number of independent pieces of information we have for modeling) approaches the total number of observations, while with ρ near 1, the effective sample size approaches the number of subjects in the study.

8.7 Binary Covariates at Level One and Level Two

8.7.1 Random Slopes and Intercepts Model

The next step in model fitting is to build a good model for predicting performance anxiety scores at Level One (within subject). We will add potentially meaningful Level One covariates—those that vary from performance-to-performance for each individual. In this case, mirroring our model from Section 8.4 we will include a binary covariate for performance type:

$$\text{LargeEns}_{ij} = \begin{cases} 1 & \text{if perf-type = Large Ensemble} \\ 0 & \text{if perf-type = Solo or Small Ensemble} \end{cases}$$

and no other Level One covariates (for now). (Note that we may later also want to include an indicator variable for "Small Ensemble" to separate the effects of `Solo` performances and `Small Ensemble` performances.) The resulting model, which we will denote as Model B, can be specified either using formulations at both levels:

- Level One:
$$Y_{ij} = a_i + b_i \text{LargeEns}_{ij} + \epsilon_{ij}$$

- Level Two:
$$a_i = \alpha_0 + u_i$$
$$b_i = \beta_0 + v_i$$

or as a composite model:

8.7 Binary Covariates at Level One and Level Two

$$Y_{ij} = [\alpha_0 + \beta_0 \text{LargeEns}_{ij}] + [u_i + v_i \text{LargeEns}_{ij} + \epsilon_{ij}]$$

where $\epsilon_{ij} \sim N(0, \sigma^2)$ and

$$\left[\begin{array}{c} u_i \\ v_i \end{array}\right] \sim N\left(\left[\begin{array}{c} 0 \\ 0 \end{array}\right], \left[\begin{array}{cc} \sigma_u^2 & \\ \rho\sigma_u\sigma_v & \sigma_v^2 \end{array}\right]\right).$$

as discussed in Section 8.5.3.

In this model, performance anxiety scores for subject i are assumed to differ (on average) for Large Ensemble performances as compared with Solos and Small Ensemble performances; the ϵ_{ij} terms capture the deviation between the true performance anxiety levels for subjects (based on performance type) and their observed anxiety levels. α_0 is then the true mean performance anxiety level for Solos and Small Ensembles, and β_0 is the true mean difference in performance anxiety for Large Ensembles compared to other performance types. As before, σ^2 quantifies the within-person variability (the scatter of points around individuals' means by performance type), while now the between-person variability is partitioned into variability in Solo and Small Ensemble scores (σ_u^2) and variability in differences with Large Ensembles (σ_v^2).

Using the composite model specification, Model B can be fit to the music performance anxiety data, producing the following output:

```
#Model B (Add large as Level 1 covariate)
model.b <- lmer(na ~ large + (large | id), data = music)
```

```
##  Groups   Name         Variance  Std.Dev.  Corr
##  id       (Intercept)  6.333     2.517
##           large        0.743     0.862     -0.76
##  Residual              21.771    4.666

##  Number of Level Two groups = 37

##                Estimate  Std. Error  t value
##  (Intercept)   16.730    0.4908      34.09
##  large         -1.676    0.5425      -3.09
```

From this output, we obtain estimates of our six model parameters (2 fixed effects and 4 variance components):

- $\hat{\alpha}_0 = 16.7 =$ the mean performance anxiety level before solos and small ensemble performances.

- $\hat{\beta}_0 = -1.7$ = the mean decrease in performance anxiety before large ensemble performances.
- $\hat{\sigma}^2 = 21.8$ = the variance in within-person deviations.
- $\hat{\sigma}_u^2 = 6.3$ = the variance in between-person deviations in performance anxiety scores before solos and small ensembles.
- $\hat{\sigma}_v^2 = 0.7$ = the variance in between-person deviations in increases (or decreases) in performance anxiety scores before large ensembles.
- $\hat{\rho}_{uv} = -0.76$ = the correlation in subjects' anxiety before solos and small ensembles and their differences in anxiety between large ensembles and other performance types.

We see that, on average, subjects had a performance anxiety level of 16.7 before solos and small ensembles, and their anxiety levels were 1.7 points lower, on average, before large ensembles, producing an average performance anxiety level before large ensembles of 15.0. According to the t-value listed in R, the difference between large ensembles and other performance types is statistically significant (t=-3.09).

This random slopes and intercepts model is illustrated in Figure 8.13. The thicker black line shows the overall trends given by our estimated fixed effects: an intercept of 16.7 and a slope of -1.7. Then, each subject is represented by a gray line. Not only do the subjects' intercepts differ (with variance 6.3), but their slopes differ as well (with variance 0.7). Additionally, subjects' slopes and intercepts are negatively associated (with correlation -0.76), so that subjects with greater intercepts tend to have steeper negative slopes. We can compare this model with the random intercepts model from Section 8.6.2, pictured in Figure 8.14. With no effect of large ensembles, each subject is represented by a gray line with identical slopes (0) but varying intercepts (with variance 5.0).

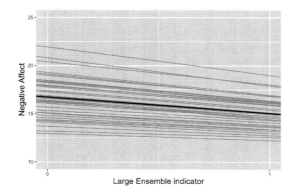

FIGURE 8.13: The random slopes and intercepts model fitted to the music performance anxiety data. Each gray line represents one subject, and the thicker black line represents the trend across all subjects.

8.7 Binary Covariates at Level One and Level Two

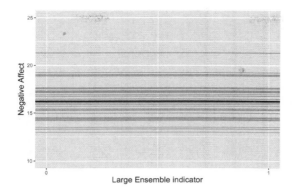

FIGURE 8.14: The random intercepts model fitted to the music performance anxiety data. Each gray line represents one subject, and the thicker black line represents the trend across all subjects.

Figures 8.13 and 8.14 use **empirical Bayes estimates** for the intercepts (a_i) and slopes (b_i) of individual subjects. Empirical Bayes estimates are sometimes called "shrinkage estimates" since they combine individual-specific information with information from all subjects, thus "shrinking" the individual estimates toward the group averages. Empirical Bayes estimates are often used when a term such as a_i involves both fixed and random components; further detail can be found in Raudenbush and Bryk [2002] and Singer and Willett [2003].

8.7.2 Pseudo R-squared Values

The estimated within-person variance $\hat{\sigma}^2$ decreased by 3.1% (from 22.5 to 21.8) from the unconditional means model, implying that only 3.1% of within-person variability in performance anxiety scores can be explained by performance type. This calculation is considered a **pseudo R-squared** value:

$$\text{Pseudo } R^2_{L1} = \frac{\hat{\sigma}^2(\text{Model A}) - \hat{\sigma}^2(\text{Model B})}{\hat{\sigma}^2(\text{Model A})} = \frac{22.5 - 21.8}{22.5} = 0.031$$

Values of $\hat{\sigma}^2_u$ and $\hat{\sigma}^2_v$ from Model B cannot be compared to between-person variability from Model A, since the inclusion of performance type has changed the interpretation of these values, although they can provide important benchmarks for evaluating more complex Level Two predictions. Finally, $\hat{\rho}_{uv} = -0.76$ indicates a strong negative relationship between a subject's performance anxiety before solos and small ensembles and their (typical) decrease in performance anxiety before large ensembles. As might be expected, subjects with higher levels of performance anxiety before solos and small ensembles tend to have smaller increases (or greater decreases) in performance anxiety before large ensembles; those with higher levels of performance anxiety before solos and small ensembles have more opportunity for decreases before large ensembles.

Pseudo R-squared values are not universally reliable as measures of model performance. Because of the complexity of estimating fixed effects and variance components at various levels of a multilevel model, it is not unusual to encounter situations in which covariates in a Level Two equation for, say, the intercept remain constant (while other aspects of the model change), yet the associated pseudo R-squared values differ or are negative. For this reason, pseudo R-squared values in multilevel models should be interpreted cautiously.

8.7.3 Adding a Covariate at Level Two

The initial two-level model described in Section 8.5.5 essentially expands upon the random slopes and intercepts model by adding a binary covariate for instrument played at Level Two. We will denote this as Model C:

- Level One:
$$Y_{ij} = a_i + b_i \text{LargeEns}_{ij} + \epsilon_{ij}$$

- Level Two:
$$a_i = \alpha_0 + \alpha_1 \text{Orch}_i + u_i$$
$$b_i = \beta_0 + \beta_1 \text{Orch}_i + v_i,$$

where $\epsilon_{ij} \sim N(0, \sigma^2)$ and

$$\begin{bmatrix} u_i \\ v_i \end{bmatrix} \sim N \left(\begin{bmatrix} 0 \\ 0 \end{bmatrix}, \begin{bmatrix} \sigma_u^2 & \\ \rho \sigma_u \sigma_v & \sigma_v^2 \end{bmatrix} \right).$$

We found that there are no highly significant fixed effects in Model C (other than the intercept). In particular, we have no significant evidence that musicians playing orchestral instruments reported different performance anxiety scores, on average, for solos and small ensembles than keyboardists and vocalists, no evidence of a difference in performance anxiety by performance type for keyboard players and vocalists, and no evidence of an instrument effect in difference between large ensembles and other types.

Since no terms were added at Level One when expanding from the random slopes and intercepts model (Model B), no discernible changes should occur in explained within-person variability (although small changes could occur due to numerical estimation procedures used in likelihood-based parameter estimates). However, Model C expanded Model B by using the instrument which a subject plays to model both intercepts and slopes at Level Two. We can use pseudo R-squared values for both intercepts and slopes to evaluate the impact on between-person variability of adding instrument to Model B.

8.7 Binary Covariates at Level One and Level Two

$$\text{Pseudo } R^2_{L2_u} = \frac{\hat{\sigma}^2_u(\text{Model B}) - \hat{\sigma}^2_u(\text{Model C})}{\hat{\sigma}^2_u(\text{Model B})} = \frac{6.33 - 5.66}{6.33} = 0.106$$

$$\text{Pseudo } R^2_{L2_v} = \frac{\hat{\sigma}^2_v(\text{Model B}) - \hat{\sigma}^2_v(\text{Model C})}{\hat{\sigma}^2_v(\text{Model B})} = \frac{0.74 - 0.45}{0.74} = 0.392$$

Pseudo $R^2_{L2_u}$ describes the improvement in Model C over Model B in explaining subject-to-subject variability in intercepts, and Pseudo $R^2_{L2_v}$ describes the improvement in Model C over Model B in explaining subject-to-subject variability in slopes. Thus, the addition of instrument at Level Two has decreased the between-person variability in mean performance anxiety before solos and small ensembles by 10.6%, and it has decreased the between-person variability in the effect of large ensembles on performance anxiety by 39.2%.

We could also run a "random intercepts" version of Model C, with no error term in the equation for the slope at Level Two (and thus no covariance between errors at Level Two as well):

- Level One:
$$Y_{ij} = a_i + b_i \text{LargeEns}_{ij} + \epsilon_{ij}$$

- Level Two:
$$a_i = \alpha_0 + \alpha_1 \text{Orch}_i + u_i$$
$$b_i = \beta_0 + \beta_1 \text{Orch}_i,$$

where $\epsilon_{ij} \sim N(0, \sigma^2)$ and $u_i \sim N(0, \sigma^2_u)$.

The output below contains REML estimates of our 6 parameters from this simplified version of Model C (which we'll call Model C2):

```
#Model C2 (Run as random intercepts model)
model.c2 <- lmer(na ~ orch + large + orch:large +
   (1|id), data = music)
```

```
##  Groups    Name         Variance Std.Dev.
##  id        (Intercept)  5.13     2.27
##  Residual               21.88    4.68

##  Number of Level Two groups = 37

##                Estimate Std. Error t value
##  (Intercept)   15.9026    0.6187   25.703
##  orch           1.7100    0.9131    1.873
##  large         -0.8918    0.8415   -1.060
##  orch:large    -1.4650    1.0880   -1.347
```

	df	AIC
model.c	8	3003
model.c2	6	2999

	df	BIC
model.c	8	3037
model.c2	6	3025

Note that parameter estimates for the remaining 6 fixed effects and variance components closely mirror the corresponding parameter estimates from Model C. In fact, removing the error term on the slope has improved (reduced) both the AIC and BIC measures of overall model performance. Instead of assuming that the large ensemble effects, after accounting for instrument played, vary by individual, we are assuming that large ensemble effect is fixed across subjects. It is not unusual to run a two-level model like this, with an error term on the intercept equation to account for subject-to-subject differences, but with no error terms on other Level Two equations unless there is an *a priori* reason to allow effects to vary by subject or if the model performs better after building in those additional error terms.

8.8 Adding Further Covariates

Recall that we are particularly interested in this study in Level Two covariates—those subject-specific variables that provide insight into why individuals react differently in anxiety-inducing situations. In Section 8.3, we saw evidence that subjects with higher baseline levels of negative emotionality tend to have higher performance anxiety levels prior to performances. Thus, in our next step in model building, we will add negative emotionality as a Level Two predictor to Model C. With this addition, our new model can be expressed as a system of Level One and Level Two models:

- Level One:
$$Y_{ij} = a_i + b_i \text{LargeEns}_{ij} + \epsilon_{ij}$$

- Level Two:
$$a_i = \alpha_0 + \alpha_1 \text{Orch}_i + \alpha_2 \text{MPQnem}_i + u_i$$
$$b_i = \beta_0 + \beta_1 \text{Orch}_i + \beta_2 \text{MPQnem}_i + v_i,$$

or as a composite model:

8.8 Adding Further Covariates

$$Y_{ij} = [\alpha_0 + \alpha_1 \text{Orch}_i + \alpha_2 \text{MPQnem}_i + \beta_0 \text{LargeEns}_{ij}$$
$$+ \beta_1 \text{Orch}_i \text{LargeEns}_{ij} + \beta_2 \text{MPQnem}_i \text{LargeEns}_{ij}]$$
$$+ [u_i + v_i \text{LargeEns}_{ij} + \epsilon_{ij}]$$

where error terms are defined as in Model C.

From the R output below, we see that, as our exploratory analyses suggested, subjects with higher baseline levels of stress reaction, alienation, and aggression (as measured by the MPQ negative emotionality scale) had significantly higher levels of performance anxiety before solos and small ensembles (t=3.893). They also had somewhat greater differences between large ensembles and other performance types, controlling for instrument (t=-0.575), although this interaction was not statistically significant.

```
# Model D (Add negative emotionality as second L2 covariate)
model.d <- lmer(na ~ orch + mpqnem + large + orch:large +
    mpqnem:large + (large | id), data = music)
```

```
##  Groups   Name         Variance Std.Dev. Corr
##  id       (Intercept)  3.286    1.813
##           large        0.557    0.746    -0.38
##  Residual              21.811   4.670

##  Number of Level Two groups =  37

##                 Estimate Std. Error t value
## (Intercept)     11.56801    1.22057  9.4775
## orch             1.00069    0.81713  1.2246
## mpqnem           0.14823    0.03808  3.8925
## large           -0.28019    1.83412 -0.1528
## orch:large      -0.94927    1.10620 -0.8581
## mpqnem:large    -0.03018    0.05246 -0.5753
```

8.8.1 Interpretation of Parameter Estimates

Compared to Model C, the directions of the effects of instrument and performance type are consistent, but the effect sizes and levels of significance are reduced because of the relative importance of the negative emotionality term. Interpretations will also change slightly to acknowledge that we have controlled for a covariate. In addition, interpretations of fixed effects involving negative emotionality must acknowledge that this covariate is a continuous measure and not binary like instrument and performance type:

- $\hat{\alpha}_0 = 11.57$. The estimated mean performance anxiety for solos and small ensembles (large=0) is 11.57 for keyboard players and vocalists (orch=0) with negative emotionality of 0 at baseline (mpqnem=0). Since the minimum negative emotionality score in this study was 11, this interpretation, while technically correct, is not practically meaningful.
- $\hat{\alpha}_1 = 1.00$. Orchestral instrument players have an estimated mean anxiety level before solos and small ensembles which is 1.00 point higher than keyboardists and vocalists, controlling for the effects of baseline negative emotionality.
- $\hat{\alpha}_2 = 0.15$. A one point increase in baseline negative emotionality is associated with an estimated 0.15 mean increase in anxiety levels before solos and small ensembles, after controlling for instrument.
- $\hat{\beta}_0 = -0.28$. Keyboard players and vocalists (orch=0) with baseline negative emotionality levels of 0 (mpqnem=0) have an estimated mean decrease in anxiety level of 0.28 points before large ensemble performances compared to other performance types.
- $\hat{\beta}_1 = -0.95$. After accounting for baseline negative emotionality, orchestral instrument players have an estimated mean anxiety level before solos and small ensembles which is 1.00 point higher than keyboardists and vocalists, while the mean anxiety of orchestral players is only .05 points higher before large ensembles (a difference of .95 points).
- $\hat{\beta}_2 = -0.03$. After accounting for instrument, a one-point increase in baseline negative emotionality is associated with an estimated 0.15 mean increase in anxiety levels before solos and small ensembles, but only an estimated 0.12 increase before large ensembles (a difference of .03 points).

Some of the detail in these parameter interpretations can be tricky—describing interaction terms, deciding if a covariate must be fixed at 0 or merely held constant, etc. Often it helps to write out models for special cases to isolate the effects of specific fixed effects. We will consider a few parameter estimates from above and see why the interpretations are written as they are.

- $\hat{\alpha}_1$. For solos and small ensembles (LargeEns=0), the following equations describe the fixed effects portion of the composite model for negative affect score for vocalists and keyboardists (Orch=0) and orchestral instrumentalists (Orch=1):

$$\text{Orch} = 0:$$
$$Y_{ij} = \alpha_0 + \alpha_2 \text{MPQnem}_i$$
$$\text{Orch} = 1:$$
$$Y_{ij} = (\alpha_0 + \alpha_1) + \alpha_2 \text{MPQnem}_i$$

Regardless of the subjects' baseline negative emotionality (MPQnem), $\hat{\alpha}_1$ represents the estimated difference in performance anxiety between those playing orchestral instruments and others. This interpretation, however, only holds for

8.8 Adding Further Covariates

solos and small ensembles. For large ensembles, the difference between those playing orchestral instruments and others is actually given by $\hat{\alpha}_1 + \hat{\beta}_1$, holding MPQnem constant (Show!).

- $\hat{\beta}_0$. Because LargeEns interacts with both Orch and MPQnem in Model C, $\hat{\beta}_0$ only describes the estimated difference between large ensembles and other performance types when both Orch=0 and MPQnem=0, thus removing the effects of the interaction terms. If, for instance, Orch=1 and MPQnem=20, then the difference between large ensembles and other performance types is given by $\hat{\beta}_0 + \hat{\beta}_1 + 20\hat{\beta}_2$.
- $\hat{\beta}_1$. As with $\hat{\alpha}_1$, we consider equations describing the fixed effects portion of the composite model for negative affect score for vocalists and keyboardists (Orch=0) and orchestral instrumentalists (Orch=1), except here we leave *LargeEns* as an unknown rather than restricting the model to solos and small ensembles:

$$\text{Orch} = 0:$$
$$Y_{ij} = \alpha_0 + \alpha_2 \text{MPQnem}_i + \beta_0 \text{LargeEns}_{ij}$$
$$+ \beta_2 \text{MPQnem}_i \text{LargeEns}_{ij}$$
$$\text{Orch} = 1:$$
$$Y_{ij} = (\alpha_0 + \alpha_1) + \alpha_2 \text{MPQnem}_i + (\beta_0 + \beta_1)\text{LargeEns}_{ij}$$
$$+ \beta_2 \text{MPQnem}_i \text{LargeEns}_{ij}$$

As long as baseline negative emotionality is held constant (at any level, not just 0), then $\hat{\beta}_1$ represents the estimated difference in the large ensemble effect between those playing orchestral instruments and others.

8.8.2 Model Comparisons

At this point, we might ask: do the two extra fixed effects terms in Model D provide a significant improvement over Model C? Nested models such as these can be tested using a **likelihood ratio test** (drop in deviance test), as we've used in Sections 4.4.4 and 6.5.4 with certain generalized linear models. Since we are comparing models nested in their fixed effects, we use full maximum likelihood methods to estimate model parameters, as discussed in Section 8.5.4. As expected, the likelihood is larger (and the log-likelihood is less negative) under the larger model (Model D); our test statistic (14.734) is then -2 times the difference in log-likelihood between Models C and D. Comparing the test statistic to a chi-square distribution with 2 degrees of freedom (signifying the number of additional terms in Model D), we obtain a p-value of .0006. Thus, Model D significantly outperforms Model C.

```
# anova() automatically uses ML for LRT tests
drop_in_dev <- anova(model.d, model.c, test = "Chisq")
```

```
          npar  AIC  BIC  logLik   dev  Chisq Df      pval
model.c      8 3007 3041   -1496  2991     NA NA        NA
model.d     10 2996 3039   -1488  2976  14.73  2 0.0006319
```

Two models, whether they are nested or not, can be compared using AIC and BIC measures, which were first seen in Chapter 1 and later used in evaluating generalized linear models. In this case, the AIC clearly favors Model D (2996.7) over Model C (3007.3), whereas the BIC favors Model D (3038.8) only slightly over Model C (3041.0) since the BIC imposes a stiffer penalty on additional terms and additional model complexity. However, the likelihood ratio test is a more reliable method for comparing nested models.

Finally, we note that Model D could be further improved by dropping the negative emotionality by large ensemble interaction term. Not only is the t-value (-0.575) associated with this term of low magnitude, but a likelihood ratio test comparing Model D to a model without `mpqnem:large` produces an insignificant p-value of 0.5534.

```
drop_in_dev <- anova(model.d, model.d1, test = "Chisq")
```

```
           npar  AIC  BIC  logLik   dev  Chisq Df   pval
model.d1      9 2995 3033   -1488  2977     NA NA     NA
model.d      10 2996 3039   -1488  2976 0.3513  1 0.5534
```

8.9 Centering Covariates

As we observed above, the addition of baseline negative emotionality in Model D did not always produce sensible interpretations of fixed effects. It makes no sense to draw conclusions about performance anxiety levels for subjects with MPQNEM scores of 0 at baseline (as in $\hat{\beta}_0$), since the minimum NEM composite score among subjects in this study was 11. In order to produce more meaningful interpretations of parameter estimates and often more stable parameter estimates, it is often wise to **center** explanatory variables. Centering involves subtracting a fixed value from each observation, where the fixed value represents a meaningful anchor value (e.g., last grade completed is 12; GPA

8.9 Centering Covariates

is 3.0). Often, when there's no pre-defined anchor value, the mean is used to represent a typical case. With this in mind, we can create a new variable

$$\text{centeredbaselineNEM} = \text{cmpqnem}$$
$$= \text{mpqnem - mean(mpqnem)}$$
$$= \text{mpqnem} - 31.63$$

and replace baseline NEM in Model D with its centered version to create Model E:

```
# Model E (Center baseline NEM in Model D)
model.e <- lmer(na ~ orch + cmpqnem + large + orch:large +
  cmpqnem:large + (large | id), REML = T, data = music)
```

```
##  Groups   Name        Variance Std.Dev. Corr
##  id       (Intercept) 3.286    1.813
##           large       0.557    0.746    -0.38
##  Residual             21.811   4.670

##  Number of Level Two groups =  37

##                 Estimate Std. Error t value
## (Intercept)     16.25679    0.54756 29.6893
## orch             1.00069    0.81713  1.2246
## cmpqnem          0.14823    0.03808  3.8925
## large           -1.23484    0.84320 -1.4645
## orch:large      -0.94927    1.10620 -0.8581
## cmpqnem:large   -0.03018    0.05246 -0.5753
```

As you compare Model D to Model E, you should notice that only two things change – $\hat{\alpha}_0$ and $\hat{\beta}_0$. All other parameter estimates—both fixed effects and variance components—remain identical; the basic model is essentially unchanged as well as the amount of variability in anxiety levels explained by the model. $\hat{\alpha}_0$ and $\hat{\beta}_0$ are the only two parameter estimates whose interpretations in Model D refer to a specific level of baseline NEM. In fact, the interpretations that held true where NEM=0 (which isn't possible) now hold true for cmpqnem=0 or when NEM is at its average value of 31.63, which is possible and quite meaningful. Now, parameter estimates using centered baseline NEM in Model E change in value from Model D and produce more useful interpretations:

- $\hat{\alpha}_0 = 16.26$. The estimated mean performance anxiety for solos and small ensembles (large=0) is 16.26 for keyboard players and vocalists (orch=0) with an average level of negative emotionality at baseline (mpqnem=31.63).

- $\hat{\beta}_0 = -1.23$. Keyboard players and vocalists (`orch=0`) with an average level of baseline negative emotionality levels (`mpqnem=31.63`) have an estimated mean decrease in anxiety level of 1.23 points before large ensemble performances compared to other performance types.

8.10 A Final Model for Music Performance Anxiety

We now begin iterating toward a "final model" for these data, on which we will base conclusions. Typical features of a "final multilevel model" include:

- fixed effects allow one to address primary research questions
- fixed effects control for important covariates at all levels
- potential interactions have been investigated
- variables are centered where interpretations can be enhanced
- important variance components have been included
- unnecessary terms have been removed
- the model tells a "persuasive story parsimoniously"

Although the process of reporting and writing up research results often demands the selection of a sensible final model, it's important to realize that (a) statisticians typically will examine and consider an entire taxonomy of models when formulating conclusions, and (b) different statisticians sometimes select different models as their "final model" for the same set of data. Choice of a "final model" depends on many factors, such as primary research questions, purpose of modeling, tradeoff between parsimony and quality of fitted model, underlying assumptions, etc. So you should be able to defend any final model you select, but you should not feel pressured to find the one and only "correct model", although most good models will lead to similar conclusions.

As we've done in previous sections, we can use (a) t-statistics for individual fixed effects when considering adding a single term to an existing model, (b) likelihood ratio tests for comparing nested models which differ by more than one parameter, and (c) model performance measures such as AIC and BIC to compare non-nested models. Below we offer one possible final model for this data—Model F:

- Level One:

$$Y_{ij} = a_i + b_i \text{previous}_{ij} + c_i \text{students}_{ij} + d_i \text{juried}_{ij} + e_i \text{public}_{ij} + f_i \text{solo}_{ij} + \epsilon_{ij}$$

- Level Two:

8.10 A Final Model for Music Performance Anxiety

$$a_i = \alpha_0 + \alpha_1 \text{mpqpem}_i + \alpha_2 \text{mpqab}_i + \alpha_3 \text{orch}_i + \alpha_4 \text{mpqnem}_i + u_i$$
$$b_i = \beta_0 + v_i,$$
$$c_i = \gamma_0 + w_i,$$
$$d_i = \delta_0 + x_i,$$
$$e_i = \varepsilon_0 + y_i,$$
$$f_i = \zeta_0 + \zeta_1 \text{mpqnem}_i + z_i,$$

where **previous** is the number of previous diary entries filled out by that individual (**diary-1**); **students**, **juried**, and **public** are indicator variables created from the **audience** categorical variable (so that "Instructor" is the reference level in this model); and, **solo** is 1 if the performance was a solo and 0 if the performance was either a small or large ensemble.

In addition, we assume the following variance-covariance structure at Level Two:

$$\begin{bmatrix} u_i \\ v_i \\ w_i \\ x_i \\ y_i \\ z_i \end{bmatrix} \sim N \left(\begin{bmatrix} 0 \\ 0 \\ 0 \\ 0 \\ 0 \\ 0 \end{bmatrix}, \begin{bmatrix} \sigma_u^2 & & & & & \\ \sigma_{uv} & \sigma_v^2 & & & & \\ \sigma_{uw} & \sigma_{vw} & \sigma_w^2 & & & \\ \sigma_{ux} & \sigma_{vx} & \sigma_{wx} & \sigma_x^2 & & \\ \sigma_{uy} & \sigma_{vy} & \sigma_{wy} & \sigma_{xy} & \sigma_y^2 & \\ \sigma_{uz} & \sigma_{vz} & \sigma_{wz} & \sigma_{xz} & \sigma_{yz} & \sigma_z^2 \end{bmatrix} \right).$$

Being able to write out these mammoth variance-covariance matrices is less important than recognizing the number of variance components that must be estimated by our intended model. In this case, we must use likelihood-based methods to obtain estimates for 6 variance terms and 15 correlation terms at Level Two, along with 1 variance term at Level One. Note that the number of correlation terms is equal to the number of unique pairs among Level Two random effects. In later sections we will consider ways to reduce the number of variance components in cases where the number of terms is exploding, or the statistical software is struggling to simultaneously find estimates for all model parameters to maximize the likelihood function.

From the signs of fixed effects estimates in the R output below, we see that performance anxiety is higher when a musician is performing in front of students, a jury, or the general public rather than their instructor, and it is lower for each additional diary the musician previously filled out. In addition, musicians with lower levels of positive emotionality and higher levels of absorption tend to experience greater performance anxiety, and those who play orchestral instruments experience more performance anxiety than those who play keyboards or sing. Addressing the researchers' primary hypothesis, after controlling for all these factors, we have significant evidence that musicians with higher levels of negative emotionality experience higher levels of performance anxiety, and that this association is even more pronounced when musicians are performing solos rather than as part of an ensemble group.

Here are how a couple of key fixed effects would be interpreted in this final model:

- $\hat{\alpha}_4 = 0.11$. A one-point increase in baseline level of negative emotionality is associated with an estimated 0.11 mean increase in performance anxiety for musicians performing in an ensemble group (solo=0), after controlling for previous diary entries, audience, positive emotionality, absorption, and instrument.
- $\hat{\zeta}_1 = 0.08$. When musicians play solos, a one-point increase in baseline level of negative emotionality is associated with an estimated 0.19 mean increase in performance anxiety, 0.08 points (73%) higher than musicians playing in ensemble groups, controlling for the effects of previous diary entries, audience, positive emotionality, absorption, and instrument.

```
# Model F (One - of many - reasonable final models)
model.f <- lmer(na ~ previous + students + juried +
    public + solo + mpqpem + mpqab + orch + mpqnem +
    mpqnem:solo + (previous + students + juried +
    public + solo | id), REML = T, data = music)
```

```
##   Groups   Name         Variance  Std.Dev.  Corr
##   id       (Intercept)  14.4802   3.805
##            previous      0.0707   0.266    -0.65
##            students      8.2151   2.866    -0.63  0.00
##            juried       18.3177   4.280    -0.64 -0.12
##            public       12.8094   3.579    -0.83  0.33
##            solo          0.7665   0.876    -0.67  0.47
##   Residual              15.2844   3.910
##
##
##
##
##      0.84
##      0.66  0.58
##      0.49  0.21  0.90
##
##   Number of Level Two groups =   37
##
##                Estimate  Std. Error  t value
##   (Intercept)   8.36883     1.91369   4.3731
```

```
## previous      -0.14303   0.06247  -2.2895
## students       3.61115   0.76796   4.7022
## juried         4.07332   1.03130   3.9497
## public         3.06453   0.89274   3.4327
## solo           0.51647   1.39635   0.3699
## mpqpem        -0.08312   0.02408  -3.4524
## mpqab          0.20377   0.04740   4.2986
## orch           1.53138   0.58384   2.6230
## mpqnem         0.11465   0.03591   3.1930
## solo:mpqnem    0.08296   0.04158   1.9951
```

8.11 Modeling Multilevel Structure: Is It Necessary?

Before going too much further, we should really consider if this multilevel structure has gained us anything over linear least squares regression. Sure, multilevel modeling seems more faithful to the inherent structure of the data—performances from the same musician should be more highly correlated than performances from different musicians—but do our conclusions change in a practical sense? Some authors have expressed doubts. For instance Robert Bickel, in his 2007 book, states, "When comparing OLS and multilevel regression results, we may find that differences among coefficient values are inconsequential, and tests of significance may lead to the same decisions. A great deal of effort seems to have yielded precious little gain" [Bickel, 2007]. Others, especially economists, advocate simply accounting for the effect of different Level Two observational units (like musicians) with a sequence of indicator variables for those observational units. We contend that (1) fitting multilevel models is a small extension of LLSR regression that is not that difficult to conceptualize and fit, especially with the software available today, and (2) using multilevel models to remain faithful to the data structure *can* lead to different coefficient estimates and *often* leads to different (and larger) standard error estimates and thus smaller test statistics. Hopefully you've seen evidence of (1) in this chapter already; the rest of this section introduces two small examples to help illustrate (2).

Figure 8.15 is based on a synthetic data set containing 10 observations from each of 4 subjects. For each subject, the relationship between previous performances and negative affect is linear and negative, with slope approximately -0.5 but different intercepts. The multilevel model (a random intercepts model as described in Section 8.7) shows an overall relationship (the dashed black line) that's consistent with the individual subjects—slope around -0.5 with an intercept that appears to average the 4 subjects. Fitting a LLSR model,

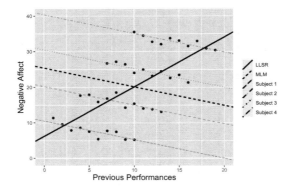

FIGURE 8.15: Hypothetical data from 4 subjects relating number of previous performances to negative affect. The dashed black line depicts the overall relationship between previous performances and negative affect as determined by a multilevel model, while the solid black line depicts the overall relationship as determined by an LLSR regression model.

however, produces an overall relationship (the solid black line) that is strongly positive. In this case, by naively fitting the 40 observations as if they were all independent and ignoring subject effects, the LLSR analysis has gotten the estimated slope of the overall relationship backwards, producing a continuous data version of Simpson's Paradox.

Our second example is based upon Model C from Section 8.7.3, with single binary predictors at both Level One and Level Two. Using the estimated fixed effects coefficients and variance components from random effects produced in Model C, we generated 1000 sets of simulated data. Each set of simulated data contained 497 observations from 37 subjects just like the original data, with relationships between negative affect and large ensembles and orchestral instruments (along with associated variability) governed by the estimated parameters from Model C. Each set of simulated data was used to fit both a multilevel model and a linear least squares regression model, and the estimated fixed effects ($\hat{\alpha}_0$, $\hat{\alpha}_1$, $\hat{\beta}_0$, and $\hat{\beta}_1$) and their standard errors were saved. Figure 8.16 shows density plots comparing the 1000 estimated values for each fixed effect from the two modeling approaches; in general, estimates from multilevel modeling and LLSR tend to agree pretty well, without noticeable bias. Based on coefficient estimates alone, there appears to be no reason to favor multilevel modeling over LLSR in this example, but Figure 8.17 tells a different story. Figure 8.17 shows density plots comparing the 1000 estimated standard errors associated with each fixed effect from the two modeling approaches; in general, standard errors are markedly larger with multilevel modeling than LLSR. This is not unusual, since LLSR assumes all 497 observations are independent, while

8.11 Modeling Multilevel Structure: Is It Necessary?

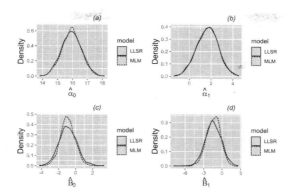

FIGURE 8.16: Density plots of parameter estimates for the four fixed effects of Model C under both a multilevel model and linear least squares regression. 1000 sets of simulated data for the 37 subjects in our study were produced using estimated fixed and random effects from Model C. For each set of simulated data, estimates of (a) α_0, (b) α_1, (c) β_0, and (d) β_1 were obtained using both a multilevel and an LLSR model. Each plot then shows a density plot for the 1000 estimates of the corresponding fixed effect using multilevel modeling vs. a similar density plot for the 1000 estimates using LLSR.

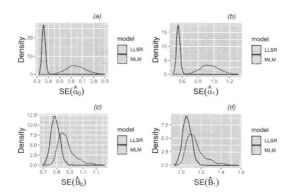

FIGURE 8.17: Density plots of standard errors of parameter estimates for the four fixed effects of Model C under both a multilevel model and linear least squares regression. 1000 sets of simulated data for the 37 subjects in our study were produced using estimated fixed and random effects from Model C. For each set of simulated data, estimates of (a) $SE(\alpha_0)$, (b) $SE(\alpha_1)$, (c) $SE(\beta_0)$, and (d) $SE(\beta_1)$ were obtained using both a multilevel and an LLSR model. Each plot then shows a density plot for the 1000 estimates of the corresponding standard error term using multilevel modeling vs. a similar density plot for the 1000 estimates using LLSR.

multilevel modeling acknowledges that, with correlated data within subject, there are fewer than 497 independent pieces of data. Therefore, linear least squares regression can overstate precision, producing t-statistics for each fixed effect that tend to be larger than they should be; the number of significant results in LLSR are then too great and not reflective of the true structure of the data.

8.12 Notes on Using R (optional)

Initial examination of the data for Case Study 8.2 shows a couple of features that must be noted. First, there are 37 unique study participants, but they are not numbered successively from 1 to 43. The majority of participants filled out 15 diaries, but several filled out fewer (with a minimum of 2); as with participant IDs, diary numbers within participant are not always successively numbered. Finally, missing data is not an issue in this data set, since researchers had already removed participants with only 1 diary entry and performances for which the type was not recorded (of which there were 11).

The R code below runs the initial multilevel model in Section 8.5.5. Multilevel model notation in R is based on the composite model formulation. Here, the response variable is `na`, while `orch`, `large`, and `orch:large` represent the fixed effects α_1, β_0, and β_1, along with the intercept α_0 which is included automatically. Note that a colon is used to denote an interaction between two variables. Error terms and their associated variance components are specified in `(large|id)`, which is equivalent to `(1+large|id)`. This specifies two error terms at Level Two (the `id` level): one corresponding to the intercept (u_i) and one corresponding to the large ensemble effect (v_i); the multilevel model will then automatically include a variance for each error term in addition to the covariance between the two error terms. A variance associated with a Level One error term is also automatically included in the multilevel model. Note that there are ways to override the automatic inclusion of certain variance components; for example, `(0+large|id)` would not include an error term for the intercept (and therefore no covariance term at Level Two either).

```
model0 <- lmer(na ~ orch + large + orch:large +
    (large | id), REML = T, data = music)
summary(model0)
```

8.13 Exercises

8.13.1 Conceptual Exercises

1. **Housing prices.** Brown and Uyar [2004] describe "A Hierarchical Linear Model Approach for Assessing the Effects of House and Neighborhood Characteristics on Housing Prices". Based on the title of their paper: (a) give the observational units at Level One and Level Two, and (b) list potential explanatory variables at both Level One and Level Two.

2. In the preceding problem, why can't we assume all houses in the data set are independent? What would be the potential implications to our analysis of assuming independence among houses?

3. In the preceding problem, for each of the following sets of predictors: (a) write out the two-level model for predicting housing prices, (b) write out the corresponding composite model, and (c) determine how many model parameters (fixed effects and variance components) must be estimated.

    - Square footage, number of bedrooms
    - Median neighborhood income, rating of neighborhood schools
    - Square footage, number of bedrooms, age of house, median neighborhood housing price
    - Square footage, median neighborhood income, rating of neighborhood schools, median neighborhood housing price

4. **Music performance anxiety.** Describe a situation in which the two plots in Figure 8.7 might tell different stories.

5. Explain the difference between a_i in Equation (8.2) and $\hat{a}_i$ in Equation (8.4).

6. Why is the contour plot for multivariate normal density in Figure 8.12(b) tilted from southwest to northeast, but the contour plot in Figure 8.12(a) is not tilted?

7. In Table 8.3, note that the standard errors associated with estimated coefficients under independence are lower than standard errors under alternative analysis methods. Why is that often the case?

8. Why is Model A (Section 8.6.2) sometimes called the "unconditional means model"? Why is it also sometimes called the "random intercepts model"? Are these two labels consistent with each other?

9. Consider adding an indicator variable in Model B (Section 8.7.1) for Small Ensemble performances.

    - Write out the two-level model for performance anxiety,
    - Write out the corresponding composite model,
    - Determine how many model parameters (fixed effects and variance components) must be estimated, and
    - Explain how the interpretation for the coefficient in front of Large Ensembles would change.

10. Give a short rule in your own words describing when an interpretation of an estimated coefficient should "hold constant" another covariate or "set to 0" that covariate (see Section 8.8.1).

11. The interpretation of $\hat{\alpha}_1$ in Section 8.8.1 claims that, "This interpretation, however, only holds for solos and small ensembles. For large ensembles, the difference between those playing orchestral instruments and others is actually given by $\hat{\alpha}_1 + \hat{\beta}_1$, holding MPQNEM constant." Show that this claim is true.

12. Explain how the interpretations of the following parameter estimates change (or don't change) as we change our model:

    - $\hat{\alpha}_0$ from Model A to B to C to D to E
    - $\hat{\beta}_1$ from Model B to C to D to E
    - $\hat{\alpha}_1$ from Model C to D to E
    - $\hat{\beta}_1$ from Model C to D to E
    - $\hat{\sigma}_u$ from Model A to B to C to D to E
    - $\hat{\sigma}_v$ from Model B to C to D to E

13. When moving from Model B to Model C in Section 8.7.3, $\hat{\sigma}_u^2$ increases slightly. Why might this have occurred?

14. Interpret other estimated parameters from Model F beyond those interpreted in Section 8.10: $\hat{\alpha}_0$, $\hat{\alpha}_2$, $\hat{\alpha}_3$, $\hat{\beta}_0$, $\hat{\gamma}_0$, $\hat{\zeta}_0$, $\hat{\rho}_{wx}$, $\hat{\sigma}^2$, $\hat{\sigma}_u^2$, and $\hat{\sigma}_z^2$.

15. Explain Figure 8.15 in your own words. Why would LLSR produce a misleading analysis in this case, but multilevel models would not?

16. Summarize Figures 8.16 and 8.17 in your own words.

8.13.2 Guided Exercises

1. **Music performance joy.** In this chapter, we studied models for predicting music performance anxiety, as measured by the negative affect scale from the PANAS instrument. Now we will examine models for predicting the happiness of musicians prior to performances, as measured by the positive affect scale from the PANAS instrument.

8.13 Exercises

To begin, run the following models:

- Model A = unconditional means model
- Model B = indicator for instructor audience type and indicator for student audience type at Level One; no Level Two predictors
- Model C = indicator for instructor audience type and indicator for student audience type at Level One; centered MPQ absorption subscale as Level Two predictor for intercept and all slope terms
- Model D = indicator for instructor audience type and indicator for student audience type at Level One; centered MPQ absorption subscale and a male indicator as Level Two predictors for intercept and all slope terms

1. Perform an exploratory data analysis by comparing positive affect (happiness) to Level One and Level Two covariates using appropriate graphs. Comment on interesting trends, supporting your comments with appropriate summary statistics.
2. Report estimated fixed effects and variance components from Model A, using proper notation from this chapter (no interpretations required). Also report and interpret an intraclass correlation coefficient.
3. Report estimated fixed effects and variance components from Model B, using proper notation from this chapter. Interpret your MLE estimates for $\hat{\alpha}_0$ (the intercept), $\hat{\beta}_1$ (the instructor indicator), and $\hat{\sigma}_u$ (the Level Two standard deviation for the intercept). Also report and interpret an appropriate pseudo R-squared value.
4. Write out Model C, using both separate Level One and Level Two models as well as a composite model. Be sure to express distributions for error terms. How many parameters must be estimated in Model C?
5. Report and interpret the following parameter estimates from Model C: $\hat{\alpha}_0$, $\hat{\alpha}_1$, $\hat{\gamma}_0$, $\hat{\beta}_1$, $\hat{\sigma}_u$, $\hat{\sigma}_v$, and $\hat{\rho}_{uv}$. Interpretations for variance components should be done in terms of standard deviations and correlation coefficients.
6. Report and interpret the same parameter estimates listed above from Model D. In each case, the new interpretation should involve a small modification of your interpretation from Model C. Use underlines or highlights to denote the part of the Model D interpretation that differs from the Model C interpretation.
7. Also report and interpret the following parameter estimates from Model D: $\hat{\alpha}_2$ and $\hat{\beta}_2$.
8. Use a drop in deviance statistic (likelihood ratio test) to compare Model C vs. Model D. Give a test statistic and p-value, then state a conclusion. Also compare Models C and D with appropriate pseudo R-squared value(s) and with AIC and BIC statistics.

8.13.3 Open-Ended Exercises

1. **Political ambiguity.** Chapp et al. [2018] explored 2014 congressional candidates' ambiguity on political issues in their paper, *Going Vague: Ambiguity and Avoidance in Online Political Messaging*. They hand-coded a random sample of 2012 congressional candidates' websites, assigning an ambiguity score. A total of 870 websites from 2014 were then automatically scored using Wordscores, a program designed for political textual analysis. In their paper, they fit a multilevel model for candidates' ambiguities with predictors at both the candidate and district levels. Some of their hypotheses include that:

   - "when incumbents do hazard issue statements, these statements will be marked by a higher degree of clarity." (Hypothesis 1b)
   - "ideological distance [from district residents] will be associated with greater ambiguity." (Hypothesis 2a)
   - "controlling for ideological distance, ideological extremity [of the candidate] should correspond to less ambiguity." (Hypothesis 2b)
   - "more variance in attitudes [among district residents] will correspond to a higher degree of ambiguity in rhetoric" (Hypothesis 3a)
   - "a more heterogeneous mix of subgroups [among district residents] will also correspond to a higher degree of ambiguity in rhetoric" (Hypothesis 3b)

   Their data can be found in `ambiguity.csv`. Variables of interest include:

   - `ambiguity` = assigned ambiguity score. Higher scores indicate greater clarity (less ambiguity)
   - `democrat` = 1 if a Democrat, 0 otherwise (Republican)
   - `incumbent` = 1 if an incumbent, 0 otherwise
   - `ideology` = a measure of the candidate's left-right orientation. Higher (positive) scores indicate more conservative candidates and lower (negative) scores indicate more liberal candidates.
   - `mismatch` = the distance between the candidate's ideology and the district's ideology (candidate ideology scores were regressed against district ideology scores; mismatch values represent the absolute value of the residual associated with each candidate)
   - `distID` = the congressional district's unique ID
   - `distLean` = the district's political leaning. Higher scores imply more conservative districts.
   - `attHeterogeneity` = a measure of the variability of ideologies within the district. Higher scores imply more attitudinal heterogeneity among voters.

8.13 Exercises

- demHeterogeneity = a measure of the demographic variability within the district. Higher scores imply more demographic heterogeneity among voters.

With this in mind, fit your own models to address these hypotheses from Chapp et al. [2018]. Be sure to use a two-level structure to account for variables at both the candidate and district levels.

2. **Airbnb in Chicago.** Trinh and Ameri [2018] collected data on 1561 Airbnb listings in Chicago from August 2016, and then they merged in information from the neighborhood (out of 43 in Chicago) where the listing was located. We can examine traits that are associated with listings that command a higher price. Conduct an EDA, build a multilevel model, and interpret model coefficients to answer questions such as: What are characteristics of a higher priced listing? Are the most influential traits associated with individual listings or entire neighborhoods? Are there intriguing interactions where the effect of one variable depends on levels of another?

The following variables can be found in airbnb.csv or derived from the variables found there:

- overall_satisfaction = rating on a 0-5 scale.
- satisfaction = 1 if overall_satisfaction is 5, 0 otherwise
- price = price for one night (in dollars)
- reviews = number of reviews posted
- room_type = Entire home/apt, Private room, or Shared room
- accommodates = number of people the unit can hold
- bedrooms = number of bedrooms
- minstay = minimum length of stay (in days)
- neighborhood = neighborhood where unit is located (1 of 43)
- district = district where unit is located (1 of 9)
- WalkScore = quality of the neighborhood for walking (0-100)
- TransitScore = quality of the neighborhood for public transit (0-100)
- BikeScore = quality of the neighborhood for biking (0-100)
- PctBlack = proportion of black residents in a neighborhood
- HighBlack = 1 if PctBlack above .60, 0 otherwise

3. **Project 5183.** The Colorado Rockies, a Major League Baseball team, instigated a radical experiment on June 20th, 2012. Hopelessly out of contention for the playoffs and struggling again with their pitching, the Rockies decided to limit their starting pitchers to 75 pitches from June 20th until the end of the year with the hope of improving a struggling starting rotation, teaching pitchers how to pitch to contact (which results in low pitch counts), and at the same time trying to conserve young arms. Data has shown that, as a game

progresses, fatigue becomes a big factor in a pitcher's performance; if a pitcher has to tweak his mechanics to try to make up for a fatigued body, injuries can often occur. In addition, pitchers often struggle as they begin facing the same batters over again later in games. The Rockies called their experiment "Project 5183" to acknowledge the altitude at Coors Field, their home ballpark, and the havoc that high altitude can wreak on pitchers.

A team of students collected 2012 data on Rockies pitchers from FanGraphs to evaluate Project 5183 [Sturtz et al., 2013]. In a successful experiment, Colorado pitchers on a strict limit of 75 pitches would throw more strikes and yet record fewer strikeouts (pitching to contact rather than throwing more pitches to attempt to strike batters out). Different theories explain whether these pitchers would throw harder (since they don't have to save themselves) or throw slower (in order to throw more strikes). But the end result the Rockies hoped to observe was that their pitchers pitch better (allow fewer runs to the opponent) with a pitch limit.

The data set `FinalRockiesdata.csv` contains information for 7 starting pitchers who started at least one game before June 20th (without a pitch limit) and at least one game after June 20th (with a limit of 75 pitches). Key response variables include:

- vFA = average fastball velocity
- K.9 = strikeouts per nine innings
- ERA = earned runs per nine innings
- Pitpct = percentage of strikes thrown

The primary explanatory variable of interest is PCL (an indicator variable for if a pitch count limit is in effect). Other potential confounding variables that may be important to control for include Coors (whether or not the game was played in Coors Field, where more runs are often scored because of the high altitude and thin air) and Age of the pitcher.

Write a short report summarizing the results of Project 5183. (You may notice a few variance components with unusual estimates, such as an estimated variance of 0 or an estimated correlation of 1. These estimates have encountered boundary constraints; we will learn how to deal with these situations in Section 10.5. For now ignore these variance components; the fixed effects coefficients are still reliable and their interpretations valid.)

4. **Replicate the Sadler and Miller paper.** Try to replicate Models 1 and 2 presented in Table 2 of Sadler and Miller [2010]. We expect small differences in parameter estimates, since they use SAS (with an unstructured covariance structure) instead of R.

8.13 Exercises

- Do the parameter estimates, SEs, AIC, and variance explained compare well?
- Explain ways in which the model equations from page 284 of Sadler and Miller do not align with Model 2 from Table 2, or with the manner in which we write out two-level models.

9

Two-Level Longitudinal Data

9.1 Learning Objectives

After finishing this chapter, you should be able to:

- Recognize longitudinal data as a special case of multilevel data, with time at Level One.
- Consider patterns of missingness and implications of that missing data on multilevel analyses.
- Apply exploratory data analysis techniques specific to longitudinal data.
- Build and understand a taxonomy of models for longitudinal data.
- Interpret model parameters in multilevel models with time at Level One.
- Compare models, both nested and not, with appropriate statistical tests and summary statistics.
- Consider different ways of modeling the variance-covariance structure in longitudinal data.
- Understand how a parametric bootstrap test of significance works and when it might be useful.

```
# Packages required for Chapter 9
library(GGally)
library(data.table)
library(Hmisc)
library(mice)
library(lattice)
library(nlme)
library(reshape2)
library(MASS)
library(mnormt)
library(lme4)
library(gridExtra)
library(knitr)
library(kableExtra)
```

```
library(broom)
library(tidyverse)
```

9.2 Case Study: Charter Schools

Charter schools were first introduced in the state of Minnesota in 1991 [U.S. Department of Education, 2018]. Since then, charter schools have begun appearing all over the United States. While publicly funded, a unique feature of charter schools is their independence from many of the regulations that are present in the public school systems of their respective city or state. Thus, charters will often extend the school days or year and tend to offer non-traditional techniques and styles of instruction and learning.

One example of this unique schedule structure is the KIPP (Knowledge Is Power Program) Stand Academy in Minneapolis, MN. KIPP stresses longer days and better partnerships with parents, and they claim that 80% of their students go to college from a population where 87% qualify for free and reduced lunch and 95% are African American or Latino [KIPP, 2018]. However, the larger question is whether or not charter schools are out-performing non-charter public schools in general. Because of the relative youthfulness of charter schools, data has just begun to be collected to evaluate the performance of charter versus non-charter schools and some of the factors that influence a school's performance. Along these lines, we will examine data collected by the Minnesota Department of Education for all Minnesota schools during the years 2008-2010.

Comparisons of student performance in charter schools versus public schools have produced conflicting results, potentially as a result of the strong differences in the structure and population of the student bodies that represent the two types of schools. A study by the Policy and Program Studies Service of five states found that charter schools are less likely to meet state performance standards than conventional public schools [Finnigan et al., 2004]. However, Witte et al. [2007] performed a statistical analysis comparing Wisconsin charter and non-charter schools and found that average achievement test scores were significantly higher in charter schools compared to non-charter schools, after controlling for demographic variables such as the percentage of white students. In addition, a study of California students who took the Stanford 9 exam from 1998 through 2002 found that charter schools, on average, were performing at the same level as conventional public schools [Buddin and Zimmer, 2005]. Although school performance is difficult to quantify with a single measure, for illustration purposes in this chapter we will focus on that aspect of school performance measured by the math portion of the Minnesota Comprehensive Assessment (MCA-II) data for 6th grade students enrolled in 618 different Minnesota schools during the years 2008, 2009, and 2010 [Minnesota Department of Education, 2018]. Similar comparisons could obviously be conducted for other grade levels or modes of assessment.

As described in Green III et al. [2003], it is very challenging to compare charter and public non-charter schools, as charter schools are often designed to target or attract specific populations of students. Without accounting for differences in student populations, comparisons lose meaning. With the assistance of multiple school-specific predictors, we will attempt to model sixth grade math MCA-II scores of Minnesota schools, focusing on the differences between charter and public non-charter school performances. In the process, we hope to answer the following research questions:

- Which factors most influence a school's performance in MCA testing?
- How do the average math MCA-II scores for 6th graders enrolled in charter schools differ from scores for students who attend non-charter public schools? Do these differences persist after accounting for differences in student populations?
- Are there differences in yearly improvement between charter and non-charter public schools?

9.3 Initial Exploratory Analyses

9.3.1 Data Organization

Key variables in `chart_wide_condense.csv` which we will examine to address the research questions above are:

- `schoolid` = includes district type, district number, and school number
- `schoolName` = name of school
- `urban` = is the school in an urban (1) or rural (0) location?
- `charter` = is the school a charter school (1) or a non-charter public school (0)?
- `schPctnonw` = proportion of non-white students in a school (based on 2010 figures)
- `schPctsped` = proportion of special education students in a school (based on 2010 figures)
- `schPctfree` = proportion of students who receive free or reduced lunches in a school (based on 2010 figures). This serves as a measure of poverty among school families.

TABLE 9.1: The first six observations in the wide data set for the Charter Schools case study.

schoolid	schoolName	urban	charter	schPctnonw	schPctsped	schPctfree	MathAvgScore.0	MathAvgScore.1	MathAvgScore.2
Dtype 1 Dnum 1 Snum 2	RIPPLESIDE ELEMENTARY	0	0	0.0000	0.1176	0.3627	652.8	656.6	652.6
Dtype 1 Dnum 100 Snum 1	WRENSHALL ELEMENTARY	0	0	0.0303	0.1515	0.4242	646.9	645.3	651.9
Dtype 1 Dnum 108 Snum 30	CENTRAL MIDDLE	0	0	0.0769	0.1231	0.2615	654.7	658.5	659.7
Dtype 1 Dnum 11 Snum 121	SANDBURG MIDDLE	1	0	0.0977	0.0827	0.2481	656.4	656.8	659.9
Dtype 1 Dnum 11 Snum 193	OAK VIEW MIDDLE	1	0	0.0538	0.0954	0.1418	657.7	658.2	659.8
Dtype 1 Dnum 11 Snum 195	ROOSEVELT MIDDLE	1	0	0.1234	0.0886	0.2405	655.9	659.1	660.3

- `MathAvgScore.0` = average MCA-II math score for all sixth grade students in a school in 2008
- `MathAvgScore.1` = average MCA-II math score for all sixth grade students in a school in 2009
- `MathAvgScore.2` = average MCA-II math score for all sixth grade students in a school in 2010

This data is stored in WIDE format, with one row per school, as illustrated in Table 9.1.

For most statistical analyses, it will be advantageous to convert WIDE format to LONG format, with one row per year per school. To make this conversion, we will have to create a time variable, which under the LONG format is very flexible—each school can have a different number of and differently-spaced time points, and they can even have predictors which vary over time. Details for making this conversion can be found in the R Markdown code for this chapter, and the form of the LONG data in this study is exhibited in the next section.

9.3.2 Missing Data

In this case, before we convert our data to LONG form, we should first address problems with missing data. Missing data is a common phenomenon in longitudinal studies. For instance, it could arise if a new school was started during the observation period, a school was shut down during the observation period, or no results were reported in a given year. Dealing with missing data in a statistical analysis is not trivial, but fortunately many multilevel packages (including the lme4 package in R) are adept at handling missing data.

First, we must understand the extent and nature of missing data in our study. Table 9.2 is a frequency table of missing data patterns, where 1 indicates presence of a variable and 0 indicates a missing value for a particular variable; this table is a helpful starting point. Among our 618 schools, 540 had complete data (all covariates and math scores for all three years), 25 were missing a math score for 2008, 35 were missing math scores in both 2008 and 2009, etc.

9.3 Initial Exploratory Analyses

TABLE 9.2: A frequency table of missing data patterns.

	charter	MathAvgScore.2	MathAvgScore.1	MathAvgScore.0	
540	1	1	1	1	0
25	1	1	1	0	1
4	1	1	0	1	1
35	1	1	0	0	2
6	1	0	1	1	1
1	1	0	1	0	2
7	1	0	0	1	2
	0	14	46	61	121

The number of schools with a particular missing data pattern are listed in the left column; the remaining columns of 0's and 1's describe the missing data pattern, with 0 indicating a missing value. Some covariates that are present for every school are not listed. The bottom row gives the number of schools with missing values for specific variables; the last entry indicates that 121 total observations were missing.

Statisticians have devised different strategies for handling missing data; a few common approaches are described briefly here:

- Include only schools with complete data. This is the cleanest approach analytically; however, ignoring data from 12.6% of the study's schools (since 78 of the 618 schools had incomplete data) means that a large amount of potentially useful data is being thrown away. In addition, this approach creates potential issues with informative missingness. Informative missingness occurs when a school's lack of scores is not a random phenomenon but provides information about the effectiveness of the school type (e.g., a school closes because of low test scores).
- Last observation carried forward. Each school's last math score is analyzed as a univariate response, whether the last measurement was taken in 2008, 2009, or 2010. With this approach, data from all schools can be used, and analyses can be conducted with traditional methods assuming independent responses. This approach is sometimes used in clinical trials because it tends to be conservative, setting a higher bar for showing that a new therapy is significantly better than a traditional therapy. Of course, we must assume that a school's 2008 score is representative of its 2010 score. In addition, information about trajectories over time is thrown away.
- Imputation of missing observations. Many methods have been developed for sensibly "filling in" missing observations, using imputation models which base imputed data on subjects with similar covariate profiles and on typical observed time trends. Once an imputed data set is created (or several imputed

TABLE 9.3: The first six observations in the long data set for the Charter Schools case study; these lines correspond to the first two observations from the wide data set illustrated in Table 9.1.

schoolName	charter	schPctsped	schPctfree	year08	MathAvgScore
RIPPLESIDE ELEMENTARY	0	0.1176	0.3627	0	652.8
RIPPLESIDE ELEMENTARY	0	0.1176	0.3627	1	656.6
RIPPLESIDE ELEMENTARY	0	0.1176	0.3627	2	652.6
WRENSHALL ELEMENTARY	0	0.1515	0.4242	0	646.9
WRENSHALL ELEMENTARY	0	0.1515	0.4242	1	645.3
WRENSHALL ELEMENTARY	0	0.1515	0.4242	2	651.9

data sets), analyses can proceed with complete data methods that are easier to apply. Risks with the imputation approach include misrepresenting missing observations and overstating precision in final results.
- Apply multilevel methods, which use available data to estimate patterns over time by school and then combine those school estimates in a way that recognizes that time trends for schools with complete data are more precise than time trends for schools with fewer measurements. Laird [1988] demonstrates that multilevel models are valid under the fairly unrestrictive condition that the probability of missingness cannot depend on any unobserved predictors or the response. This is the approach we will follow in the remainder of the text.

Now, we are ready to create our LONG data set. Fortunately, many packages (including R) have built-in functions for easing this conversion, and the functions are improving constantly. The resulting LONG data set is shown in Table 9.3, where `year08` measures the number of years since 2008.

9.3.3 Exploratory Analyses for General Multilevel Models

Notice the **longitudinal** structure of our data—we have up to three measurements of test scores at different time points for each of our 618 schools. With this structure, we can address questions at two levels:

- Within school—changes over time
- Between schools—effects of school-specific covariates (charter or non-charter, urban or rural, percent free and reduced lunch, percent special education, and percent non-white) on 2008 math scores and rate of change between 2008 and 2010.

As with any statistical analysis, it is vitally important to begin with graphical and numerical summaries of important variables and relationships between variables. We'll begin with initial exploratory analyses that we introduced in the previous chapter, noting that we have no Level One covariates other

9.3 Initial Exploratory Analyses

than time at this point (potential covariates at this level may have included measures of the number of students tested or funds available per student). We will, however, consider the Level Two variables of charter or non-charter, urban or rural, percent free and reduced lunch, percent special education, and percent non-white. Although covariates such as percent free and reduced lunch may vary slightly from year to year within a school, the larger and more important differences tend to occur between schools, so we used percent free and reduced lunch for a school in 2010 as a Level Two variable.

As in Chapter 8, we can conduct initial investigations of relationships between Level Two covariates and test scores in two ways. First, we can use all 1733 observations to investigate relationships of Level Two covariates with test scores. Although these plots will contain dependent points, since each school is represented by up to three years of test score data, general patterns exhibited in these plots tend to be real. Second, we can calculate mean scores across all years for each of the 618 schools. While we lose some information with this approach, we can more easily consider each plotted point to be independent. Typically, both types of exploratory plots illustrate similar relationships, and in this case, both approaches are so similar that we will only show plots using the second approach, with one observation per school.

Figure 9.1 shows the distribution of MCA math test scores as somewhat left-skewed. MCA test scores for sixth graders are scaled to fall between 600 and 700, where scores above 650 for individual students indicate "meeting standards". Thus, schools with averages below 650 will often have increased incentive to improve their scores the following year. When we refer to the "math score" for a particular school in a particular year, we will assume that score represents the average for all sixth graders at that school. In Figure 9.2, we see that test scores are generally higher for both schools in rural areas and for public non-charter schools. Note that in this data set there are 237 schools in rural areas and 381 schools in urban areas, as well as 545 public non-charter schools and 73 charter schools. In addition, we can see in Figure 9.3 that schools tend to have lower math scores if they have higher percentages of students with free and reduced lunch, with special education needs, or who are non-white.

9.3.4 Exploratory Analyses for Longitudinal Data

In addition to the initial exploratory analyses above, longitudinal data— multilevel data with time at Level One—calls for further plots and summaries that describe time trends within and across individuals. For example, we can examine trends over time within individual schools. Figure 9.4 provides a **lattice plot** illustrating trends over time for the first 24 schools in the data set. We note differences among schools in starting point (test scores in 2008), slope

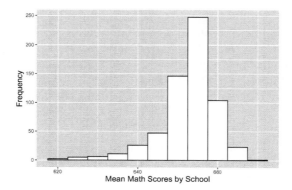

FIGURE 9.1: Histogram of mean sixth grade MCA math test scores over the years 2008-2010 for 618 Minnesota schools.

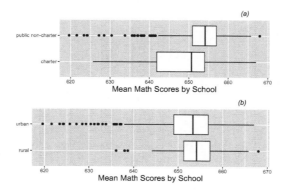

FIGURE 9.2: Boxplots of categorical Level Two covariates vs. average MCA math scores. Plot (a) shows charter vs. public non-charter schools, while plot (b) shows urban vs. rural schools.

(change in test scores over the three-year period), and form of the relationship. These differences among schools are nicely illustrated in so-called **spaghetti plots** such as Figure 9.5, which overlays the individual schools' time trends (for the math test scores) from Figure 9.4 on a single set of axes. In order to illustrate the overall time trend without making global assumptions about the form of the relationship, we overlaid in bold a non-parametric fitted curve through a **loess smoother**. LOESS comes from "locally estimated scatterplot smoother", in which a low-degree polynomial is fit to each data point using weighted regression techniques, where nearby points receive greater weight. LOESS is a computationally intensive method which performs especially well with larger sets of data, although ideally there would be a greater diversity of

9.3 Initial Exploratory Analyses

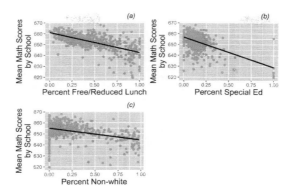

FIGURE 9.3: Scatterplots of average MCA math scores by (a) percent free and reduced lunch, (b) percent special education, and (c) percent non-white in a school.

x-values than the three time points we have. In this case, the loess smoother follows very closely to a linear trend, indicating that assuming a linear increase in test scores over the three-year period is probably a reasonable simplifying assumption. To further examine the hypothesis that linearity would provide a reasonable approximation to the form of the individual time trends in most cases, Figure 9.6 shows a lattice plot containing linear fits through ordinary least squares rather than connected time points as in Figure 9.4.

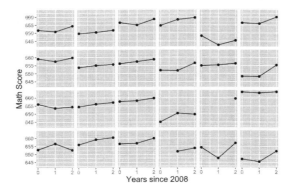

FIGURE 9.4: Lattice plot by school of math scores over time for the first 24 schools in the data set.

Just as we explored the relationship between our response (average math scores) and important covariates in Section 9.3.3, we can now examine the relationships between time trends by school and important covariates. For instance, Figure 9.7 shows that charter schools had math scores that were lower on average than public non-charter schools and more variable. This type of plot

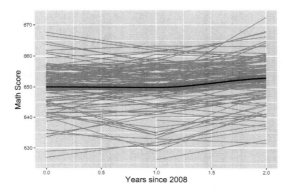

FIGURE 9.5: Spaghetti plot of math scores over time by school, for all the charter schools and a random sample of public non-charter schools, with overall fit using loess (bold).

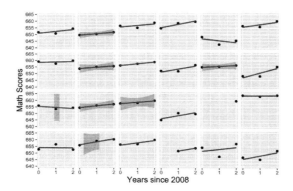

FIGURE 9.6: Lattice plot by school of math scores over time with linear fit for the first 24 schools in the data set.

is sometimes called a **trellis graph**, since it displays a grid of smaller charts with consistent scales, where each smaller chart represents a condition—an item in a category. Trends over time by school type are denoted by bold loess curves. Public non-charter schools have higher scores across all years; both school types show little growth between 2008 and 2009, but greater growth between 2009 and 2010, especially charter schools. Exploratory analyses like this can be repeated for other covariates, such as percent free and reduced lunch in Figure 9.8. The trellis plot automatically divides schools into four groups based on quartiles of their percent free and reduced lunch, and we see that schools with lower percentages of free and reduced lunch students tend to have higher math scores and less variability. Across all levels of free and reduced lunch, we see greater gains between 2009 and 2010 than between 2008 and 2009.

9.4 Preliminary Two-Stage Modeling

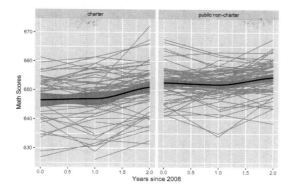

FIGURE 9.7: Spaghetti plots showing time trends for each school by school type, for a random sample of charter schools (left) and public non-charter schools (right), with overall fits using loess (bold).

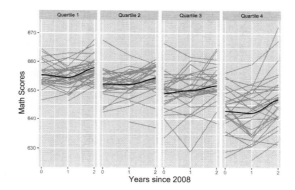

FIGURE 9.8: Spaghetti plots showing time trends for each school by quartiles of percent free and reduced lunch, with loess fits.

9.4 Preliminary Two-Stage Modeling

9.4.1 Linear Trends Within Schools

Even though we know that every school's math test scores were not strictly linearly increasing or decreasing over the observation period, a linear model for individual time trends is often a simple but reasonable way to model data. One advantage of using a linear model within school is that each school's data points can be summarized with two summary statistics—an intercept and a slope (obviously, this is an even bigger advantage when there are more observations over time per school). For instance, we see in Figure 9.6 that sixth graders from the school depicted in the top right slot slowly increased math scores over the three-year observation period, while students from the school depicted in the fourth column of the top row generally experienced decreasing math scores over the same period. As a whole, the linear model fits individual trends pretty well, and many schools appear to have slowly increasing math scores over time, as researchers in this study may have hypothesized.

Another advantage of assuming a linear trend at Level One (within schools) is that we can examine summary statistics across schools. Both the intercept and slope are meaningful for each school: the *intercept* conveys the school's math score in 2008, while the *slope* conveys the school's average yearly increase or decrease in math scores over the three-year period. Figure 9.9 shows that point estimates and uncertainty surrounding individual estimates of intercepts and slopes vary considerably. In addition, we can generate summary statistics and histograms for the 618 intercepts and slopes produced by fitting linear regression models at Level One, in addition to R-squared values which describe the strength of fit of the linear model for each school (Figure 9.10). For our 618 schools, the mean math score for 2008 was 651.4 (SD=7.28), and the mean yearly rate of change in math scores over the three-year period was 1.30 (SD=2.51). We can further examine the relationship between schools' intercepts and slopes. Figure 9.11 shows a general decreasing trend, suggesting that schools with lower 2008 test scores tend to have greater growth in scores between 2008 and 2010 (potentially because those schools have more room for improvement); this trend is supported with a correlation coefficient of -0.32 between fitted intercepts and slopes. Note that, with only 3 or fewer observations for each school, extreme or intractable values for the slope and R-squared are possible. For example, slopes cannot be estimated for those schools with just a single test score, R-squared values cannot be calculated for those schools with no variability in test scores between 2008 and 2010, and R-squared values must be 1 for those schools with only two test scores.

9.4.2 Effects of Level Two Covariates on Linear Time Trends

Summarizing trends over time within schools is typically only a start, however. Most of the primary research questions from this study involve comparisons among schools, such as: (a) are there significant differences between charter schools and public non-charter schools, and (b) do any differences between charter schools and public schools change with percent free and reduced lunch, percent special education, or location? These are Level Two questions, and we can begin to explore these questions by graphically examining the effects of school-level variables on schools' linear time trends. By school-level variables, we are referring to those covariates that differ by school but are not dependent on time. For example, school type (charter or public non-charter), urban or rural location, percent non-white, percent special education, and percent free

9.4 Preliminary Two-Stage Modeling

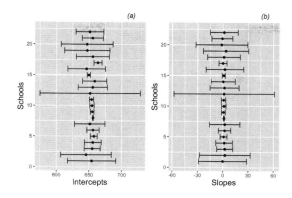

FIGURE 9.9: Point estimates and 95% confidence intervals for (a) intercepts and (b) slopes by school, for the first 24 schools in the data set.

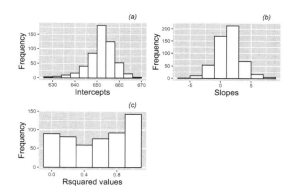

FIGURE 9.10: Histograms for (a) intercepts, (b) slopes, and (c) R-squared values from fitted regression lines by school.

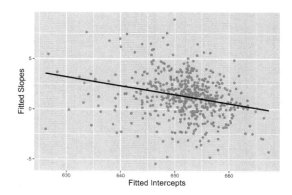

FIGURE 9.11: Scatterplot showing the relationship between intercepts and slopes from fitted regression lines by school.

and reduced lunch are all variables which differ by school but which don't change over time, at least as they were assessed in this study. Variables which would be time-dependent include quantities such as per pupil funding and reading scores.

Figure 9.12 shows differences in the average time trends by school type, using estimated intercepts and slopes to support observations from the spaghetti plots in Figure 9.7. Based on intercepts, charter schools have lower math scores, on average, in 2008 than public non-charter schools. Based on slopes, however, charter schools tend to improve their math scores at a slightly faster rate than public schools, especially at the 75th percentile and above. By the end of the three-year observation period, we would nevertheless expect charter schools to have lower average math scores than public schools. For another exploratory perspective on school type comparisons, we can examine differences between school types with respect to math scores in 2008 and math scores in 2010. As expected, boxplots by school type (Figure 9.13) show clearly lower math scores for charter schools in 2008, but differences are slightly less dramatic in 2010.

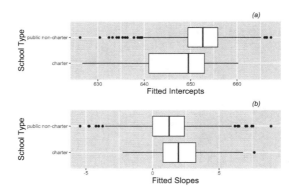

FIGURE 9.12: Boxplots of (a) intercepts and (b) slopes by school type (charter vs. public non-charter).

Any initial exploratory analyses should also investigate effects of potential confounding variables such as school demographics and location. If we discover, for instance, that those schools with higher levels of poverty (measured by the percentage of students receiving free and reduced lunch) display lower test scores in 2008 but greater improvements between 2008 and 2010, then we might be able to use percentage of free and reduced lunch in statistical modeling of intercepts and slopes, leading to more precise estimates of the charter school effects on these two outcomes. In addition, we should also look for any interaction with school type—any evidence that the difference between charter and non-charter schools changes based on the level of a confounding

9.4 Preliminary Two-Stage Modeling

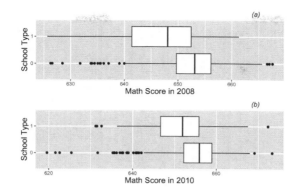

FIGURE 9.13: Boxplots of (a) 2008 and (b) 2010 math scores by school type (charter (1) vs. public non-charter (0)).

variable. For example, do charter schools perform better relative to non-charter schools when there is a large percentage of non-white students at a school?

With a confounding variable such as percentage of free and reduced lunch, we will treat this variable as continuous to produce the most powerful exploratory analyses. We can begin by examining boxplots of free and reduced lunch percentage against school type (Figure 9.14). We observe that charter schools tend to have greater percentages of free and reduced lunch students as well as greater school-to-school variability. Next, we can use scatterplots to graphically illustrate the relationships between free and reduced lunch percentages and significant outcomes such as intercept and slope (also Figure 9.14). In this study, it appears that schools with higher levels of free and reduced lunch (i.e., greater poverty) tend to have lower math scores in 2008, but there is little evidence of a relationship between levels of free and reduced lunch and improvements in test scores between 2008 and 2010. These observations are supported with correlation coefficients between percent free and reduced lunch and intercepts (r=-0.61) and slopes (r=-0.06).

A less powerful but occasionally informative way to look at the effect of a continuous confounder on an outcome variable is by creating a categorical variable out of the confounder. For instance, we could classify any school with a percentage of free and reduced lunch students above the median as having a high percentage of free and reduced lunch students, and all other schools as having a low percentage of free and reduced lunch students. Then we could examine a possible interaction between percent free and reduced lunch and school type through a series of four boxplots (Figure 9.15). In fact, these boxplots suggest that the gap between charter and public non-charter schools in 2008 was greater in schools with a high percentage of free and reduced lunch students, while the difference in rate of change in test scores between charter and public non-charter schools appeared similar for high and low levels of free and reduced lunch. We will investigate these trends more thoroughly with statistical modeling.

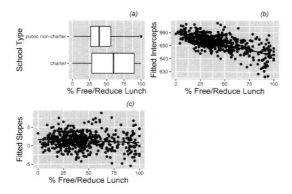

FIGURE 9.14: (a) Boxplot of percent free and reduced lunch by school type (charter vs. public non-charter), along with scatterplots of (b) intercepts and (c) slopes from fitted regression lines by school vs. percent free and reduced lunch.

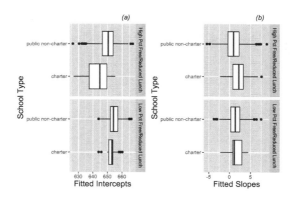

FIGURE 9.15: Boxplots of (a) intercepts and (b) slopes from fitted regression lines by school vs. school type (charter vs. public non-charter), separated by high and low levels of percent free and reduced lunch.

The effect of other confounding variables (e.g., percent non-white, percent special education, urban or rural location) can be investigated in a similar fashion to free and reduced lunch percentage, both in terms of main effect (variability in outcomes such as slope and intercept which can be explained by the confounding variable) and interaction with school type (ability of the confounding variable to explain differences between charter and public non-charter schools). We leave these explorations as an exercise.

9.4.3 Error Structure Within Schools

Finally, with longitudinal data it is important to investigate the error variance-covariance structure of data collected within a school (the Level Two observational unit). In multilevel data, as in the examples we introduced in Chapter 7, we suspect observations within group (like a school) to be correlated, and we strive to model that correlation. When the data within group is collected over time, we often see distinct patterns in the residuals that can be modeled—correlations which decrease systematically as the time interval increases, variances that change over time, correlation structure that depends on a covariate, etc. A first step in modeling the error variance-covariance structure is the production of an exploratory plot such as Figure 9.16. To generate this plot, we begin by modeling MCA math score as a linear function of time using all 1733 observations and ignoring the school variable. This population (marginal) trend is illustrated in Figure 9.5 and is given by:

$$\hat{Y}_{ij} = 651.69 + 1.20 \text{Time}_{ij},$$

where $\hat{Y}_{ij}$ is the predicted math score of the ith school at time j, where time j is the number of years since 2008. In this model, the predicted math score will be identical for all schools at a given time point j. Residuals $Y_{ij} - \hat{Y}_{ij}$ are then calculated for each observation, measuring the difference between actual math score and the average overall time trend. Figure 9.16 then combines three pieces of information: the upper right triangle contains correlation coefficients for residuals between pairs of years, the diagonal contains histograms of residuals at each time point, and the lower left triangle contains scatterplots of residuals from two different years. In our case, we see that correlation between residuals from adjacent years is strongly positive (0.81-0.83) and does not drop off greatly as the time interval between years increases.

9.5 Initial Models

Throughout the exploratory analysis phase, our original research questions have guided our work, and now with modeling we return to familiar questions such as:

- are differences between charter and public non-charter schools (in intercept, in slope, in 2010 math score) statistically significant?
- are differences between school types statistically significant, even after accounting for school demographics and location?
- do charter schools offer any measurable benefit over non-charter public schools,

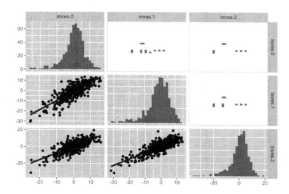

FIGURE 9.16: Correlation structure within school. The upper right contains correlation coefficients between residuals at pairs of time points, the lower left contains scatterplots of the residuals at time point pairs, and the diagonal contains histograms of residuals at each of the three time points.

either overall or within certain subgroups of schools based on demographics or location?

As you might expect, answers to these questions will arise from proper consideration of variability and properly identified statistical models. As in Chapter 8, we will begin model fitting with some simple, preliminary models, in part to establish a baseline for evaluating larger models. Then, we can build toward a final model for inference by attempting to add important covariates, centering certain variables, and checking assumptions.

9.5.1 Unconditional Means Model

In the multilevel context, we almost always begin with the **unconditional means model**, in which there are no predictors at any level. The purpose of the unconditional means model is to assess the amount of variation at each level, and to compare variability within school to variability between schools. Define Y_{ij} as the MCA-II math score from school i and year j. Using the composite model specification from Chapter 8:

$$Y_{ij} = \alpha_0 + u_i + \epsilon_{ij} \text{ with } u_i \sim N(0, \sigma_u^2) \text{ and } \epsilon_{ij} \sim N(0, \sigma^2)$$

the unconditional means model can be fit to the MCA-II data:

9.5 Initial Models

```
#Model A (Unconditional means model)
model.a <- lmer(MathAvgScore~ 1 + (1|schoolid),
                REML=T, data=chart.long)
```

```
##  Groups     Name         Variance  Std.Dev.
##  schoolid   (Intercept)  41.9      6.47
##  Residual                10.6      3.25

##  Number of Level Two groups = 618

##                 Estimate  Std. Error  t value
##  (Intercept)    652.7     0.2726      2395
```

From this output, we obtain estimates of our three model parameters:

- $\hat{\alpha}_0 = 652.7 =$ the mean math score across all schools and all years
- $\hat{\sigma}^2 = 10.6 =$ the variance in within-school deviations between individual yearly scores and the school mean across all years
- $\hat{\sigma}_u^2 = 41.9 =$ the variance in between-school deviations between school means and the overall mean across all schools and all years

Based on the intraclass correlation coefficient:

$$\hat{\rho} = \frac{\hat{\sigma}_u^2}{\hat{\sigma}_u^2 + \hat{\sigma}^2} = \frac{41.869}{41.869 + 10.571} = 0.798$$

79.8% of the total variation in math scores is attributable to differences among schools rather than changes over time within schools. We can also say that the average correlation for any pair of responses from the same school is 0.798.

9.5.2 Unconditional Growth Model

The second model in most multilevel contexts introduces a covariate at Level One (see Model B in Chapter 8). With longitudinal data, this second model introduces time as a predictor at Level One, but there are still no predictors at Level Two. This model is then called the **unconditional growth model**. The unconditional growth model allows us to assess how much of the within-school variability can be attributed to systematic changes over time.

At the lowest level, we can consider building individual growth models over time for each of the 618 schools in our study. First, we must decide upon a form for each of our 618 growth curves. Based on our initial exploratory analyses, assuming that an individual school's MCA-II math scores follow a linear trend seems like a reasonable starting point. Under the assumption of linearity, we must estimate an intercept and a slope for each school, based on their 1-3 test scores over a period of three years. Compared to time series analyses of economic data, most longitudinal data analyses have relatively few time periods for each subject (or school), and the basic patterns within subject are often reasonably described by simpler functional forms.

Let Y_{ij} be the math score of the ith school in year j. Then we can model the linear change in math test scores over time for School i according to Model B:

$$Y_{ij} = a_i + b_i \text{Year08}_{ij} + \epsilon_{ij} \text{ where } \epsilon_{ij} \sim N(0, \sigma^2)$$

The parameters in this model (a_i, b_i, and σ^2) can be estimated through LLSR methods. a_i represents the true intercept for School i—i.e., the expected test score level for School i when time is zero (2008)—while b_i represents the true slope for School i—i.e., the expected yearly rate of change in math score for School i over the three-year observation period. Here we use Roman letters rather than Greek for model parameters since models by school will eventually be a conceptual first step in a multilevel model. The ϵ_{ij} terms represent the deviation of School i's actual test scores from the expected results under linear growth—the part of school i's test score at time j that is not explained by linear changes over time. The variability in these deviations from the linear model is given by σ^2. In Figure 9.17, which illustrates a linear growth model for Norwood Central Middle School, a_i is estimated by the y-intercept of the fitted regression line, b_i is estimated by the slope of the fitted regression line, and σ^2 is estimated by the variability in the vertical distances between each point (the actual math score in year j) and the line (the predicted math score in year j).

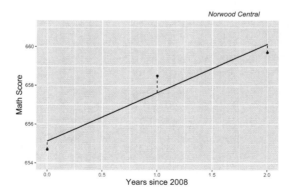

FIGURE 9.17: Linear growth model for Norwood Central Middle School.

9.5 Initial Models

In a multilevel model, we let intercepts (a_i) and slopes (b_i) vary by school and build models for these intercepts and slopes using school-level variables at Level Two. An unconditional growth model features no predictors at Level Two and can be specified either using formulations at both levels:

- Level One:
$$Y_{ij} = a_i + b_i \text{Year08}_{ij} + \epsilon_{ij}$$

- Level Two:
$$a_i = \alpha_0 + u_i$$
$$b_i = \beta_0 + v_i$$

or as a composite model:

$$Y_{ij} = \alpha_0 + \beta_0 \text{Year08}_{ij} + u_i + v_i \text{Year08}_{ij} + \epsilon_{ij}$$

where $\epsilon_{ij} \sim N(0, \sigma^2)$ and

$$\left[\begin{array}{c} u_i \\ v_i \end{array} \right] \sim N \left(\left[\begin{array}{c} 0 \\ 0 \end{array} \right], \left[\begin{array}{cc} \sigma_u^2 & \\ \sigma_{uv} & \sigma_v^2 \end{array} \right] \right).$$

As before, σ^2 quantifies the within-school variability (the scatter of points around schools' linear growth trajectories), while now the between-school variability is partitioned into variability in initial status (σ_u^2) and variability in rates of change (σ_v^2).

Using the composite model specification, the unconditional growth model can be fit to the MCA-II test data:

```
#Model B (Unconditional growth)
model.b <- lmer(MathAvgScore~ year08 + (year08|schoolid),
   REML=T, data=chart.long)
```

```
##  Groups   Name        Variance Std.Dev. Corr
##  schoolid (Intercept) 39.441   6.280
##           year08      0.111    0.332    0.72
##  Residual             8.820    2.970

##  Number of Level Two groups =  618

##                Estimate Std. Error t value
##  (Intercept)   651.408  0.27934    2331.96
##  year08        1.265    0.08997    14.06

##   AIC =   10352 ;  BIC =   10384
```

From this output, we obtain estimates of our six model parameters:

- $\hat{\alpha}_0 = 651.4 =$ the mean math score for the population of schools in 2008.
- $\hat{\beta}_0 = 1.26 =$ the mean yearly change in math test scores for the population during the three-year observation period.
- $\hat{\sigma}^2 = 8.82 =$ the variance in within-school deviations.
- $\hat{\sigma}_u^2 = 39.4 =$ the variance between schools in 2008 scores.
- $\hat{\sigma}_v^2 = 0.11 =$ the variance between schools in rates of change in math test scores during the three-year observation period.
- $\hat{\rho}_{uv} = 0.72 =$ the correlation in schools' 2008 math score and their rate of change in scores between 2008 and 2010.

We see that schools had a mean math test score of 651.4 in 2008 and their mean test scores tended to increase by 1.26 points per year over the three-year observation period, producing a mean test score at the end of three years of 653.9. According to the t-value (14.1), the increase in mean test scores noted during the three-year observation period is statistically significant.

The estimated within-school variance $\hat{\sigma}^2$ decreased by about 17% from the unconditional means model, implying that 17% of within-school variability in test scores can be explained by a linear increase over time:

$$\text{Pseudo } R^2_{L1} = \frac{\hat{\sigma}^2(\text{uncond means}) - \hat{\sigma}^2(\text{uncond growth})}{\hat{\sigma}^2(\text{uncond means})}$$
$$= \frac{10.571 - 8.820}{10.571} = 0.17$$

9.5.3 Modeling Other Trends over Time

While modeling linear trends over time is often a good approximation of reality, it is by no means the only way to model the effect of time. One alternative is to model the quadratic effect of time, which implies adding terms for both time and the square of time. Typically, to reduce the correlation between the linear and quadratic components of the time effect, the time variable is often centered first; we have already "centered" on 2008. Modifying Model B to produce an **unconditional quadratic growth model** would take the following form:

- Level One:
$$Y_{ij} = a_i + b_i \text{Year08}_{ij} + c_i \text{Year08}_{ij}^2 + \epsilon_{ij}$$

- Level Two:
$$a_i = \alpha_0 + u_i$$
$$b_i = \beta_0 + v_i$$
$$c_i = \gamma_0 + w_i$$

where $\epsilon_{ij} \sim N(0, \sigma^2)$ and

9.5 Initial Models

$$\begin{bmatrix} u_i \\ v_i \\ w_i \end{bmatrix} \sim N \left(\begin{bmatrix} 0 \\ 0 \\ 0 \end{bmatrix}, \begin{bmatrix} \sigma_u^2 & & \\ \sigma_{uv} & \sigma_v^2 & \\ \sigma_{uw} & \sigma_{vw} & \sigma_w^2 \end{bmatrix} \right).$$

With the extra term at Level One for the quadratic effect, we now have 3 equations at Level Two, and 6 variance components at Level Two (3 variance terms and 3 covariance terms). However, with only a maximum of 3 observations per school, we lack the data for fitting 3 equations with error terms at Level Two. Instead, we could model the quadratic time effect with fewer variance components—for instance, by only using an error term on the intercept at Level Two:

$$a_i = \alpha_0 + u_i$$
$$b_i = \beta_0$$
$$c_i = \gamma_0$$

where $u_i \sim N(0, \sigma_u^2)$. Models like this are frequently used in practice—they allow for a separate overall effect on test scores for each school, while minimizing parameters that must be estimated. The tradeoff is that this model does not allow linear and quadratic effects to differ by school, but we have little choice here without more observations per school. Thus, using the composite model specification, the unconditional quadratic growth model with random intercept for each school can be fit to the MCA-II test data:

```
# Modeling quadratic time trend
model.b2 <- lmer(MathAvgScore~ yearc + yearc2 + (1|schoolid),
  REML=T, data=chart.long)
```

```
##  Groups   Name        Variance Std.Dev.
##  schoolid (Intercept) 43.05    6.56
##  Residual             8.52     2.92

##  Number of Level Two groups =  618

##              Estimate Std. Error  t value
##  (Intercept) 651.942  0.29229     2230.448
##  yearc         1.270  0.08758       14.501
##  yearc2        1.068  0.15046        7.101

##   AIC =  10308 ;  BIC =  10335
```

From this output, we see that the quadratic effect is positive and significant (t=7.1), in this case indicating that increases in test scores are greater between 2009 and 2010 than between 2008 and 2009. Based on AIC and BIC values, the quadratic growth model outperforms the linear growth model with random intercepts only at level Two (AIC: 10308 vs. 10354; BIC: 10335 vs. 10375).

Another frequently used approach to modeling time effects is the **piecewise linear model**. In this model, the complete time span of the study is divided into two or more segments, with a separate slope relating time to the response in each segment. In our case study there is only one piecewise option—fitting separate slopes in 2008-09 and 2009-10. With only 3 time points, creating a piecewise linear model is a bit simplified, but this idea can be generalized to segments with more than two years each.

The performance of this model is very similar to the quadratic growth model by AIC and BIC measures, and the story told by fixed effects estimates is also very similar. While the mean yearly increase in math scores was 0.2 points between 2008 and 2009, it was 2.3 points between 2009 and 2010.

Despite the good performances of the quadratic growth and piecewise linear models on our three-year window of data, we will continue to use linear growth assumptions in the remainder of this chapter. Not only is a linear model easier to interpret and explain, but it's probably a more reasonable assumption in years beyond 2010. Predicting future performance is more risky by assuming a steep one-year rise or a non-linear rise will continue, rather than by using the average increase over two years.

9.6 Building to a Final Model

9.6.1 Uncontrolled Effects of School Type

Initially, we can consider whether or not there are significant differences in individual school growth parameters (intercepts and slopes) based on school type. From a modeling perspective, we would build a system of two Level Two models:

$$a_i = \alpha_0 + \alpha_1 \text{Charter}_i + u_i$$
$$b_i = \beta_0 + \beta_1 \text{Charter}_i + v_i$$

where $\text{Charter}_i = 1$ if School i is a charter school, and $\text{Charter}_i = 0$ if School i is a non-charter public school. In addition, the error terms at Level Two are assumed to follow a multivariate normal distribution:

9.6 Building to a Final Model

$$\begin{bmatrix} u_i \\ v_i \end{bmatrix} \sim N\left(\begin{bmatrix} 0 \\ 0 \end{bmatrix}, \begin{bmatrix} \sigma_u^2 & \\ \sigma_{uv} & \sigma_v^2 \end{bmatrix}\right).$$

With a binary predictor at Level Two such as school type, we can write out what our Level Two model looks like for public non-charter schools and charter schools.

- Public schools

$$a_i = \alpha_0 + u_i$$
$$b_i = \beta_0 + v_i,$$

- Charter schools

$$a_i = (\alpha_0 + \alpha_1) + u_i$$
$$b_i = (\beta_0 + \beta_1) + v_i$$

Writing the Level Two model in this manner helps us interpret the model parameters from our two-level model. We can use statistical software (such as the lmer() function from the lme4 package in R) to obtain parameter estimates using our 1733 observations, after first converting our Level One and Level Two models into a composite model (Model C) with fixed effects and random effects separated:

$$\begin{aligned} Y_{ij} &= a_i + b_i \text{Year08}_{ij} + \epsilon_{ij} \\ &= (\alpha_0 + \alpha_1 \text{Charter}_i + u_i) + (\beta_0 + \beta_1 \text{Charter}_i + v_i)\text{Year08}_{ij} + \epsilon_{ij} \\ &= [\alpha_0 + \beta_0 \text{Year08}_{ij} + \alpha_1 \text{Charter}_i + \beta_1 \text{Charter}_i \text{Year08}_{ij}] + \\ &\quad [u_i + v_i \text{Year08}_{ij} + \epsilon_{ij}] \end{aligned}$$

```
#Model C (uncontrolled effects of school type on
#   intercept and slope)
model.c <- lmer(MathAvgScore~ charter + year08 +
  charter:year08 + (year08|schoolid),
  REML=T, data=chart.long)
```

```
## Groups     Name         Variance Std.Dev. Corr
## schoolid  (Intercept)   35.832   5.986
##           year08         0.131   0.362    0.88
## Residual                 8.784   2.964

## Number of Level Two groups =  618
```

```
##                 Estimate Std. Error  t value
## (Intercept)     652.0584   0.28449 2291.996
## charter          -6.0184   0.86562   -6.953
## year08            1.1971   0.09427   12.698
## charter:year08    0.8557   0.31430    2.723
```

```
## AIC =  10308 ;  BIC =  10351
```

Armed with our parameter estimates, we can offer concrete interpretations:

- Fixed effects:

    - $\hat{\alpha}_0 = 652.1$. The estimated mean test score for 2008 for non-charter public schools is 652.1.
    - $\hat{\alpha}_1 = -6.02$. Charter schools have an estimated test score in 2008 which is 6.02 points lower than public non-charter schools.
    - $\hat{\beta}_0 = 1.20$. Public non-charter schools have an estimated mean increase in test scores of 1.20 points per year.
    - $\hat{\beta}_1 = 0.86$. Charter schools have an estimated mean increase in test scores of 2.06 points per year over the three-year observation period, 0.86 points higher than the mean yearly increase among public non-charter schools.

- Variance components:

    - $\hat{\sigma}_u = 5.99$. The estimated standard deviation of 2008 test scores is 5.99 points, after controlling for school type.
    - $\hat{\sigma}_v = 0.36$. The estimated standard deviation of yearly changes in test scores during the three-year observation period is 0.36 points, after controlling for school type.
    - $\hat{\rho}_{uv} = 0.88$. The estimated correlation between 2008 test scores and yearly changes in test scores is 0.88, after controlling for school type.
    - $\hat{\sigma} = 2.96$. The estimated standard deviation in residuals for the individual growth curves is 2.96 points.

Based on t-values reported by R, the effects of year08 and charter both appear to be statistically significant, and there is also significant evidence of an interaction between year08 and charter. Public schools had a significantly higher mean math score in 2008, while charter schools had significantly greater improvement in scores between 2008 and 2010 (although the mean score of charter schools still lagged behind that of public schools in 2010, as indicated in the graphical comparison of Models B and C in Figure 9.18). Based on pseudo

9.6 Building to a Final Model

R-squared values, the addition of a charter school indicator to the unconditional growth model has decreased unexplained school-to-school variability in 2008 math scores by 4.7%, while unexplained variability in yearly improvement actually increased slightly. Obviously, it makes little sense that introducing an additional predictor would *reduce* the amount of variability in test scores explained, but this is an example of the limitations in the pseudo R-squared values discussed in Section 8.7.2.

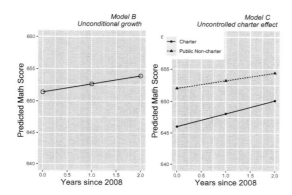

FIGURE 9.18: Fitted growth curves for Models B and C.

9.6.2 Add Percent Free and Reduced Lunch as a Covariate

Although we will still be primarily interested in the effect of school type on both 2008 test scores and rate of change in test scores (as we observed in Model C), we can try to improve our estimates of school type effects through the introduction of meaningful covariates. In this study, we are particularly interested in Level Two covariates—those variables which differ by school but which remain basically constant for a given school over time—such as urban or rural location, percentage of special education students, and percentage of students with free and reduced lunch. In Section 9.4, we investigated the relationship between percent free and reduced lunch and a school's test score in 2008 and their rate of change from 2008 to 2010.

Based on these analyses, we will begin by adding percent free and reduced lunch as a Level Two predictor for both intercept and slope (Model D):

- Level One:
$$Y_{ij} = a_i + b_i \text{Year08}_{ij} + \epsilon_{ij}$$

- Level Two:
$$a_i = \alpha_0 + \alpha_1 \text{Charter}_i + \alpha_2 \text{schpctfree}_i + u_i$$
$$b_i = \beta_0 + \beta_1 \text{Charter}_i + \beta_2 \text{schpctfree}_i + v_i$$

The composite model is then:

$$Y_{ij} = [\alpha_0 + \alpha_1 \text{Charter}_i + \alpha_2 \text{schpctfree}_i + \beta_0 \text{Year08}_{ij}$$
$$+ \beta_1 \text{Charter}_i \text{Year08}_{ij} + \beta_2 \text{schpctfree}_i \text{Year08}_{ij}]$$
$$+ [u_i + v_i \text{Year08}_{ij} + \epsilon_{ij}]$$

where error terms are defined as in Model C.

```
#Model D2 (Introduce SchPctFree at level 2)
model.d2 <- lmer(MathAvgScore~ charter + SchPctFree + year08 +
  charter:year08 + SchPctFree:year08 + (year08|schoolid),
  REML=T, data=chart.long)
```

```
##  Groups    Name         Variance  Std.Dev. Corr
##  schoolid  (Intercept)  19.13     4.37
##            year08        0.16     0.40     0.51
##  Residual                8.80     2.97

##  Number of Level Two groups =  618

##                     Estimate  Std. Error  t value
##  (Intercept)        659.27848 0.444690    1482.558
##  charter             -3.43994 0.712836      -4.826
##  SchPctFree          -0.16654 0.008907     -18.697
##  year08               1.64137 0.189499       8.662
##  charter:year08       0.98076 0.318583       3.078
##  SchPctFree:year08   -0.01041 0.003839      -2.711

##  AIC =  9988 ;  BIC =   10043
```

```
drop_in_dev <- anova(model.d2, model.c, test = "Chisq")
```

```
          npar  AIC    BIC    logLik  dev   Chisq Df
model.c      8 10305  10348  -5144   10289   NA   NA
model.d2    10  9967  10022  -4974    9947  341.5  2
          pval
model.c     NA
model.d2  7.158e-75
```

9.6 Building to a Final Model

Compared to Model C, the introduction of school-level poverty based on percentage of students receiving free and reduced lunch in Model D leads to similar conclusions about the significance of the charter school effect on both the intercept and the slope, although the magnitude of these estimates changes after controlling for poverty levels. The estimated gap in test scores between charter and non-charter schools in 2008 is smaller in Model D, while estimates of improvement between 2008 and 2010 increase for both types of schools. Inclusion of free and reduced lunch reduces the unexplained variability between schools in 2008 math scores by 27%, while unexplained variability in rates of change between schools again increases slightly based on pseudo R-squared values. A **likelihood ratio test** using maximum likelihood estimates illustrates that adding free and reduced lunch as a Level Two covariate significantly improves our model ($\chi^2 = 341.5, df = 2, p < .001$). Specific fixed effect parameter estimates are given below:

- $\hat{\alpha}_0 = 659.3$. The estimated mean math test score for 2008 is 659.3 for non-charter public schools with no students receiving free and reduced lunch.

- $\hat{\alpha}_1 = -3.44$. Charter schools have an estimated mean math test score in 2008 which is 3.44 points lower than non-charter public schools, controlling for effects of school-level poverty.

- $\hat{\alpha}_2 = -0.17$. Each 10% increase in the percentage of students at a school receiving free and reduced lunch is associated with a 1.7 point decrease in mean math test scores for 2008, after controlling for school type.

- $\hat{\beta}_0 = 1.64$. Public non-charter schools with no students receiving free and reduced lunch have an estimated mean increase in math test score of 1.64 points per year during the three years of observation.

- $\hat{\beta}_1 = 0.98$. Charter schools have an estimated mean yearly increase in math test scores over the three-year observation period of 2.62, which is 0.98 points higher than the annual increase for public non-charter schools, after controlling for school-level poverty.

- $\hat{\beta}_2 = -0.010$. Each 10% increase in the percentage of students at a school receiving free and reduced lunch is associated with a 0.10 point decrease in rate of change over the three years of observation, after controlling for school type.

9.6.3 A Final Model with Three Level Two Covariates

We now begin iterating toward a "final model" for these data, on which we will base conclusions. Being cognizant of typical features of a "final model" as outlined in Chapter 8, we offer one possible final model for this data—Model F:

- Level One:

$$Y_{ij} = a_i + b_i \text{Year08}_{ij} + \epsilon_{ij}$$

- Level Two:

$$a_i = \alpha_0 + \alpha_1 \text{Charter}_i + \alpha_2 \text{urban}_i + \alpha_3 \text{schpctsped}_i + \alpha_4 \text{schpctfree}_i + u_i$$
$$b_i = \beta_0 + \beta_1 \text{Charter}_i + \beta_2 \text{urban}_i + \beta_3 \text{schpctsped}_i + v_i$$

where we find the effect of charter schools on 2008 test scores after adjusting for urban or rural location, percentage of special education students, and percentage of students that receive free or reduced lunch, and the effect of charter schools on yearly change between 2008 and 2010 after adjusting for urban or rural location and percentage of special education students. We can use AIC and BIC criteria to compare Model F with Model D, since the two models are not nested. By both criteria, Model F is significantly better than Model D: AIC of 9885 vs. 9988, and BIC of 9956 vs. 10043. Based on the R output below, we offer interpretations for estimates of model fixed effects:

```
model.f2 <- lmer(MathAvgScore ~ charter + urban + SchPctFree +
  SchPctSped + charter:year08 + urban:year08 +
  SchPctSped:year08 + year08 +
  (year08|schoolid), REML=T, data=chart.long)
```

```
##  Groups   Name         Variance Std.Dev. Corr
##  schoolid (Intercept)  16.94756 4.1167
##           year08        0.00475 0.0689   0.85
##  Residual               8.82197 2.9702

##  Number of Level Two groups =  618

##                     Estimate Std. Error  t value
## (Intercept)         661.01042   0.512888 1288.800
## charter              -3.22286   0.698547   -4.614
## urban                -1.11383   0.427566   -2.605
## SchPctFree           -0.15281   0.008096  -18.874
## SchPctSped           -0.11770   0.020612   -5.710
## year08                2.14430   0.200867   10.675
## charter:year08        1.03087   0.315159    3.271
## urban:year08         -0.52749   0.186480   -2.829
## SchPctSped:year08    -0.04674   0.010166   -4.598

##   AIC =  9885 ;  BIC =   9956
```

9.6 Building to a Final Model

- $\hat{\alpha}_0 = 661.0$. The estimated mean math test score for 2008 is 661.0 for public schools in rural areas with no students qualifying for special education or free and reduced lunch.
- $\hat{\alpha}_1 = -3.22$. Charter schools have an estimated mean math test score in 2008 which is 3.22 points lower than non-charter public schools, after controlling for urban or rural location, percent special education, and percent free and reduced lunch.
- $\hat{\alpha}_2 = -1.11$. Schools in urban areas have an estimated mean math score in 2008 which is 1.11 points lower than schools in rural areas, after controlling for school type, percent special education, and percent free and reduced lunch.
- $\hat{\alpha}_3 = -0.118$. A 10% increase in special education students at a school is associated with a 1.18 point decrease in estimated mean math score for 2008, after controlling for school type, urban or rural location, and percent free and reduced lunch.
- $\hat{\alpha}_4 = -0.153$. A 10% increase in free and reduced lunch students at a school is associated with a 1.53 point decrease in estimated mean math score for 2008, after controlling for school type, urban or rural location, and percent special education.
- $\hat{\beta}_0 = 2.14$. Public non-charter schools in rural areas with no students qualifying for special education have an estimated increase in mean math test score of 2.14 points per year over the three-year observation period, after controlling for percent of students receiving free and reduced lunch.
- $\hat{\beta}_1 = 1.03$. Charter schools have an estimated mean annual increase in math score that is 1.03 points higher than public non-charter schools over the three-year observation period, after controlling for urban or rural location, percent special education, and percent free and reduced lunch.
- $\hat{\beta}_2 = -0.53$. Schools in urban areas have an estimated mean annual increase in math score that is 0.53 points lower than schools from rural areas over the three-year observation period, after controlling for school type, percent special education, and percent free and reduced lunch.
- $\hat{\beta}_3 = -0.047$. A 10% increase in special education students at a school is associated with an estimated mean annual increase in math score that is 0.47 points lower over the three-year observation period, after controlling for school type, urban or rural location, and percent free and reduced lunch.

From this model, we again see that 2008 sixth grade math test scores from charter schools were significantly lower than similar scores from public non-charter schools, after controlling for school location and demographics. However, charter schools showed significantly greater improvement between 2008 and 2010 compared to public non-charter schools, although charter school test scores were still lower than public school scores in 2010, on average. We also tested several interactions between Level Two covariates and charter schools and found none to be significant, indicating that the 2008 gap between charter schools and public non-charter schools was consistent across demographic subgroups. The faster improvement between 2008 and 2010 for charter schools was also consistent across demographic subgroups (found by testing three-way interactions). Controlling for school location and demographic variables provided more reliable and nuanced estimates of the effects of charter schools, while also providing interesting insights. For example, schools in rural areas not only had higher test scores than schools in urban areas in 2008, but the gap grew larger over the study period given fixed levels of percent special education, percent free and reduced lunch, and school type. In addition, schools with higher levels of poverty lagged behind other schools and showed no signs of closing the gap, and schools with higher levels of special education students had both lower test scores in 2008 and slower rates of improvement during the study period, again given fixed levels of other covariates.

As we demonstrated in this case study, applying multilevel methods to two-level longitudinal data yields valuable insights about our original research questions while properly accounting for the structure of the data.

9.6.4 Parametric Bootstrap Testing

We could further examine whether or not a simplified version of Model F, with fewer random effects, might be preferable. For instance, consider testing whether we could remove v_i in Model F in Section 9.6.3, to create Model F0. Removing v_i is equivalent to setting $\sigma_v^2 = 0$ and $\rho_{uv} = 0$. We begin by comparing Model F (full model) to Model F0 (reduced model) using a likelihood ratio test:

```
#Model F0 (remove 2 variance components from Model F)
model.f0 <- lmer(MathAvgScore ~ charter + urban + SchPctFree +
  SchPctSped + charter:year08 + urban:year08 +
  SchPctSped:year08 + year08 +
  (1|schoolid), REML=T, data=chart.long)
```

```
##  Groups   Name        Variance  Std.Dev.
##  schoolid (Intercept) 17.46     4.18
##  Residual              8.83     2.97

##  Number of Level Two groups =  618

##                    Estimate  Std. Error  t value
##  (Intercept)       661.02599  0.51691    1278.813
##  charter            -3.21918  0.70543      -4.563
```

9.6 Building to a Final Model

```
## urban                   -1.11852    0.43204   -2.589
## SchPctFree              -0.15302    0.00810  -18.892
## SchPctSped              -0.11813    0.02080   -5.680
## year08                   2.13924    0.20097   10.644
## charter:year08           1.03157    0.31551    3.270
## urban:year08            -0.52330    0.18651   -2.806
## SchPctSped:year08       -0.04645    0.01017   -4.568

## AIC =  9881 ;  BIC =  9941
```

```
drop_in_dev <- anova(model.f2, model.f0, test = "Chisq")
```

```
         npar  AIC  BIC logLik  dev Chisq Df    pval
model.f0   11 9854 9914  -4916 9832    NA NA      NA
model.f2   13 9857 9928  -4916 9831 0.3376  2  0.8447
```

When testing random effects at the boundary (such as $\sigma_v^2 = 0$) or those with restricted ranges (such as $\rho_{uv} = 0$), using a chi-square distribution to conduct a likelihood ratio test is not appropriate. In fact, this will produce a conservative test, with p-values that are too large and not rejected enough (Raudenbush and Bryk [2002], Singer and Willett [2003], Faraway [2005]). For example, we should suspect that the p-value (.8447) produced by the likelihood ratio test comparing Models F and F0 is too large, that the real probability of getting a likelihood ratio test statistic of 0.3376 or greater when Model F0 is true is smaller than .8447.

Researchers often use methods like the **parametric bootstrap** to better approximate the distribution of the likelihood test statistic and produce more accurate p-values by simulating data under the null hypothesis. Here are the basic steps for running a parametric bootstrap procedure to compare Model F0 with Model F (see associated diagram in Figure 9.19):

- Fit Model F0 (the null model) to obtain estimated fixed effects and variance components (this is the "parametric" part.)
- Use the estimated fixed effects and variance components from the null model to regenerate a new set of math test scores with the same sample size ($n = 1733$) and associated covariates for each observation as the original data (this is the "bootstrap" part.)
- Fit both Model F0 (the reduced model) and Model F (the full model) to the new data
- Compute a likelihood ratio statistic comparing Models F0 and F

- Repeat the previous 3 steps many times (e.g., 1000)
- Produce a histogram of likelihood ratio statistics to illustrate its behavior when the null hypothesis is true
- Calculate a p-value by finding the proportion of times the bootstrapped test statistic is greater than our observed test statistic

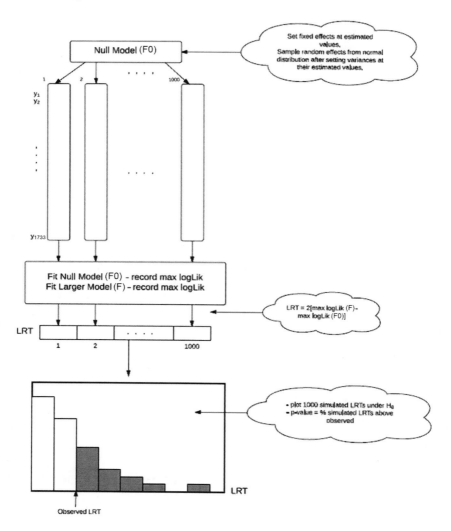

FIGURE 9.19: The steps in conducting a parametric bootstrap test comparing Models F and F0.

Let's see how new test scores are generated under the parametric bootstrap. Consider, for instance, $i = 1$ and $j = 1, 2, 3$; that is, consider test scores for

9.6 Building to a Final Model

TABLE 9.4: Original data for Rippleside Elementary (School 1).

schoolName	urban	charter	schPctsped	schPctfree	year08	MathAvgScore
RIPPLESIDE ELEMENTARY	0	0	0.1176	0.3627	0	652.8
RIPPLESIDE ELEMENTARY	0	0	0.1176	0.3627	1	656.6
RIPPLESIDE ELEMENTARY	0	0	0.1176	0.3627	2	652.6

School #1 (Rippleside Elementary) across all three years (2008, 2009, and 2010). Table 9.4 shows the original data for Rippleside Elementary.

Level Two

One way to see the data generation process under the null model (Model F0) is to start with Level Two and work backwards to Level One. Recall that our Level Two models for a_i and b_i, the true intercept and slope for school i, in Model F0 are:

$$a_i = \alpha_0 + \alpha_1 \text{Charter}_i + \alpha_2 \text{urban}_i + \alpha_3 \text{schpctsped}_i + \alpha_4 \text{schpctfree}_i + u_i$$
$$b_i = \beta_0 + \beta_1 \text{Charter}_i + \beta_2 \text{urban}_i + \beta_3 \text{schpctsped}_i$$

All the α and β terms will be fixed at their estimated values, so the one term that will change for each bootstrapped data set is u_i. As we obtain a numeric value for u_i for each school, we will fix the subscript. For example, if u_i is set to -5.92 for School #1, then we would denote this by $u_1 = -5.92$. Similarly, in the context of Model F0, a_1 represents the 2008 math test score for School #1, where u_1 quantifies how School #1's 2008 score differs from the average 2008 score across all schools with the same attributes: charter status, urban or rural location, percent of special education students, and percent of free and reduced lunch students.

According to Model F0, each u_i is sampled from a normal distribution with mean 0 and standard deviation 4.18. That is, a random component to the intercept for School #1 (u_1) would be sampled from a normal distribution with mean 0 and SD 4.18; say, for instance, $u_1 = -5.92$. We would sample $u_2, ..., u_{618}$ in a similar manner for all 618 schools. Then we can produce a model-based intercept and slope for School #1:

$$a_1 = 661.03 - 3.22(0) - 1.12(0) - 0.12(11.8) - 0.15(36.3) - 5.92 = 648.2$$
$$b_1 = 2.14 + 1.03(0) - 0.52(0) - .046(11.8) = 1.60$$

Notice a couple of features of the above derivations. First, all of the coefficients from the above equations ($\alpha_0 = 661.03$, $\alpha_1 = -3.22$, etc.) come from the estimated fixed effects from Model F0. Second, "public non-charter" is the reference level for charter and "rural" is the reference level for urban, so both of those predictors are 0 for Rippleside Elementary. Third, the mean intercept (2008 test scores) for schools like Rippleside that are rural and public non-charter, with 11.8% special education students and 36.3% free and reduced lunch students, is 661.03 - 0.12(11.8) - 0.15(36.3) = 654.2. The mean yearly improvement in test scores for rural, public non-charter schools with 11.8% special education students is then 1.60 points per year (2.14 - .046*11.8). School #1 (Rippleside) therefore has a 2008 test score that is 5.92 points below the mean for all similar schools, but every such school is assumed to have the same improvement rate in test scores of 1.60 points per year because of our assumption that there is no school-to-school variability in yearly rate of change (i.e., $v_i = 0$).

Level One

We next proceed to Level One, where the scores from Rippleside are modeled as a linear function of year ($654.2 + 1.60 \text{Year08}_{ij}$) with a normally distributed residual ϵ_{1k} at each time point k. Three residuals (one for each year) are sampled independently from a normal distribution with mean 0 and standard deviation 2.97 – the standard deviation again coming from parameter estimates from fitting Model F0 to the actual data. Suppose we obtain residuals of $\epsilon_{11} = -3.11$, $\epsilon_{12} = 1.19$, and $\epsilon_{13} = 2.41$. In that case, our parametrically generated data for Rippleside Elementary (School #1) would look like:

$$\begin{aligned} Y_{11} &= 654.2 + 1.60(0) - 3.11 &= 651.1 \\ Y_{12} &= 654.2 + 1.60(1) + 1.19 &= 657.0 \\ Y_{13} &= 654.2 + 1.60(2) + 2.41 &= 659.8 \end{aligned}$$

We would next turn to School #2 ($i = 2$)—Wrenshall Elementary. Fixed effects would remain the same but covariates would change, as Wrenshall has 15.2% special education students and 42.4% free and reduced lunch students. We would, however, sample a new residual u_2 at Level Two, producing a different intercept a_2 than observed for School #1. Three new independent residuals ϵ_{2k} would also be selected at Level One, from the same normal distribution as before with mean 0 and standard deviation 2.97.

Once an entire set of simulated scores for every school and year have been generated based on Model F0, two models are fit to this data:

- Model F0 – the correct (null) model that was actually used to generate the responses
- Model F – the incorrect (full) model that contains two extra variance components – σ_v^2 and σ_{uv} – that were not actually used when generating the responses

9.6 Building to a Final Model

```
# Generate 1 set of bootstrapped data and run chi-square test
# (will also work if use REML models, but may take longer)
set.seed(3333)
d <- drop(simulate(model.f0ml))
m2 <-refit(model.f2ml, newresp=d)
m1 <-refit(model.f0ml, newresp=d)
drop_in_dev <- anova(m2, m1, test = "Chisq")
```

	npar	AIC	BIC	logLik	dev	Chisq	Df	pval
m1	11	9891	9951	-4935	9869	NA	NA	NA
m2	13	9891	9962	-4932	9865	4.581	2	0.1012

A likelihood ratio test statistic is calculated comparing Model F0 to Model F. For example, after continuing as above to generate new Y_{ij} values corresponding to all 1733 score observations, we fit both models to the "bootstrapped" data. Since the data was generated using Model F0, we would expect the two extra terms in Model F (σ_v^2 and σ_{uv}) to contribute very little to the quality of the fit; Model F will have a slightly larger likelihood and loglikelihood since it contains every parameter from Model F0 plus two more, but the difference in the likelihoods should be due to chance. In fact, that is what the output above shows. Model F does have a larger loglikelihood than Model F0 (-4932 vs. -4935), but this small difference is not statistically significant based on a chi-square test with 2 degrees of freedom (p=.1012).

However, we are really only interested in saving the likelihood ratio test statistic from this bootstrapped sample $(2 * (-4932) - (-4935) = 4.581)$. By generating ("bootstrapping") many sets of responses based on estimated parameters from Model F0 and calculating many likelihood ratio test statistics, we can observe how this test statistic behaves under the null hypothesis of $\sigma_v^2 = \sigma_{uv} = 0$, rather than making the (dubious) assumption that its behavior is described by a chi-square distribution with 2 degrees of freedom. Figure 9.20 illustrates the null distribution of the likelihood ratio test statistic derived by the parametric bootstrap procedure as compared to a chi-square distribution. A p-value for comparing our full and reduced models can be approximated by finding the proportion of likelihood ratio test statistics generated under the null model which exceed our observed likelihood ratio test (0.3376). The parametric bootstrap provides a more reliable p-value in this case (.578 from table below); a chi-square distribution puts too much mass in the tail and not enough near 0, leading to overestimation of the p-value. Based on this test, we would still choose our simpler Model F0.

```
bootstrapAnova(mA=model.f2ml, m0=model.f0ml, B=1000)
```

```
        npar  logLik  dev  Chisq  Df  pval_boot
m0        11   -4916 9832     NA  NA         NA
mA        13   -4916 9831 0.3376   2      0.578
```

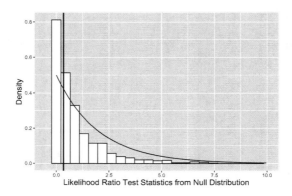

FIGURE 9.20: Null distribution of likelihood ratio test statistic derived using parametric bootstrap (histogram) compared to a chi-square distribution with 2 degrees of freedom (smooth curve). The vertical line represents the observed likelihood ratio test statistic.

Another way of examining whether or not we should stick with the reduced model or reject it in favor of the larger model is by generating parametric bootstrap samples, and then using those samples to produce 95% confidence intervals for both ρ_{uv} and σ_v.

```
bootciF = confint(model.f2, method="boot", oldNames=F)
bootciF
```

```
##                                       2.5 %      97.5 %
## sd_(Intercept)|schoolid             3.801826    4.50866
## cor_year08.(Intercept)|schoolid    -1.000000    1.00000
## sd_year08|schoolid                  0.009203    0.91060
## sigma                               2.779393    3.07776
## (Intercept)                       660.071996  662.04728
## charter                            -4.588611   -2.07372
## urban                              -2.031600   -0.29152
## SchPctFree                         -0.169426   -0.13738
## SchPctSped                         -0.156065   -0.07548
## year08                              1.722458    2.56106
## charter:year08                      0.449139    1.65928
## urban:year08                       -0.905941   -0.17156
## SchPctSped:year08                  -0.066985   -0.02617
```

From the output above, the 95% bootstrapped confidence interval for ρ_{uv} (-1, 1) contains 0, and the interval for σ_v (0.0092, 0.9106) nearly contains 0, providing further evidence that the larger model is not needed.

In this section, we have offered the parametric bootstrap as a noticeable improvement over the likelihood ratio test with an approximate chi-square distribution for testing random effects, especially those near a boundary. Typically when we conduct hypothesis tests involving variance terms we are testing at the boundary, since we are asking if the variance term is really necessary (i.e., $H_0 : \sigma^2 = 0$ vs. $H_A : \sigma^2 > 0$). However, what if we are conducting a hypothesis test about a fixed effect? For the typical test of whether or not a fixed effect is significant – e.g., $H_0 : \alpha_i = 0$ vs. $H_A : \alpha_i \neq 0$ – we are *not* testing at the boundary, since most fixed effects have no bounds on allowable values. We have often used a likelihood ratio test with an approximate chi-square distribution in these settings. Does that provide accurate p-values? Although some research (e.g., Faraway [2005]) shows that p-values of fixed effects from likelihood ratio tests can tend to be anti-conservative (too low), in general the approximation is not bad. We will continue to use the likelihood ratio test with a chi-square distribution for fixed effects, but you could always check your p-values using a parametric bootstrap approach.

9.7 Covariance Structure among Observations

Part of our motivation for framing our model for multilevel data was to account for the correlation among observations made on the same school (the Level Two observational unit). Our two-level model, through error terms on both Level One and Level Two variables, actually implies a specific within-school covariance structure among observations, yet we have not (until now) focused on this imposed structure. For example:

- What does our two-level model say about the relative variability of 2008 and 2010 scores from the same school?
- What does it say about the correlation between 2008 and 2009 scores from the same school?

In this section, we will describe the within-school covariance structure imposed by our two-level model and offer alternative covariance structures that we might consider, especially in the context of longitudinal data. In short, we will discuss how we might decide if our implicit covariance structure in our two-level model is satisfactory for the data at hand. Then, in the succeeding optional section, we provide derivations of the imposed within-school covariance structure for our standard two-level model using results from probability theory.

9.7.1 Standard Covariance Structure

We will use Model C (uncontrolled effects of school type) to illustrate covariance structure within subjects. Recall that, in composite form, Model C is:

$$\begin{aligned}
Y_{ij} &= a_i + b_i \text{Year08}_{ij} + \epsilon_{ij} \\
&= (\alpha_0 + \alpha_1 \text{Charter}_i + u_i) + (\beta_0 + \beta_1 \text{Charter}_i + v_i)\text{Year08}_{ij} + \epsilon_{ij} \\
&= [\alpha_0 + \alpha_1 \text{Charter}_i + \beta_0 \text{Year08}_{ij} + \beta_1 \text{Charter}_i \text{Year08}_{ij}] + [u_i \\
&\quad + v_i \text{Year08}_{ij} + \epsilon_{ij}]
\end{aligned}$$

where $\epsilon_{ij} \sim N(0, \sigma^2)$ and

$$\begin{bmatrix} u_i \\ v_i \end{bmatrix} \sim N\left(\begin{bmatrix} 0 \\ 0 \end{bmatrix}, \begin{bmatrix} \sigma_u^2 & \\ \sigma_{uv} & \sigma_v^2 \end{bmatrix} \right).$$

For School i, the covariance structure for the three time points has general form:

$$Cov(\mathbf{Y}_i) = \begin{bmatrix} Var(Y_{i1}) & Cov(Y_{i1}, Y_{i2}) & Cov(Y_{i1}, Y_{i3}) \\ Cov(Y_{i1}, Y_{i2}) & Var(Y_{i2}) & Cov(Y_{i2}, Y_{i3}) \\ Cov(Y_{i1}, Y_{i3}) & Cov(Y_{i2}, Y_{i3}) & Var(Y_{i3}) \end{bmatrix}$$

where, for instance, $Var(Y_{i1})$ is the variability in 2008 test scores (time $j = 1$), $Cov(Y_{i1}, Y_{i2})$ is the covariance between 2008 and 2009 test scores (times $j = 1$ and $j = 2$), etc. Since covariance measures the tendency of two variables to move together, we expect positive values for all three covariance terms in $Cov(\mathbf{Y}_i)$, since schools with relatively high test scores in 2008 are likely to also have relatively high test scores in 2009 or 2010. The correlation between two variables then scales covariance terms to values between -1 and 1, so by the same rationale, we expect correlation coefficients between two years to be near 1. If observations within school were independent—that is, knowing a school had relatively high scores in 2008 tells nothing about whether that school will have relatively high scores in 2009 or 2010—then we would expect covariance and correlation values near 0.

It is important to notice that the error structure at Level Two is *not* the same as the within-school covariance structure among observations. That is, the relationship between u_i and v_i from the Level Two equations is not the same as the relationship between test scores from different years at the same school (e.g., the relationship between Y_{i1} and Y_{i2}). In other words,

9.7 Covariance Structure among Observations

$$Cov(\mathbf{Y}_i) \neq \begin{bmatrix} u_i \\ v_i \end{bmatrix} \sim N \left(\begin{bmatrix} 0 \\ 0 \end{bmatrix}, \begin{bmatrix} \sigma_u^2 & \\ \sigma_{uv} & \sigma_v^2 \end{bmatrix} \right).$$

Yet, the error structure and the covariance structure *are* connected to each other, as we will now explore.

Using results from probability theory (see Section 9.7.5), we can show that:

$$Var(Y_{ij}) = \sigma_u^2 + t_{ij}^2 \sigma_v^2 + \sigma^2 + 2t_{ij}\sigma_{uv},$$
$$Cov(Y_{ij}, Y_{ik}) = \sigma_u^2 + t_{ij}t_{ik}\sigma_v^2 + (t_{ij} + t_{ik})\sigma_{uv}$$

for all i, where our time variable (year08) has values $t_{i1} = 0$, $t_{i2} = 1$, and $t_{i3} = 2$ for every School i. Intuitively, these formulas are sensible. For instance, $Var(Y_{i1})$, the uncertainty (variability) around a school's score in 2008, increases as the uncertainty in intercepts and slopes increases, as the uncertainty around that school's linear time trend increases, and as the covariance between intercept and slope residuals increases (since if one is off, the other one is likely off as well). Also, $Cov(Y_{i1}, Y_{i2})$, the covariance between 2008 and 2009 scores, does not depend on Level One error. Thus, in the 3-by-3 within-school covariance structure of the charter schools case study, our standard two-level model determines all 6 covariance matrix elements through the estimation of four parameters $(\sigma_u^2, \sigma_{uv}, \sigma_v^2, \sigma^2)$ and the imposition of a specific structure related to time.

To obtain estimated variances for individual observations and covariances between two time points from the same school, we can simply plug estimated variance components from our two-level model along with time points from our data collection into the equations above. For instance, in Section 9.6.1, we obtained the following estimates of variance components: $\hat{\sigma}^2 = 8.784$, $\hat{\sigma}_u^2 = 35.832$, $\hat{\sigma}_v^2 = 0.131$, and $\hat{\sigma}_{uv} = \hat{\rho}\hat{\sigma}_u\hat{\sigma}_v = 1.907$. Therefore, our estimated within-school variances for the three time points would be:

$$\hat{Var}(Y_{i1}) = 35.832 + 0^2 0.131 + 8.784 + 2(0)1.907 = 44.62$$
$$\hat{Var}(Y_{i2}) = 35.832 + 1^2 0.131 + 8.784 + 2(1)1.907 = 48.56$$
$$\hat{Var}(Y_{i3}) = 35.832 + 2^2 0.131 + 8.784 + 2(2)1.907 = 52.77$$

and our estimated within-school covariances between different time points would be:

$$\hat{Cov}(Y_{i1}, Y_{i2}) = 35.832 + (0)(1)0.131 + (0+1)1.907 = 37.74$$
$$\hat{Cov}(Y_{i1}, Y_{i3}) = 35.832 + (0)(2)0.131 + (0+2)1.907 = 39.65$$
$$\hat{Cov}(Y_{i2}, Y_{i3}) = 35.832 + (1)(2)0.131 + (1+2)1.907 = 41.81$$

In fact, these values will be identical for every School i, since scores were assessed at the same three time points. Thus, we will drop the subscript i moving forward.

Written in matrix form, our two-level model implicitly imposes this estimated covariance structure on within-school observations for any specific School i:

$$\hat{Cov}(\mathbf{Y}) = \begin{bmatrix} 44.62 & & \\ 37.74 & 48.56 & \\ 39.65 & 41.81 & 52.77 \end{bmatrix}$$

and this estimated covariance matrix can be converted into an estimated within-school correlation matrix using the identity $Corr(Y_1, Y_2) = \frac{Cov(Y_1, Y_2)}{\sqrt{Var(Y_1)Var(Y_2)}}$:

$$\hat{Corr}(\mathbf{Y}) = \begin{bmatrix} 1 & & \\ .811 & 1 & \\ .817 & .826 & 1 \end{bmatrix}$$

A couple of features of these two matrices can be highlighted that offer insights into implications of our standard two-level model on the covariance structure among observations at Level One from the same school:

- Many longitudinal data sets show higher correlation for observations that are closer in time. In this case, we see that correlation is very consistent between all pairs of observations from the same school; the correlation between test scores separated by two years (.817) is approximately the same as the correlation between test scores separated by a single year (.811 for 2008 and 2009 scores; .826 for 2009 and 2010 scores).
- Many longitudinal data sets show similar variability at all time points. In this case, the variability in 2010 (52.77) is about 18% greater than the variability in 2008 (44.62), while the variability in 2009 is in between (48.56).
- Our two-level model actually imposes a quadratic structure on the relationship between variance and time; note that the equation for $Var(Y_j)$ contains both t_j^2 and t_j. The variance is therefore minimized at $t = \frac{-\sigma_{uv}}{\sigma_v^2}$. With the charter school data, the variance in test scores is minimized when $t = \frac{-\sigma_{uv}}{\sigma_v^2} = \frac{-1.907}{0.131} = -14.6$; that is, the smallest within-school variance in test scores is expected 14.6 years prior to 2008 (i.e., about 1994), and the variance increases parabolically from there. In general, cases in which σ_v^2 and σ_{uv} are relatively small have little curvature and fairly consistent variability over time.
- There is no requirement that time points within school need to be evenly spaced or even that each school has an equal number of measurements over time, which makes the two-level model structure nicely flexible.

9.7.2 Alternative Covariance Structures

The standard covariance structure that's implied by our multilevel modeling structure provides a useful model in a wide variety of situations—it provides a reasonable model for Level One variability with a relatively small number of parameters, and it has sufficient flexibility to accommodate irregular time intervals as well as subjects with a different number of observations over time. However, there may be cases in which a better fitting model requires additional parameters, or when a simpler model with fewer parameters still provides a good fit to the data. Here is an outline of a few alternative error structures:

- *Unstructured* - Every variance and covariance term for observations within a school is a separate parameter and is therefore estimated uniquely; no patterns among variances or correlations are assumed. This structure offers maximum flexibility but is most costly in terms of parameters estimated.
- *Compound symmetry* - Assume variance is constant across all time points and correlation is constant across all pairs of time points. This structure is highly restrictive but least costly in terms of parameters estimated.
- *Autoregressive* - Assume variance is constant across all time points, but correlation drops off in a systematic fashion as the gap in time increases. Autoregressive models expand compound symmetry by allowing for a common structure where points closest in time are most highly correlated.
- *Toeplitz* - Toeplitz is similar to the autoregressive model, except that it does not impose any structure on the decline in correlation as time gaps increase. Thus, it requires more parameters to be estimated than the autoregressive model while providing additional flexibility.
- *Heterogeneous variances* - The assumption that variances are equal across time points found in the compound symmetry, autoregressive, and Toeplitz models can be relaxed by introducing additional parameters to allow unequal (heterogeneous) variances.

When the focus of an analysis is on stochastic parameters (variance components) rather than fixed effects, parameter estimates are typically based on restricted maximum likelihood (REML) methods; model performance statistics then reflect only the stochastic portion of the model. Models with the same fixed effects but different covariance structures can be compared as usual—with AIC and BIC measures when models are not nested and with likelihood ratio tests when models are nested. However, using a chi-square distribution to conduct a likelihood ratio test in these cases can often produce a conservative test, with p-values that are too large and not rejected enough (Raudenbush and Bryk [2002]; Singer and Willett [2003]; Faraway [2005]). In Section 9.6.4, we introduced the parametric bootstrap as a potentially better way of testing models nested in their random effects.

9.7.3 Non-longitudinal Multilevel Models

Careful modeling and estimation of the Level One covariance matrix is especially important and valuable for longitudinal data (with time at Level One) and as we've seen, our standard two-level model has several nice properties for this purpose. The standard model is also often appropriate for non-longitudinal multilevel models as discussed in Chapter 8, although we must remain aware of the covariance structure implicitly imposed. In other words, the ideas in this section generalize even if time isn't a Level One covariate.

As an example, in Case Study 8.2 where Level One observational units are musical performances rather than time points, the standard model implies the following covariance structure for Musician i in Model C, which uses an indicator for large ensembles as a Level One predictor:

$$Var(Y_{ij}) = \sigma_u^2 + \text{Large}_{ij}^2 \sigma_v^2 + \sigma^2 + 2\text{Large}_{ij}\sigma_{uv}$$
$$= \begin{cases} \sigma^2 + \sigma_u^2 & \text{if Large}_{ij} = 0 \\ \sigma^2 + \sigma_u^2 + \sigma_v^2 + 2\sigma_{uv} & \text{if Large}_{ij} = 1 \end{cases}$$

and

$$Cov(Y_{ij}, Y_{ik}) = \sigma_u^2 + \text{Large}_{ij}\text{Large}_{ik}\sigma_v^2 + (\text{Large}_{ij} + \text{Large}_{ik})\sigma_{uv}$$
$$= \begin{cases} \sigma_u^2 & \text{if Large}_{ij} = \text{Large}_{ik} = 0 \\ \sigma_u^2 + \sigma_{uv} & \text{if Large}_{ij} = 0, \text{Large}_{ik} = 1 \text{ or vice versa} \\ \sigma_u^2 + \sigma_v^2 + 2\sigma_{uv} & \text{if Large}_{ij} = \text{Large}_{ik} = 1 \end{cases}$$

Note that, in the Music Performance Anxiety case study, each subject will have a unique Level One variance-covariance structure, since each subject has a different number of performances and a different mix of large ensemble and small ensemble or solo performances.

9.7.4 Final Thoughts Regarding Covariance Structures

In the charter school example, as is often true in multilevel models, the choice of covariance matrix does not greatly affect estimates of fixed effects. The choice of covariance structure could potentially impact the standard errors of fixed effects, and thus the associated test statistics, but the impact appears minimal in this particular case study. In fact, the standard model typically works very well. So is it worth the time and effort to accurately model the covariance structure? If primary interest is in inference regarding fixed effects, and if the standard errors for the fixed effects appear robust to choice of covariance structure, then extensive time spent modeling the covariance structure is not advised. However, if researchers are interested in predicted random effects and estimated variance components in addition to estimated fixed effects,

9.7 Covariance Structure among Observations

then choice of covariance structure can make a big difference. For instance, if researchers are interested in drawing conclusions about particular schools rather than charter schools in general, they may more carefully model the covariance structure in this study.

9.7.5 Details of Covariance Structures (optional)

Using Model C as specified in Section 9.7.1, we specified the general covariance structure for School i as:

$$Cov(\mathbf{Y}_i) = \begin{bmatrix} Var(Y_{i1}) & Cov(Y_{i1}, Y_{i2}) & Cov(Y_{i1}, Y_{i3}) \\ Cov(Y_{i1}, Y_{i2}) & Var(Y_{i2}) & Cov(Y_{i2}, Y_{i3}) \\ Cov(Y_{i1}, Y_{i3}) & Cov(Y_{i2}, Y_{i3}) & Var(Y_{i3}) \end{bmatrix}$$

If $Y_1 = a_1 X_1 + a_2 X_2 + a_3$ and $Y_2 = b_1 X_1 + b_2 X_2 + b_3$ where X_1 and X_2 are random variables and a_i and b_i are constants for $i = 1, 2, 3$, then we know from probability theory that:

$$Var(Y_1) = a_1^2 Var(X_1) + a_2^2 Var(X_2) + 2a_1 a_2 Cov(X_1, X_2)$$
$$Cov(Y_1, Y_2) = a_1 b_1 Var(X_1) + a_2 b_2 Var(X_2) + (a_1 b_2 + a_2 b_1) Cov(X_1, X_2)$$

Applying these identities to Model C, we first see that we can ignore all fixed effects, since they do not contribute to the variability. Thus,

$$\begin{aligned} Var(Y_{ij}) &= Var(u_i + v_i \text{Year08}_{ij} + \epsilon_{ij}) \\ &= Var(u_i) + \text{Year08}_{ij}^2 Var(v_i) + Var(\epsilon_{ij}) + 2\text{Year08}_{ij} Cov(u_i, v_i) \\ &= \sigma_u^2 + \text{Year08}_{ij}^2 \sigma_v^2 + \sigma^2 + 2\text{Year08}_{ij} \sigma_{uv} \\ &= \sigma_u^2 + t_j^2 \sigma_v^2 + \sigma^2 + 2 t_j \sigma_{uv} \end{aligned}$$

where the last line reflects the fact that observations were taken at the same time points for all schools. We can derive the covariance terms in a similar fashion:

$$\begin{aligned} Cov(Y_{ij}, Y_{ik}) &= Cov(u_i + v_i \text{Year08}_{ij} + \epsilon_{ij}, u_i + v_i \text{Year08}_{ik} + \epsilon_{ik}) \\ &= Var(u_i) + \text{Year08}_{ij} \text{Year08}_{ik} Var(v_i) + \\ &\quad (\text{Year08}_{ij} + \text{Year08}_{ik}) Cov(u_i, v_i) \\ &= \sigma_u^2 + t_j t_k \sigma_v^2 + (t_j + t_k) \sigma_{uv} \end{aligned}$$

In Model C, we obtained the following estimates of variance components: $\hat{\sigma}^2 = 8.784$, $\hat{\sigma}_u^2 = 35.832$, $\hat{\sigma}_v^2 = 0.131$, and $\hat{\sigma}_{uv} = \hat{\rho} \hat{\sigma}_u \hat{\sigma}_v = 1.907$. Therefore, our two-level model implicitly imposes this covariance structure on within-subject observations:

$$Cov(\mathbf{Y}_i) = \begin{bmatrix} 44.62 & & \\ 37.74 & 48.56 & \\ 39.65 & 41.81 & 52.77 \end{bmatrix}$$

and this covariance matrix can be converted into a within-subject correlation matrix:

$$Corr(\mathbf{Y}_i) = \begin{bmatrix} 1 & & \\ .811 & 1 & \\ .817 & .826 & 1 \end{bmatrix}$$

9.8 Notes on Using R (optional)

The model below is our final model with σ_{uv} set to 0—i.e., we have added the restriction that Level Two error terms are uncorrelated. Motivation for this restriction came from repeated estimates of correlation in different versions of the final model near 1, when empirically a slightly negative correlation might be expected. As we will describe in Chapter 10, inclusion of the Level Two correlation as a model parameter appears to lead to boundary constraints—maximum likelihood parameter estimates near the maximum or minimum allowable value for a parameter. A likelihood ratio test using full maximum likelihood estimates confirms that the inclusion of a correlation term does not lead to an improved model (LRT test statistic = .223 on 1 df, $p = .637$); a parametric bootstrap test provides a similar result and is more trustworthy when testing a hypothesis about a variance component. Estimates of fixed effects and their standard errors are extremely consistent with the full model in Section 9.6.3; only the estimate of the variability in σ_1 is noticeably higher.

```
# Modified final model
model.f2a <- lmer(MathAvgScore ~ charter + urban + SchPctFree +
   SchPctSped + charter:year08 + urban:year08 +
   SchPctSped:year08 + year08 +
   (1|schoolid) + (0+year08|schoolid), REML=T, data=chart.long)
```

```
##  Groups     Name        Variance Std.Dev.
##  schoolid   (Intercept) 17.355   4.166
##  schoolid.1 year08       0.114   0.337
##  Residual                8.716   2.952

##  Number of Level Two groups =  618

##                       Estimate Std. Error  t value
```

9.9 Exercises

```
## (Intercept)         661.01770    0.515461  1282.381
## charter              -3.22468    0.703174    -4.586
## urban                -1.11663    0.430422    -2.594
## SchPctFree           -0.15295    0.008096   -18.890
## SchPctSped           -0.11777    0.020739    -5.679
## year08                2.14271    0.202090    10.603
## charter:year08        1.03341    0.317174     3.258
## urban:year08         -0.52442    0.187678    -2.794
## SchPctSped:year08    -0.04672    0.010219    -4.572
## 
##   AIC =  9883 ;  BIC =  9948
```

```
# LRT comparing final model in chapter (model.f2ml) with maximum
#   likelihood estimates to modified final model (model.f2aml)
#   with uncorrelated Level Two errors.
drop_in_dev <- anova(model.f2ml, model.f2aml, test = "Chisq")
```

```
             npar  AIC  BIC  logLik   dev  Chisq Df
model.f2aml    12 9855 9921  -4916   9831    NA  NA
model.f2ml     13 9857 9928  -4916   9831  0.2231  1
              pval
model.f2aml    NA
model.f2ml   0.6367
```

9.9 Exercises

9.9.1 Conceptual Exercises

1. **Parenting and gang activity.** Walker-Barnes and Mason [2001] describe, "Ethnic differences in the effect of parenting on gang involvement and gang delinquency: a longitudinal, hierarchical linear modeling perspective". In this study, 300 ninth graders from one high school in an urban southeastern city were assessed at the beginning of the school year about their gang activity, the gang activity of their peers, behavior of their parents, and their ethnic and cultural heritage. Then, information about their gang activity was collected at 7 additional occasions during the school year. For this study: (a) give the observational units at Level One and Level Two, and (b) list potential explanatory variables at both Level One and Level Two.

2. Describe the difference between the wide and long formats for longitudinal data in this study.

3. Describe scenarios or research questions in which a lattice plot would be more informative than a spaghetti plot, and other scenarios or research questions in which a spaghetti plot would be preferable to a lattice plot.

4. Walker-Barnes and Mason summarize their analytic approach in the following way, where HLM = hierarchical linear models, a synonym for multilevel models:

   The first series [of analyses] tested whether there was overall change and/or significant individual variability in gang [activity] over time, regardless of parenting behavior, peer behavior, or ethnic and cultural heritage. Second, given the well documented relation between peer and adolescent behavior . . . HLM analyses were conducted examining the effect of peer gang [activity] on [initial gang activity and] changes in gang [activity] over time. Finally, four pairs of analyses were conducted examining the role of each of the four parenting variables on [initial gang activity and] changes in gang [activity].

   The last series of analyses controlled for peer gang activity and ethnic and cultural heritage, in addition to examining interactions between parenting and ethnic and cultural heritage.

   Although the authors examined four parenting behaviors—behavioral control, lax control, psychological control, and parental warmth—they did so one at a time, using four separate multilevel models. Based on their description, write out a sample model from each of the three steps in the series. For each model, (a) write out the two-level model for predicting gang activity, (b) write out the corresponding composite model, and (c) determine how many model parameters (fixed effects and variance components) must be estimated.

5. Table 9.5 shows a portion of Table 2: Results of Hierarchical Linear Modeling Analyses Modeling Gang Involvement from Walker-Barnes and Mason [2001]. Provide interpretations of significant coefficients in context.

6. **Charter schools.** Differences exist in both sets of boxplots in Figure 9.12. What do these differences imply for multilevel modeling?

7. What implications do the scatterplots in Figures 9.14 (b) and (c) have for multilevel modeling? What implications does the boxplot in Figure 9.14 (a) have?

TABLE 9.5: A portion of Table 2: Results of Hierarchical Linear Modeling Analyses Modeling Gang Involvement from Walker-Barnes and Mason (2001).

Predictor	Coefficient	SE
Intercept (initial status)		
Base (intercept for predicting int term)	-.219	.160
Peer behavior	.252**	.026
Black ethnicity	.671*	.289
White/Other ethnicity	.149	.252
Parenting	.076	.050
Black ethnicity X parenting	-.161+	.088
White/Other ethnicity X parenting	-.026	.082
Slope (change)		
Base (intercept for predicting slope term)	.028	.030
Peer behavior	-.011*	.005
Black ethnicity	-.132*	.054
White/Other ethnicity	-.059	.046
Parenting	-.015+	.009
Black ethnicity X parenting	-.048**	.017
White/Other ethnicity X parenting	.016	.015

These columns focus on the parenting behavior of psychological control. Table reports values for coefficients in the final model with all variables entered. * p<.05; ** p<.01; + p<.10

8. What are the implications of Figure 9.15 for multilevel modeling?

9. Sketch a set of boxplots to indicate an obvious interaction between percent special education and percent non-white in modeling 2008 math scores. Where would this interaction appear in the multilevel model?

10. In Model A, σ^2 is defined as the variance in within-school deviations and σ_u^2 is defined as the variance in between-school deviations. Give potential sources of within-school and between-school deviations.

11. In Chapter 8 Model B is called the "random slopes and intercepts model", while in this chapter Model B is called the "unconditional growth model". Are these models essentially the same or systematically different? Explain.

12. In Section 9.5.2, why don't we examine the pseudo R-squared value for Level Two?

13. If we have test score data from 2001-2010, explain how we'd create new variables to fit a piecewise model.

14. In Section 9.6.2, could we have used percent free and reduced lunch as a Level One covariate rather than 2010 percent free and reduced lunch as a Level Two covariate? If so, explain how interpretations would have changed. What if we had used average percent free and reduced lunch over all three years or 2008 percent free and reduced lunch instead of 2010 percent free and reduced lunch. How would this have changed the interpretation of this term?

15. In Section 9.6.2, why do we look at a 10% increase in the percentage of students receiving free and reduced lunch when interpreting $\hat{\alpha}_2$?

16. In Section 9.6.3, if the gap in 2008 math scores between charter and non-charter schools differed for schools of different poverty levels (as measured by percent free and reduced lunch), how would the final model have differed?

17. Explain in your own words why "the error structure at Level Two is *not* the same as the within-school covariance structure among observations".

18. Here is the estimated unstructured covariance matrix for Model C:

$$Cov(\mathbf{Y}_i) = \begin{bmatrix} 41.87 & & \\ 36.46 & 48.18 & \\ 35.20 & 39.84 & 45.77 \end{bmatrix}$$

Explain why this matrix cannot represent an estimated covariance matrix with a compound symmetry, autoregressive, or Toeplitz structure. Also explain why it cannot represent our standard two-level model.

9.9.2 Guided Exercises

1. **Teen alcohol use.** Curran et al. [1997] collected data on 82 adolescents at three time points starting at age 14 to assess factors that affect teen drinking behavior. Key variables in the data set alcohol.csv (accessed via Singer and Willett [2003]) are as follows:
    - id = numerical identifier for subject
    - age = 14, 15, or 16
    - coa = 1 if the teen is a child of an alcoholic parent; 0 otherwise
    - male = 1 if male; 0 if female
    - peer = a measure of peer alcohol use, taken when each subject was 14. This is the square root of the sum of two 6-point

items about the proportion of friends who drink occasionally or regularly.
- `alcuse` = the primary response. Four items—(a) drank beer or wine, (b) drank hard liquor, (c) 5 or more drinks in a row, and (d) got drunk—were each scored on an 8-point scale, from 0="not at all" to 7="every day". Then `alcuse` is the square root of the sum of these four items.

Primary research questions included: Do trajectories of alcohol use differ by parental alcoholism? Do trajectories of alcohol use differ by peer alcohol use?

a. Identify Level One and Level Two predictors.

b. Perform a quick EDA. What can you say about the shape of `alcuse`, and the relationship between `alcuse` and `coa`, `male`, and `peer`? Appeal to plots and summary statistics in making your statements.

c. Generate a plot as in Figure 9.4 with alcohol use over time for all 82 subjects. Comment.

d. Generate three spaghetti plots with loess fits similar to Figure 9.7 (one for `coa`, one for `male`, and one after creating a binary variable from `peer`). Comment on what you can conclude from each plot.

e. Fit a linear trend to the data from each of the 82 subjects using `age` as the time variable. Generate histograms as in Figure 9.10 showing the results of these 82 linear regression lines, and generate pairs of boxplots as in Figure 9.12 for `coa` and `male`. No commentary necessary. [Hint: to produce Figure 9.12, you will need a data frame with one observation per subject.]

f. Repeat (e), using centered age (`age14 = age - 14`) as the time variable. Also generate a pair of scatterplots as in Figure 9.14 for peer alcohol use. Comment on trends you observe in these plots. [Hint: after forming `age14`, append it to your current data frame.]

g. Discuss similarities and differences between (e) and (f). Why does using `age14` as the time variable make more sense in this example?

h. (Model A) Run an unconditional means model. Report and interpret the intraclass correlation coefficient.

i. (Model B) Run an unconditional growth model with `age14` as the time variable at Level One. Report and interpret estimated fixed effects, using proper notation. Also report and interpret a pseudo R-squared value.

j. (Model C) Build upon the unconditional growth model by adding the effects of having an alcoholic parent and peer alcohol use in both Level Two equations. Report and interpret all estimated fixed effects, using proper notation.

k. (Model D) Remove the child of an alcoholic indicator variable as a predictor of slope in Model C (it will still be a predictor of intercept). Write out Model D as both a two-level and a composite model using proper notation (including error distributions); how many parameters (fixed effects and variance components) must be estimated? Compare Model D to Model C using an appropriate method and state a conclusion.

2. **Ambulance diversions**. One response to emergency department overcrowding is "ambulance diversion"—closing its doors and forcing ambulances to bring patients to alternative hospitals. The California Office of Statewide Health Planning and Development collected data on how often hospitals enacted "diversion status", enabling researchers to investigate factors associated with increasing amounts of ambulance diversions. An award-winning[1] student project [Fisher et al., 2019] examined a data set (ambulance3.csv) which contains the following variables from 184 California hospitals over a 3-year period (2013-2015):

   - diverthours = number of hours of diversion status over the year (response)
   - year2013 = year (centered at 2013)
   - totalvisits1 = total number of patient visits to the emergency department over the year (in 1000s)
   - ems_basic = 1 if the emergency department can only handle a basic level of severity; 0 if the emergency department can handle higher levels of severity
   - stations = number of emergency department stations available for patients (fixed over 3 years)

   a. State the observational units at Level One and Level Two in this study, then state the explanatory variables at each level from the list above.

   b. Create latticed spaghetti plots that illustrate the relationship between diversion hours and (i) EMS level, and (ii) number of stations (divided into "high" and "low"). Describe terms that might be worth testing in your final model based on these plots.

   c. Write out an unconditional growth model, where Y_{ij} is the number of diversion hours for the ith hospital in the jth year.

---

[1] https://www.causeweb.org/usproc/usclap/2019/spring/winners

Interpret both a_i and v_i in the context of this problem (using words – no numbers necessary).

d. In Model E (see R code at the end of these problems), focus on the `ems_basic:year2013` interaction term.
   - provide a careful interpretation in context.
   - why are there no p-values for testing significance in the `lmer()` output?
   - confidence intervals can be formed for parameters in Model E using two different methods. What can we conclude about the significance of the interaction from the CIs? Be sure to make a statement about significance in context; no need to interpret the CI itself.

e. Write out Model G in terms of its Level 1 and Level 2 equations (see R code at the end of these problems). Be sure to use proper subscripts everywhere, and be sure to also provide expressions for any assumptions made about error terms. How many total parameters must be estimated?

f. In Model G, provide careful interpretations in context for the coefficient estimates of `year2013` and `stations`.

g. We wish to compare Models D and D0.
   - Write out null and alternative hypotheses in terms of model parameters.
   - State a conclusion based on a likelihood ratio test.
   - State a conclusion based on a parametric bootstrap.
   - Generate a plot that compares the null distributions and p-values for the likelihood ratio test and parametric bootstrap.
   - Why might we consider using a parametric bootstrap p-value rather than a likelihood ratio test p-value?
   - Show how you would produce a bootstrapped value for Y_{11}, the first row of `ambulance3.csv`. Show all calculations with as many specific values filled in as possible. If you need to select a random value from a normal distribution, identify the mean and SD for the normal distribution you'd like to sample from.

```
modelD <- lmer(diverthours ~ year2013 + ems_basic +
  (year2013 | id), data = ambulance3)

modelD0 <- lmer(diverthours ~ year2013 + ems_basic +
  (1 | id), data = ambulance3)
```

```
modelE <- lmer(diverthours ~ year2013 + ems_basic +
  ems_basic:year2013 + (year2013 | id), data = ambulance3)

modelG <- lmer(diverthours ~ year2013 + totalvisits1 +
  ems_basic + stations + ems_basic:year2013 +
  stations:year2013 + (year2013 | id), data = ambulance3)
```

9.9.3 Open-Ended Exercises

1. **UCLA nurse blood pressure study.** A study by Goldstein and Shapiro [2000] collected information from 203 registered nurses in the Los Angeles area between 24 and 50 years of age on blood pressure (BP) and potential factors that contribute to hypertension. This information includes family history, and whether the subject had one or two hypertensive parents, as well as a wide range of measures of the physical and emotional condition of each nurse throughout the day. Researchers sought to study the links between BP and family history, personality, mood changes, working status, and menstrual phase.

   Data from this study provided by Weiss [2005] includes observations (40-60 per nurse) repeatedly taken on the 203 nurses over the course of a single day. The first BP measurement was taken half an hour before the subject's normal start of work, and BP was then measured approximately every 20 minutes for the rest of the day. At each BP reading, the nurses also rate their mood on several dimensions, including how stressed they feel at the moment the BP is taken. In addition, the activity of each subject during the 10 minutes before each reading was measured using an actigraph worn on the waist. Each of the variables in nursebp.csv is described below:
   - SNUM: subject identification number
   - SYS: systolic blood pressure (mmHg)
   - DIA: diastolic blood pressure (mmHg)
   - HRT: heart rate (beats per minute)
   - MNACT5: activity level (frequency of movements in 1-minute intervals, over a 10-minute period)
   - PHASE: menstrual phase (follicular—beginning with the end of menstruation and ending with ovulation, or luteal—beginning with ovulation and ending with pregnancy or menstruation)
   - DAY: workday or non-workday
   - POSTURE: position during BP measurement—either sitting, standing, or reclining

9.9 Exercises

- **STR, HAP, TIR**: self-ratings by each nurse of their level of stress, happiness and tiredness at the time of each BP measurement on a 5-point scale, with 5 being the strongest sensation of that feeling and 1 the weakest
- **AGE**: age in years
- **FH123**: coded as either NO (no family history of hypertension), YES (1 hypertensive parent), or YESYES (both parents hypertensive)
- **time**: in minutes from midnight
- **timept**: number of the measurement that day (approximately 50 for each subject)
- **timepass**: time in minutes beginning with 0 at time point 1

Using systolic blood pressure as the primary response, write a short report detailing factors that are significantly associated with higher systolic blood pressure. Be sure to support your conclusions with appropriate exploratory plots and multilevel models. In particular, how are work conditions—activity level, mood, and work status—related to trends in BP levels? As an appendix to your report, describe your modeling process—how did you arrive at your final model, which covariates are Level One or Level Two, what did you learn from exploratory plots, etc.?

Potential alternative directions: consider diastolic blood pressure or heart rate as the primary response variable, or even try modeling emotion rating using a multilevel model.

2. **Completion rates at U.S. colleges.** Education researchers wonder which factors most affect the completion rates at U.S. colleges. Using the IPEDS database containing data from 1310 institutions over the years 2002-2009 [National Center for Education Statistics, 2018], the following variables were assembled in colleges.csv:

- **id** = unique identification number for each college or university

Response:

- **rate** = completion rate (number of degrees awarded per 100 students enrolled)

Level 1 predictors:

- **year**
- **instpct** = percentage of students who receive an institutional grant
- **instamt** = typical amount of an institutional grant among recipients (in $1000s)

Level 2 predictors:

- **faculty** = mean number of full-time faculty per 100 students during 2002-2009
- **tuition** = mean yearly tuition during 2002-2009 (in $1000s)

Perform exploratory analyses and run multilevel models to determine significant predictors of baseline (2002) completion rates and changes in completion rates between 2002 and 2009. In particular, is the percentage of grant recipients or the average institutional grant awarded related to completion rate?

3. **Beating the Blues.** Depression is a common mental disorder affecting approximately 121 million people worldwide, making it one of the leading causes of disability. Evidence has shown that cognitive behavioral therapy (CBT) can be an effective treatment, but delivery of the usual face-to-face treatment is expensive and dependent on the availability of trained therapists. As a result, Proudfoot et al. [2003] developed and studied an interactive multimedia program of CBT called Beating the Blues (BtheB). In their study, 167 participants suffering from anxiety and/or depression were randomly allocated to receive BtheB therapy or treatment as usual (TAU). BtheB consisted of 8, 50-minute computerized weekly sessions with "homework" projects between sessions, while treatment as usual consisted of whatever treatment the patient's general practitioner (GP) prescribed, including drug treatment or referral to a counselor. Subjects in the BtheB group could also receive pharmacotherapy if prescribed by their GP (who reviewed a computer-generated progress report after each subject's session), but they could not receive face-to-face counseling. The primary response was the Beck Depression Inventory (BDI), measured prior to treatment, at the end of treatment (2 months later), and at 2, 4, and 6 months post-treatment follow-up. Researchers wished to examine the effect of treatment on depression levels, controlling for potential explanatory variables such as baseline BDI, if the patient took anti-depressant drugs, and the length of the current episode of depression (more or less than 6 months). Was treatment effective in both the active treatment phase and the post-treatment follow-up?

Data from the BtheB study can be found in BtheB.csv; it is also part of the HSAUR package [Everitt and Hothorn, 2006] in R. Examination of the data reveals the following variables:

- **drug** = Was the subject prescribed concomitant drug therapy?
- **length** = Was the current episode of depression (at study entry) longer or shorter than 6 months?
- **treatment** = TAU or BtheB

9.9 Exercises

- **bdi.pre** = Baseline BDI at time of study entry (before treatment began)
- **bdi.2m** = BDI level after 2 months (at the end of treatment phase)
- **bdi.4m** = BDI level after 4 months (or 2 months after treatment ended)
- **bdi.6m** = BDI level after 6 months (or 4 months after treatment ended)
- **bdi.8m** = BDI level after 8 months (or 6 months after treatment ended)

Things to consider when analyzing data from this case study:

- Examine patterns of missing data.
- Convert to LONG form (and eliminate subjects with no post-baseline data).
- Exploratory data analyses, including lattice, spaghetti, and correlation plots.
- Set time 0 to be 2 months into the study (then the intercept represents BDI level at the end of active treatment, while the slope represents change in BDI level over the posttreatment follow-up).
- Note that treatment is the variable of primary interest, while baseline BDI, concomitant drug use, and length of previous episode are confounding variables.

10
Multilevel Data With More Than Two Levels

10.1 Learning Objectives

After finishing this chapter, you should be able to:

- Extend the standard multilevel model to cases with more than two levels.
- Apply exploratory data analysis techniques specific to data from more than two levels.
- Formulate multilevel models including the variance-covariance structure.
- Build and understand a taxonomy of models for data with more than two levels.
- Interpret parameters in models with more than two levels.
- Develop strategies for handling an exploding number of parameters in multilevel models.
- Recognize when a fitted model has encountered boundary constraints and understand strategies for moving forward.
- Apply a parametric bootstrap test of significance to appropriate situations with more than two levels.

```
# Packages required for Chapter 10
library(knitr)
library(gridExtra)
library(GGally)
library(mice)
library(nlme)
library(lme4)
library(mnormt)
library(boot)
library(HLMdiag)
library(kableExtra)
library(pander)
library(tidyverse)
```

10.2 Case Studies: Seed Germination

It is estimated that 82-99% of historic tallgrass prairie ecosystems have been converted to agricultural use [Baer et al., 2002]. A prime example of this large scale conversion of native prairie to agricultural purposes can be seen in Minnesota, where less than 1% of the prairies that once existed in the state remain [Camill et al., 2004]. Such large-scale alteration of prairie communities has been associated with numerous problems. For example, erosion and decomposition that readily take place in cultivated soils have increased atmospheric CO_2 levels and increased nitrogen inputs to adjacent waterways (Baer et al. [2002], Camill et al. [2004], Knops and Tilman [2000]). In addition, cultivation practices are known to affect rhizosphere composition as tilling can disrupt networks of soil microbes [Allison et al., 2005]. The rhizosphere is the narrow region of soil that is directly influenced by root secretions and associated soil microorganisms; much of the nutrient cycling and disease suppression needed by plants occur immediately adjacent to roots. It is important to note that microbial communities in prairie soils have been implicated with plant diversity and overall ecosystem function by controlling carbon and nitrogen cycling in the soils [Zak et al., 2003].

There have been many responses to these claims, but one response in recent years is reconstruction of the native prairie community. These reconstruction projects provide a new habitat for a variety of native prairie species, yet it is important to know as much as possible about the outcomes of prairie reconstruction projects in order to ensure that a functioning prairie community is established. The ecological repercussions resulting from prairie reconstruction are not well known. For example, all of the aforementioned changes associated with cultivation practices are known to affect the subsequent reconstructed prairie community (Baer et al. [2002], Camill et al. [2004]), yet there are few explanations for this phenomenon. For instance, prairies reconstructed in different years (using the same seed combinations and dispersal techniques) have yielded disparate prairie communities.

Researchers at a small midwestern college decided to experimentally explore the underlying causes of variation in reconstruction projects in order to make future projects more effective. Introductory ecology classes were organized to collect longitudinal data on native plant species grown in a greenhouse setting, using soil samples from surrounding lands [Angell, 2010]. We will examine their data to compare germination and growth of two species of prairie plants—leadplants (*Amorpha canescens*) and coneflowers (*Ratibida pinnata*)—in soils taken from a remnant (natural) prairie, a cultivated (agricultural) field, and a restored (reconstructed) prairie. Additionally, half of the sampled soil was sterilized to determine if rhizosphere differences were responsible for the observed variation, so we will examine the effects of sterilization as well.

10.3 Initial Exploratory Analyses

The data we'll examine was collected through an experiment run using a 3x2x2 factorial design, with 3 levels of soil type (remnant, cultivated, and restored), 2 levels of sterilization (yes or no), and 2 levels of species (leadplant and coneflower). Each of the 12 treatments (unique combinations of factor levels) was replicated in 6 pots, for a total of 72 pots. Six seeds were planted in each pot (although a few pots had 7 or 8 seeds), and initially student researchers recorded days to germination (defined as when two leaves are visible), if germination occurred. In addition, the height of each germinated plant (in mm) was measured at 13, 18, 23, and 28 days after planting. The study design is illustrated in Figure 10.1.

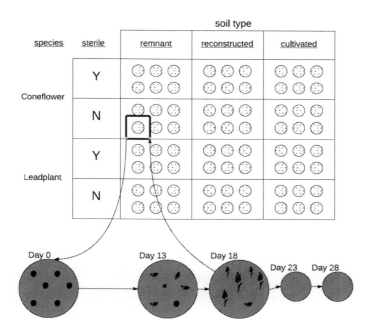

FIGURE 10.1: The design of the seed germination study.

10.3 Initial Exploratory Analyses

10.3.1 Data Organization

Data for Case Study 10.2 in `seeds2.csv` contains the following variables:

TABLE 10.1: A snapshot of data (Plants 231-246) from the Seed Germination case study in wide format.

	pot	plant	soil	sterile	species	germin	hgt13	hgt18	hgt23	hgt28
135	23	231	CULT	N	C	Y	1.1	1.4	1.6	1.7
136	23	232	CULT	N	C	Y	1.3	2.2	2.5	2.7
137	23	233	CULT	N	C	Y	0.5	1.4	2.0	2.3
138	23	234	CULT	N	C	Y	0.3	0.4	1.2	1.7
139	23	235	CULT	N	C	Y	0.5	0.5	0.8	2.0
140	23	236	CULT	N	C	Y	0.1	NA	NA	NA
141	24	241	STP	Y	L	Y	1.8	2.6	3.9	4.2
142	24	242	STP	Y	L	Y	1.3	1.7	2.8	3.7
143	24	243	STP	Y	L	Y	1.5	1.6	3.9	3.9
144	24	244	STP	Y	L	Y	NA	1.0	2.3	3.8
145	24	245	STP	Y	L	N	NA	NA	NA	NA
146	24	246	STP	Y	L	N	NA	NA	NA	NA

- pot = Pot plant was grown in (1-72)
- plant = Unique plant identification number
- species = L for leadplant and C for coneflower
- soil = STP for reconstructed prairie, REM for remnant prairie, and CULT for cultivated land
- sterile = Y for yes and N for no
- germin = Y if plant germinated, N if not
- hgt13 = height of plant (in mm) 13 days after seeds planted
- hgt18 = height of plant (in mm) 18 days after seeds planted
- hgt23 = height of plant (in mm) 23 days after seeds planted
- hgt28 = height of plant (in mm) 28 days after seeds planted

This data is stored in wide format, with one row per plant (see 12 sample plants in Table 10.1). As we have done in previous multilevel analyses, we will convert to long format (one observation per plant-time combination) after examining the missing data pattern and removing any plants with no growth data. In this case, we are almost assuredly losing information by removing plants with no height data at all four time points, since these plants did not germinate, and there may well be differences between species, soil type, and sterilization with respect to germination rates. We will handle this possibility by analyzing germination rates separately (see Chapter 11); the analysis in this chapter will focus on effects of species, soil type, and sterilization on initial growth and growth rate among plants that germinate.

Although the experimental design called for $72 * 6 = 432$ plants, the wide data set has 437 plants because a few pots had more than six plants (likely because two of the microscopically small seeds stuck together when planted). Of those

10.3 Initial Exploratory Analyses

TABLE 10.2: A snapshot of data (Plants 236-242) from the Seed Germination case study in long format.

pot	plant	soil	sterile	species	germin	hgt	time13
23	236	CULT	N	C	Y	0.1	0
23	236	CULT	N	C	Y	NA	5
23	236	CULT	N	C	Y	NA	10
23	236	CULT	N	C	Y	NA	15
24	241	STP	Y	L	Y	1.8	0
24	241	STP	Y	L	Y	2.6	5
24	241	STP	Y	L	Y	3.9	10
24	241	STP	Y	L	Y	4.2	15
24	242	STP	Y	L	Y	1.3	0
24	242	STP	Y	L	Y	1.7	5
24	242	STP	Y	L	Y	2.8	10
24	242	STP	Y	L	Y	3.7	15

437 plants, 154 had no height data (did not germinate by the 28th day) and were removed from analysis (for example, see rows 145-146 in Table 10.1). A total of 248 plants had complete height data (e.g., rows 135-139 and 141-143), 13 germinated later than the 13th day but had complete heights once they germinated (e.g., row 144), and 22 germinated and had measurable height on the 13th day but died before the 28th day (e.g., row 140). Ultimately, the long data set contains 1132 unique observations where plant heights were recorded; representation of plants 236-242 in the long data set can be seen in Table 10.2.

Notice the **three-level structure** of this data. Treatments (levels of the three experimental factors) were assigned at the pot level, then multiple plants were grown in each pot, and multiple measurements were taken over time for each plant. Our multilevel analysis must therefore account for pot-to-pot variability in height measurements (which could result from factor effects), plant-to-plant variability in height within a single pot, and variability over time in height for individual plants. In order to fit such a three-level model, we must extend the two-level model which we have used thus far.

10.3.2 Exploratory Analyses

We start by taking an initial look at the effect of Level Three covariates (factors applied at the pot level: species, soil type, and sterilization) on plant height, pooling observations across pot, across plant, and across time of measurement within plant. First, we observe that the initial balance which existed after randomization of pot to treatment no longer holds. After removing plants that did not germinate (and therefore had no height data), more height measurements exist for coneflowers (n=704, compared to 428 for leadplants), soil from restored prairies (n=524, compared to 288 for cultivated land and 320 for remnant prairies), and unsterilized soil (n=612, compared to 520 for sterilized soil). This imbalance indicates possible factor effects on germination rate; we will take up those hypotheses in Chapter 11. In this chapter, we will focus on the effects of species, soil type, and sterilization on the growth patterns of plants that germinate.

Because we suspect that height measurements over time for a single plant are highly correlated, while height measurements from different plants from the same pot are less correlated, we calculate mean height per plant (over all available time points) before generating exploratory plots investigating Level Three factors. Figure 10.2 then examines the effects of soil type and sterilization separately by species. Sterilization seems to have a bigger benefit for coneflowers, while soil from remnant prairies seems to lead to smaller leadplants and taller coneflowers.

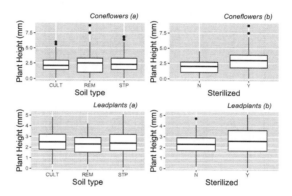

FIGURE 10.2: Plant height comparisons of (a) soil type and (b) sterilization within species. Each plant is represented by the mean height over all measurements at all time points for that plant.

We also use spaghetti plots to examine time trends within species to see (a) if it is reasonable to assume linear growth between Day 13 and Day 28 after planting, and (b) if initial height and rate of growth is similar in the two species. Figure 10.3 illustrates differences between species. While both species have similar average heights 13 days after planting, coneflowers appear to have faster early growth which slows later, while leadplants have a more linear growth rate which culminates in greater average heights 28 days after planting. Coneflowers also appear to have greater variability in initial height and growth rate, although there are more coneflowers with height data.

Exploratory analyses such as these confirm the suspicions of biology researchers that leadplants and coneflowers should be analyzed separately. Because of

10.3 Initial Exploratory Analyses

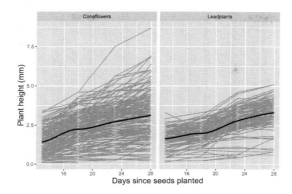

FIGURE 10.3: Spaghetti plot by species with loess fit. Each line represents one plant.

biological differences, it is expected that these two species will show different growth patterns and respond differently to treatments such as fertilization. Coneflowers are members of the aster family, growing up to 4 feet tall with their distinctive gray seed heads and drooping yellow petals. Leadplants, on the other hand, are members of the bean family, with purple flowers, a height of 1 to 3 feet, and compound grayish green leaves which look to be dusted with white lead. Leadplants have deep root systems and are symbiotic N-fixers, which means they might experience stifled growth in sterilized soil compared with other species. For the remainder of this chapter, we will focus on **leadplants** and how their growth patterns are affected by soil type and sterilization. You will have a chance to analyze coneflower data later in the Exercises section.

Lattice plots, illustrating several observational units simultaneously, each with fitted lines where appropriate, are also valuable to examine during the exploratory analysis phase. Figure 10.4 shows height over time for 24 randomly selected leadplants that germinated in this study, with a fitted linear regression line. Linearity appears reasonable in most cases, although there is some variability in the intercepts and a good deal of variability in the slopes of the fitted lines. These intercepts and slopes by plant, of course, will be potential parameters in a multilevel model which we will fit to this data. Given the three-level nature of this data, it is also useful to examine a spaghetti plot by pot (Figure 10.5). While linearity appears to reasonably model the average trend over time within pot, we see differences in the plant-to-plant variability within pot, but some consistency in intercept and slope from pot to pot.

Spaghetti plots can also be an effective tool for examining the potential effects of soil type and sterilization on growth patterns of leadplants. Figure 10.6 and Figure 10.7 illustrate how the growth patterns of leadplants depend on soil type and sterilization. In general, we observe slower growth in soil from remnant prairies and soil that has not been sterilized.

We can further explore the variability in linear growth among plants and among pots by fitting regression lines and examining the estimated intercepts

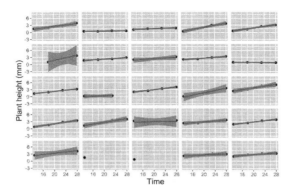

FIGURE 10.4: Lattice plot of linear trends fit to 24 randomly selected leadplants. Two plants with only a single height measurement have no associated regression line.

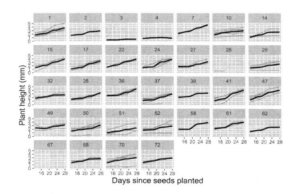

FIGURE 10.5: Spaghetti plot for leadplants by pot with loess fit.

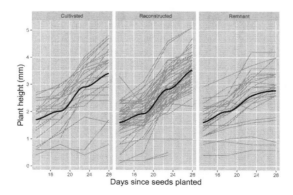

FIGURE 10.6: Spaghetti plot for leadplants by soil type with loess fit.

10.3 Initial Exploratory Analyses

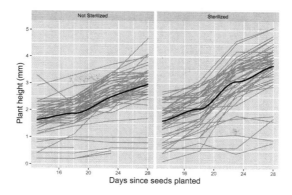

FIGURE 10.7: Spaghetti plot for leadplants by sterilization with loess fit.

and slopes, as well as the corresponding R^2 values. Figures 10.8 and 10.9 provide just such an analysis, where Figure 10.8 shows results of fitting lines by plant, and Figure 10.9 shows results of fitting lines by pot. Certain caveats accompany these summaries. In the case of fitted lines by plant, each plant is given equal weight regardless of the number of observations (2-4) for a given plant, and in the case of fitted lines by pot, a line is estimated by simply pooling all observations from a given pot, ignoring the plant from which the observations came, and equally weighting pots regardless of how many plants germinated and survived to Day 28. Nevertheless, the summaries of fitted lines provide useful information. When fitting regression lines by plant, we see a mean intercept of 1.52 (SD=0.66), indicating an estimated average height at 13 days of 1.5 mm, and a mean slope of 0.114 mm per day of growth from Days 13 to 28 (SD=0.059). Most R-squared values were strong (e.g., 84% were above 0.8). Summaries of fitted regression lines by pot show similar mean intercepts (1.50) and slopes (0.107), but somewhat less variability pot-to-pot than we observed plant-to-plant (SD=0.46 for intercepts and SD=0.050 for slopes).

Another way to examine variability due to plant vs. variability due to pot is through summary statistics. Plant-to-plant variability can be estimated by averaging standard deviations from each pot (.489 for intercepts and .039 for slopes), while pot-to-pot variability can be estimated by finding the standard deviation of average intercept (.478) or slope (.051) within pot. Based on these rough measurements, variability due to plants and pots is comparable.

Fitted lines by plant and pot are modeled using a centered time variable (`time13`), adjusted so that the first day of height measurements (13 days after planting) corresponds to `time13`=0. This centering has two primary advantages. First, the estimated intercept becomes more interpretable. Rather than representing height on the day of planting (which should be 0 mm, but which represents a hefty extrapolation from our observed range of days 13 to

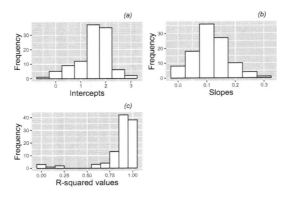

FIGURE 10.8: Histograms of (a) intercepts, (b) slopes, and (c) R-squared values for linear fits across all leadplants.

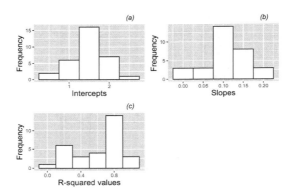

FIGURE 10.9: Histograms of (a) intercepts, (b) slopes, and (c) R-squared values for linear fits across all pots with leadplants.

28), the intercept now represents height on Day 13. Second, the intercept and slope are much less correlated (r=-0.16) than when uncentered time is used, which improves the stability of future models.

Fitted intercepts and slopes by plant can be used for an additional exploratory examination of factor effects to complement those from the earlier spaghetti plots. Figure 10.10 complements Figure 10.3, again showing differences between species—coneflowers tend to start smaller and have slower growth rates, although they have much more variability in growth patterns than leadplants. Returning to our focus on leadplants, Figure 10.11 shows that plants grown in soil from cultivated fields tend to be taller at Day 13, and plants grown in soil from remnant prairies tend to grow more slowly than plants grown in other soil types. Figure 10.12 shows the strong tendency for plants grown in

10.4 Initial Exploratory Analyses

sterilized soil to grow faster than plants grown in non-sterilized soil. We will soon see if our fitted multilevel models support these observed trends.

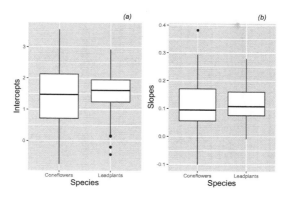

FIGURE 10.10: Boxplots of (a) intercepts and (b) slopes for all plants by species, based on a linear fit to height data from each plant.

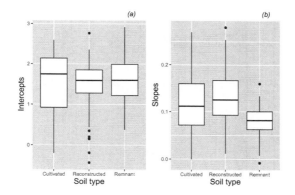

FIGURE 10.11: Boxplots of (a) intercepts and (b) slopes for all leadplants by soil type, based on a linear fit to height data from each plant.

Since we have time at Level One, any exploratory analysis of Case Study 10.2 should contain an investigation of the variance-covariance structure within plant. Figure 10.13 shows the potential for an autocorrelation structure in which the correlation between observations from the same plant diminishes as the time between measurements increases. Residuals five days apart have correlations ranging from .77 to .91, while measurements ten days apart have correlations of .62 and .70, and measurements fifteen days apart have correlation of .58.

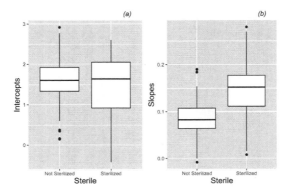

FIGURE 10.12: Boxplots of (a) intercepts and (b) slopes for all leadplants by sterilization, based on a linear fit to height data from each plant.

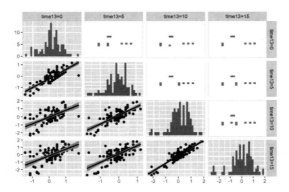

FIGURE 10.13: Correlation structure within plant. The upper right contains correlation coefficients between residuals at pairs of time points, the lower left contains scatterplots of the residuals at time point pairs, and the diagonal contains histograms of residuals at each of the four time points.

10.4 Initial Models

The structure and notation for three level models will closely resemble the structure and notation for two-level models, just with extra subscripts. Therein lies some of the power of multilevel models—extensions are relatively easy and allow you to control for many sources of variability, obtaining more precise estimates of important parameters. However, the number of variance component parameters to estimate can quickly mushroom as covariates are added at lower

10.4 Initial Models

levels, so implementing simplifying restrictions will often become necessary (see Section 10.7).

10.4.1 Unconditional Means

We once again begin with the **unconditional means model**, in which there are no predictors at any level, in order to assess the amount of variation at each level. Here, Level Three is pot, Level Two is plant within pot, and Level One is time within plant. Using model formulations at each of the three levels, the unconditional means three-level model can be expressed as:

- Level One (timepoint within plant):

$$Y_{ijk} = a_{ij} + \epsilon_{ijk} \text{ where } \epsilon_{ijk} \sim N(0, \sigma^2)$$

- Level Two (plant within pot):

$$a_{ij} = a_i + u_{ij} \text{ where } u_{ij} \sim N(0, \sigma_u^2)$$

- Level Three (pot):

$$a_i = \alpha_0 + \tilde{u}_i \text{ where } \tilde{u}_i \sim N(0, \sigma_{\tilde{u}}^2)$$

where the heights of plants from different pots are considered independent, but plants from the same pot are correlated as well as measurements at different times from the same plant.

Keeping track of all the model terms, especially with three subscripts, is not a trivial task, but it's worth spending time thinking it through. Here is a quick guide to the meaning of terms found in our three-level model:

- Y_{ijk} is the height (in mm) of plant j from pot i at time k
- a_{ij} is the true mean height for plant j from pot i across all time points. This is not considered a model parameter, since we further model a_{ij} at Level Two.
- a_i is the true mean height for pot i across all plants from that pot and all time points. This is also not considered a model parameter, since we further model a_i at Level Three.
- α_0 is a fixed effects model parameter representing the true mean height across all pots, plants, and time points.

- ϵ_{ijk} describes how far an observed height Y_{ijk} is from the mean height for plant j from pot i.
- u_{ij} describe how far the mean height of plant j from pot i is from the mean height of all plants from pot i.
- $\tilde{u}_i$ describes how far the mean height of all observations from pot i is from the overall mean height across all pots, plants, and time points. None of the error terms ($\epsilon, u, \tilde{u}$) are considered model parameters; they simply account for differences between the observed data and expected values under our model.
- σ^2 is a variance component (random effects model parameter) that describes within-plant variability over time.
- σ_u^2 is the variance component describing plant-to-plant variability within pot.
- $\sigma_{\tilde{u}}^2$ is the variance component describing pot-to-pot variability.

The three-level unconditional means model can also be expressed as a composite model:

$$Y_{ijk} = \alpha_0 + \tilde{u}_i + u_{ij} + \epsilon_{ijk}$$

and this composite model can be fit using statistical software:

```
# Model A - unconditional means
modela1 = lmer(hgt ~ 1 + (1|plant) + (1|pot),
               REML=T, data=leaddata)
```

```
##  Groups   Name        Variance Std.Dev.
##  plant    (Intercept) 0.2782   0.527
##  pot      (Intercept) 0.0487   0.221
##  Residual             0.7278   0.853

## Number of Level Two groups =   107
## Number of Level Three groups = 32

##             Estimate Std. Error t value
## (Intercept)   2.388    0.07887   30.28
```

From this output, we obtain estimates of our four model parameters:

- $\hat{\alpha}_0 = 2.39 =$ the mean height (in mm) across all time points, plants, and pots.
- $\hat{\sigma}^2 = 0.728 =$ the variance over time within plants.

10.4 Initial Models

- $\hat{\sigma}_u^2 = 0.278 =$ the variance between plants from the same pot.
- $\hat{\sigma}_{\tilde{u}}^2 = 0.049 =$ the variance between pots.

From the estimates of variance components, 69.0% of total variability in height measurements is due to differences over time for each plant, 26.4% of total variability is due to differences between plants from the same pot, and only 4.6% of total variability is due to difference between pots. Accordingly, we will next explore whether the incorporation of time as a linear predictor at Level One can reduce the unexplained variability within plant.

10.4.2 Unconditional Growth

The **unconditional growth model** introduces time as a predictor at Level One, but there are still no predictors at Levels Two or Three. The unconditional growth model allows us to assess how much of the within-plant variability (the variability among height measurements from the same plant at different time points) can be attributed to linear changes over time, while also determining how much variability we see in the intercept (Day 13 height) and slope (daily growth rate) from plant-to-plant and pot-to-pot. Later, we can model plant-to-plant and pot-to-pot differences in intercepts and slopes with Level Two and Three covariates.

The three-level unconditional growth model (Model B) can be specified either using formulations at each level:

- Level One (timepoint within plant):

$$Y_{ijk} = a_{ij} + b_{ij}\text{time}_{ijk} + \epsilon_{ijk}$$

- Level Two (plant within pot):

$$a_{ij} = a_i + u_{ij}$$
$$b_{ij} = b_i + v_{ij}$$

- Level Three (pot):

$$a_i = \alpha_0 + \tilde{u}_i$$
$$b_i = \beta_0 + \tilde{v}_i$$

or as a composite model:

$$Y_{ijk} = [\alpha_0 + \beta_0 \text{time}_{ijk}] + [\tilde{u}_i + v_{ij} + \epsilon_{ijk} + (\tilde{v}_i + v_{ij})\text{time}_{ijk}]$$

where $\epsilon_{ijk} \sim N(0, \sigma^2)$,

$$\begin{bmatrix} u_{ij} \\ v_{ij} \end{bmatrix} \sim N\left(\begin{bmatrix} 0 \\ 0 \end{bmatrix}, \begin{bmatrix} \sigma_u^2 & \\ \sigma_{uv} & \sigma_v^2 \end{bmatrix} \right),$$

and

$$\begin{bmatrix} \tilde{u}_i \\ \tilde{v}_i \end{bmatrix} \sim N\left(\begin{bmatrix} 0 \\ 0 \end{bmatrix}, \begin{bmatrix} \sigma_{\tilde{u}}^2 & \\ \sigma_{\tilde{u}\tilde{v}} & \sigma_{\tilde{v}}^2 \end{bmatrix} \right).$$

In this model, at Level One the trajectory for plant j from pot i is assumed to be linear, with intercept a_{ij} (height on Day 13) and slope b_{ij} (daily growth rate between Days 13 and 28); the ϵ_{ijk} terms capture the deviation between the true growth trajectory of plant j from pot i and its observed heights. At Level Two, a_i represents the true mean intercept and b_i represents the true mean slope for all plants from pot i, while u_{ij} and v_{ij} capture the deviation between plant j's true growth trajectory and the mean intercept and slope for pot i. The deviations in intercept and slope at Level Two are allowed to be correlated through the covariance parameter σ_{uv}. Finally, α_0 is the true mean intercept and β_0 is the true mean daily growth rate over the entire population of leadplants, while $\tilde{u}_i$ and $\tilde{v}_i$ capture the deviation between pot i's true overall growth trajectory and the population mean intercept and slope. Note that between-plant and between-pot variability are both partitioned now into variability in initial status (σ_u^2 and $\sigma_{\tilde{u}}^2$) and variability in rates of change (σ_v^2 and $\sigma_{\tilde{v}}^2$).

Using the composite model specification, the unconditional growth model can be fit to the seed germination data:

```
# Model B - unconditional growth
modelbl = lmer(hgt ~ time13 + (time13|plant) + (time13|pot),
          REML=T, data=leaddata)
```

```
##  Groups    Name         Variance Std.Dev. Corr
##  plant     (Intercept)  0.29914  0.5469
##            time13       0.00119  0.0346   0.28
##  pot       (Intercept)  0.04423  0.2103
##            time13       0.00126  0.0355   -0.61
##  Residual               0.08216  0.2866

##  Number of Level Two groups =   107
##  Number of Level Three groups = 32
```

```
##                 Estimate Std. Error t value
## (Intercept)     1.5377   0.070305   21.87
## time13          0.1121   0.007924   14.15
```

From this output, we obtain estimates of our nine model parameters (two fixed effects and seven variance components):

- $\hat{\alpha}_0 = 1.538 =$ the mean height of leadplants 13 days after planting.
- $\hat{\beta}_0 = 0.112 =$ the mean daily change in height of leadplants from 13 to 28 days after planting.
- $\hat{\sigma} = .287 =$ the standard deviation in within-plant residuals after accounting for time.
- $\hat{\sigma}_u = .547 =$ the standard deviation in Day 13 heights between plants from the same pot.
- $\hat{\sigma}_v = .0346 =$ the standard deviation in rates of change in height between plants from the same pot.
- $\hat{\rho}_{uv} = .280 =$ the correlation in plants' Day 13 height and their rate of change in height.
- $\hat{\sigma}_{\tilde{u}} = .210 =$ the standard deviation in Day 13 heights between pots.
- $\hat{\sigma}_{\tilde{v}} = .0355 =$ the standard deviation in rates of change in height between pots.
- $\hat{\rho}_{\tilde{u}\tilde{v}} = -.610 =$ the correlation in pots' Day 13 height and their rate of change in height.

We see that, on average, leadplants have a height of 1.54 mm 13 days after planting (pooled across pots and treatment groups), and their heights tend to grow by 0.11 mm per day, producing an average height at the end of the study (Day 28) of 3.22 mm. According to the t-values listed in R, both the Day 13 height and the growth rate are statistically significant. The estimated within-plant variance $\hat{\sigma}^2$ decreased by 88.7% from the unconditional means model (from 0.728 to 0.082), implying that 88.7% of within-plant variability in height can be explained by linear growth over time.

10.5 Encountering Boundary Constraints

Typically, with models consisting of three or more levels, the next step after adding covariates at Level One (such as time) is considering covariates at Level Two. In the seed germination experiment, however, there are no Level Two covariates of interest, and the treatments being studied were applied to pots (Level Three). We are primarily interested in the effects of soil type and sterilization on the growth of leadplants. Since soil type is a categorical factor with three levels, we can represent soil type in our model with indicator variables for cultivated lands (`cult`) and remnant prairies (`rem`), using reconstructed prairies as the reference level. For sterilization, we create a single indicator variable (`strl`) which takes on the value 1 for sterilized soil.

Our Level One and Level Two models will look identical to those from Model B; our Level Three models will contain the new covariates for soil type (`cult` and `rem`) and sterilization (`strl`):

$$a_i = \alpha_0 + \alpha_1 \text{strl}_i + \alpha_2 \text{cult}_i + \alpha_3 \text{rem}_i + \tilde{u}_i$$
$$b_i = \beta_0 + \beta_1 \text{strl}_i + \beta_2 \text{cult}_i + \beta_3 \text{rem}_i + \tilde{v}_i$$

where the error terms at Level Three follow the same multivariate normal distribution as in Model B. In our case, the composite model can be written as:

$$Y_{ijk} = (\alpha_0 + \alpha_1 \text{strl}_i + \alpha_2 \text{cult}_i + \alpha_3 \text{rem}_i + \tilde{u}_i + u_{ij}) +$$
$$(\beta_0 + \beta_1 \text{strl}_i + \beta_2 \text{cult}_i + \beta_3 \text{rem}_i + \tilde{v}_i + v_{ij}) \text{time}_{ijk} + \epsilon_{ijk}$$

which, after combining fixed effects and random effects, can be rewritten as:

$$Y_{ijk} = [\alpha_0 + \alpha_1 \text{strl}_i + \alpha_2 \text{cult}_i + \alpha_3 \text{rem}_i + \beta_0 \text{time}_{ijk} +$$
$$\beta_1 \text{strl}_i \text{time}_{ijk} + \beta_2 \text{cult}_i \text{time}_{ijk} + \beta_3 \text{rem}_i \text{time}_{ijk}] +$$
$$[\tilde{u}_i + u_{ij} + \epsilon_{ijk} + \tilde{v}_i \text{time}_{ijk} + v_{ij} \text{time}_{ijk}]$$

From the output below, the addition of Level Three covariates in Model C (`cult`, `rem`, `strl`, and their interactions with `time`) appears to provide a significant improvement (likelihood ratio test statistic = 32.2 on 6 df, $p < .001$) to the unconditional growth model (Model B).

```
# Model C - add covariates at pot level
modelcl = lmer(hgt ~ time13 + strl + cult + rem +
    time13:strl + time13:cult + time13:rem + (time13|plant) +
    (time13|pot), REML=T, data=leaddata)
```

```
##  Groups   Name        Variance Std.Dev. Corr
##  plant    (Intercept) 0.298087 0.5460
##           time13      0.001208 0.0348   0.28
##  pot      (Intercept) 0.053118 0.2305
##           time13      0.000132 0.0115   -1.00
##  Residual             0.082073 0.2865
```

10.5 Encountering Boundary Constraints

```
## Number of Level Two groups =  107
## Number of Level Three groups =  32

##              Estimate Std. Error t value
## (Intercept)   1.50289   0.126992 11.8345
## time13        0.10107   0.008292 12.1889
## strl         -0.07655   0.151361 -0.5058
## cult          0.13002   0.182715  0.7116
## rem           0.13840   0.176189  0.7855
## time13:strl   0.05892   0.010282  5.7300
## time13:cult  -0.02977   0.012263 -2.4273
## time13:rem   -0.03586   0.011978 -2.9939
```

```
drop_in_dev <- anova(modelbl, modelcl, test = "Chisq")
```

```
         npar   AIC   BIC logLik   dev Chisq Df
modelbl     9 603.8 640.0 -292.9 585.8    NA NA
modelcl    15 583.6 643.9 -276.8 553.6  32.2  6
              pval
modelbl         NA
modelcl 1.493e-05
```

However, Model C has encountered a **boundary constraint** with an estimated Level 3 correlation between the intercept and slope error terms of -1. "Allowable" values of correlation coefficients run from -1 to 1; by definition, it is impossible to have a correlation between two error terms below -1. Thus, our estimate of -1 is right on the boundary of the allowable values. But how did this happen, and why is it potentially problematic?

Consider a model in which we have two parameters that must be estimated: β_0 and σ^2. As the intercept, β_0 can take on any value; any real number is "allowable". But, by definition, variance terms such as σ^2 must be non-negative; that is, $\sigma^2 \geq 0$. Under the Principle of Maximum Likelihood, maximum likelihood estimators for β_0 and σ^2 will be chosen to maximize the likelihood of observing our given data. The left plot in Figure 10.14 shows hypothetical contours of the likelihood function $L(\beta_0, \sigma^2)$; the likelihood is clearly maximized at $(\hat{\beta}_0, \hat{\sigma}^2) = (4, -2)$. However, variance terms cannot be negative! A more sensible approach would have been to perform a constrained search for MLEs, considering any potential values for β_0 but only non-negative values for σ^2. This constrained search is illustrated in the right plot in Figure 10.14. In this case, the likelihood is maximized at $(\hat{\beta}_0, \hat{\sigma}^2) = (4, 0)$. Note that the estimated intercept did not change, but the estimated variance is simply set at the smallest allowable value – at the **boundary constraint**.

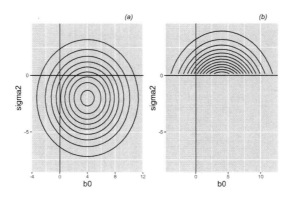

FIGURE 10.14: Left (a): hypothetical contours of the likelihood function $L(\beta_0, \sigma^2)$ with no restrictions on σ^2; the likelihood function is maximized at $(\hat{\beta}_0, \hat{\sigma}^2) = (4, -2)$. Right (b): hypothetical contours of the likelihood function $L(\beta_0, \sigma^2)$ with the restriction that $\sigma^2 \geq 0$; the constrained likelihood function is maximized at $(\hat{\beta}_0, \hat{\sigma}^2) = (4, 0)$.

Graphically, in this simple illustration, the effect of the boundary constraint is to alter the likelihood function from a nice hill (in the left plot in Figure 10.14) with a single peak at $(4, -2)$, to a hill with a huge cliff face where $\sigma^2 = 0$. The highest point overlooking this cliff is at $(4, 0)$, straight down the hill from the original peak.

In general, then, boundary constraints occur when the maximum likelihood estimator of at least one model parameter occurs at the limits of allowable values (such as estimated correlation coefficients of -1 or 1, or estimated variances of 0). Maximum likelihood estimates at the boundary tend to indicate that the likelihood function would be maximized at non-allowable values of that parameter, if an unconstrained search for MLEs was conducted. Most software packages, however, will only report maximum likelihood estimates with allowable values. Therefore, boundary constraints would ideally be avoided, if possible.

What should you do if you encounter boundary constraints? Often, boundary constraints signal that your model needs to be reparameterized, i.e., you should alter your model to feature different parameters or ones that are interpreted differently. This can be accomplished in several ways:

- remove parameters, especially those variance and correlation terms which are being estimated on their boundaries.
- fix the values of certain parameters; for instance, you could set two variance terms equal to each other, thereby reducing the number of unknown parameters to estimate by one.

10.5 Encountering Boundary Constraints

- transform covariates. Centering variables, standardizing variables, or changing units can all help stabilize a model. Numerical procedures for searching for and finding maximum likelihood estimates can encounter difficulties when variables have very high or low values, extreme ranges, outliers, or are highly correlated.

Although it is worthwhile attempting to reparameterize models to remove boundary constraints, sometimes they can be tolerated if (a) you are not interested in estimates of those parameters encountering boundary issues, and (b) removing those parameters does not affect conclusions about parameters of interest. For example, in the output below we explore the implications of simply removing the correlation between error terms at the pot level (i.e., assume $\rho_{\tilde{u}\tilde{v}} = 0$ rather than accepting the (constrained) maximum likelihood estimate of $\hat{\rho}_{\tilde{u}\tilde{v}} = -1$ that we saw in Model C).

```
# Try Model C without correlation between L2 errors
modelcl.noL2corr <- lmer(hgt ~ time13 + strl + cult + rem +
    time13:strl + time13:cult + time13:rem + (time13|plant) +
    (1|pot) + (0+time13|pot), REML=T, data=leaddata)
```

```
##  Groups   Name        Variance Std.Dev. Corr
##  plant    (Intercept) 0.294076 0.5423
##           time13      0.001206 0.0347   0.22
##  pot      (Intercept) 0.059273 0.2435
##  pot.1    time13      0.000138 0.0117
##  Residual             0.082170 0.2867

##  Number of Level Two groups =   107
##  Number of Level Three groups =  32

##               Estimate Std. Error t value
## (Intercept)    1.51209   0.129695 11.6588
## time13         0.10158   0.008361 12.1490
## strl          -0.08742   0.154113 -0.5673
## cult           0.13233   0.185687  0.7126
## rem            0.10657   0.179363  0.5942
## time13:strl    0.05869   0.010382  5.6529
## time13:cult   -0.03065   0.012337 -2.4843
## time13:rem    -0.03810   0.012096 -3.1500
```

Note that the estimated variance components are all very similar to Model C, and the estimated fixed effects and their associated t-statistics are also very similar to Model C. Therefore, in this case we could consider simply reporting the results of Model C despite the boundary constraint.

However, when it is possible to remove boundary constraints through reasonable model reparameterizations, that is typically the preferred route. In this case, one option we might consider is simplifying Model C by setting $\sigma_{\tilde{v}}^2 = \sigma_{\tilde{u}\tilde{v}} = 0$. We can then write our new model (Model C.1) in level-by-level formulation:

- Level One (timepoint within plant):

$$Y_{ijk} = a_{ij} + b_{ij}\text{time}_{ijk} + \epsilon_{ijk}$$

- Level Two (plant within pot):

$$a_{ij} = a_i + u_{ij}$$
$$b_{ij} = b_i + v_{ij}$$

- Level Three (pot):

$$a_i = \alpha_0 + \alpha_1\text{strl}_i + \alpha_2\text{cult}_i + \alpha_3\text{rem}_i + \tilde{u}_i$$
$$b_i = \beta_0 + \beta_1\text{strl}_i + \beta_2\text{cult}_i + \beta_3\text{rem}_i$$

Note that there is no longer an error term associated with the model for mean growth rate b_i at the pot level. The growth rate for pot i is assumed to be fixed, after accounting for soil type and sterilization; all pots with the same soil type and sterilization are assumed to have the same growth rate. As a result, our error assumption at Level Three is no longer bivariate normal, but rather univariate normal: $\tilde{u}_i \sim N(0, \sigma_{\tilde{u}}^2)$. By removing one of our two Level Three error terms ($\tilde{v}_i$), we effectively removed two parameters: the variance for $\tilde{v}_i$ and the correlation between $\tilde{u}_i$ and $\tilde{v}_i$. Fixed effects remain similar, as can be seen in the output below:

```
# Try Model C without time at pot level
modelc10 <- lmer(hgt ~ time13 + strl + cult + rem +
    time13:strl + time13:cult + time13:rem + (time13|plant) +
    (1|pot), REML=T, data=leaddata)
```

```
## Groups    Name         Variance Std.Dev. Corr
## plant     (Intercept)  0.29471  0.5429
##           time13       0.00133  0.0364   0.19
## pot       (Intercept)  0.05768  0.2402
## Residual               0.08222  0.2867

## Number of Level Two groups   = 107
## Number of Level Three groups = 32

##               Estimate Std. Error t value
## (Intercept)    1.51223   0.129016 11.7212
## time13         0.10109   0.007452 13.5648
## strl          -0.08752   0.153451 -0.5703
## cult           0.13287   0.184898  0.7186
## rem            0.10653   0.178612  0.5964
## time13:strl    0.05926   0.009492  6.2437
## time13:cult   -0.03082   0.011353 -2.7151
## time13:rem    -0.03624   0.011101 -3.2649
```

We now have a more stable model, free of boundary constraints. In fact, we can attempt to determine whether or not removing the two variance component parameters for Model C.1 provides a significant reduction in performance. Based on a likelihood ratio test (see below), we do not have significant evidence (chi-square test statistic=2.089 on 2 df, p=0.3519) that $\sigma_{\tilde{v}}^2$ or $\sigma_{\tilde{u}\tilde{v}}$ is non-zero, so it is advisable to use the simpler Model C.1. However, Section 10.6 describes why this test may be misleading and prescribes a potentially better approach.

```
drop_in_dev <- anova(modelcl0, modelcl, test = "Chisq")
```

```
         npar   AIC   BIC logLik    dev Chisq Df   pval
modelcl0   13 581.6 633.9 -277.8  555.6    NA NA     NA
modelcl    15 583.6 643.9 -276.8  553.6 2.089  2 0.3519
```

10.6 Parametric Bootstrap Testing

As in Section 9.6.4, when testing variance components at the boundary (such as $\sigma_{\tilde{v}}^2 = 0$), a method like the **parametric bootstrap** should be used, since using a chi-square distribution to conduct a likelihood ratio test produces p-values that tend to be too large.

TABLE 10.3: Original data for Plants 11 and 12 from Pot 1.

pot	plant	soil	sterile	species	germin	hgt13	hgt18	hgt23	hgt28
1	11	STP	Y	L	Y	2.3	2.9	4.5	5.1
1	12	STP	Y	L	Y	1.9	2.0	2.6	3.5

Under the parametric bootstrap, we must simulate data under the null hypothesis many times. Here are the basic steps for running a parametric bootstrap procedure to compare Model C.1 with Model C:

- Fit Model C.1 (the null model) to obtain estimated fixed effects and variance components (this is the "parametric" part).
- Use the estimated fixed effects and variance components from the null model to regenerate a new set of plant heights with the same sample size ($n = 413$) and associated covariates for each observation as the original data (this is the "bootstrap" part).
- Fit both Model C.1 (the reduced model) and Model C (the full model) to the new data.
- Compute a likelihood ratio statistic comparing Models C.1 and C.
- Repeat the previous 3 steps many times (e.g., 1000).
- Produce a histogram of likelihood ratio statistics to illustrate its behavior when the null hypothesis is true.
- Calculate a p-value by finding the proportion of times the bootstrapped test statistic is greater than our observed test statistic.

Let's see how new plant heights are generated under the parametric bootstrap. Consider, for instance, $i = 1$ and $j = 1, 2$. That is, consider Plants #11 and #12 as shown in Table 10.3. These plants are found in Pot #1, which was randomly assigned to contain sterilized soil from a restored prairie (STP):

Level Three

One way to see the data generation process under the null model (Model C.1) is to start with Level Three and work backwards to Level One. Recall that our Level Three models for a_i and b_i, the true intercept and slope from Pot i, in Model C.1 are:

$$a_i = \alpha_0 + \alpha_1 \text{strl}_i + \alpha_2 \text{cult}_i + \alpha_3 \text{rem}_i + \tilde{u}_i$$
$$b_i = \beta_0 + \beta_1 \text{strl}_i + \beta_2 \text{cult}_i + \beta_3 \text{rem}_i$$

All the α and β terms will be fixed at their estimated values, so the one term that will change for each bootstrapped data set is $\tilde{u}_i$. As we obtain a numeric value for $\tilde{u}_i$ for each pot, we will fix the subscript. For example, if $\tilde{u}_i$ is set to -.192 for Pot #1, then we would denote this by $\tilde{u}_1 = -.192$. Similarly, in the context of Model C.1, a_1 represents the mean height at Day 13 across all

10.6 Parametric Bootstrap Testing

plants in Pot #1, where $\tilde{u}_1$ quantifies how Pot #1's Day 13 height relates to other pots with the same sterilization and soil type.

According to Model C.1, each $\tilde{u}_i$ is sampled from a normal distribution with mean 0 and standard deviation .240 (note that the standard deviation σ_u is also fixed at its estimated value from Model C.1, given in Section 10.5). That is, a random component to the intercept for Pot #1 ($\tilde{u}_1$) would be sampled from a normal distribution with mean 0 and SD .240; say, for instance, $\tilde{u}_1 = -.192$. We would sample $\tilde{u}_2, ..., \tilde{u}_{72}$ in a similar manner. Then we can produce a model-based intercept and slope for Pot #1:

$$a_1 = 1.512 - .088(1) + .133(0) + .107(0) - .192 = 1.232$$
$$b_1 = .101 + .059(1) - .031(0) - .036(0) = .160$$

Notice a couple of features of the above derivations. First, all of the coefficients from the above equations ($\alpha_0 = 1.512$, $\alpha_1 = -.088$, etc.) come from the estimated fixed effects from Model C.1 reported in Section 10.5. Second, "restored prairie" is the reference level for soil type, so that indicators for "cultivated land" and "remnant prairie" are both 0. Third, the mean intercept (Day 13 height) for observations from sterilized restored prairie soil is $1.512 - 0.088 = 1.424$ mm across all pots, while the mean daily growth is .160 mm. Pot #1 therefore has mean Day 13 height that is .192 mm below the mean for all pots with sterilized restored prairie soil, but every such pot is assumed to have the same growth rate of .160 mm/day because of our assumption that there is no pot-to-pot variability in growth rate (i.e., $\tilde{v}_i = 0$).

Level Two

We next proceed to Level Two, where our equations for Model C.1 are:

$$a_{ij} = a_i + u_{ij}$$
$$b_{ij} = b_i + v_{ij}$$

We will initially focus on Plant #11 from Pot #1. Notice that the intercept (Day 13 height = a_{11}) for Plant #11 has two components: the mean Day 13 height for Pot #1 (a_1) which we specified at Level Three, and an error term (u_{11}) which indicates how the Day 13 height for Plant #11 differs from the overall average for all plants from Pot #1. The slope (daily growth rate = b_{11}) for Plant #11 similarly has two components. Since both a_1 and b_1 were determined at Level Three, at this point we need to find the two error terms for Plant #11: u_{11} and v_{11}. According to our multilevel model, we can sample u_{11} and v_{11} from a bivariate normal distribution with means both equal to 0, standard deviation for the intercept of .543, standard deviation for the slope of .036, and correlation between the intercept and slope of .194.

For instance, suppose we sample $u_{11} = .336$ and $v_{11} = .029$. Then we can produce a model-based intercept and slope for Plant #11:

$$a_{11} = 1.232 + .336 = 1.568$$
$$b_{11} = .160 + .029 = .189$$

Although plants from Pot #1 have a mean Day 13 height of 1.232 mm, Plant #11's mean Day 13 height is .336 mm above that. Similarly, although plants from Pot #1 have a mean growth rate of .160 mm/day (just like every other pot with sterilized restored prairie soil), Plant #11's growth rate is .029 mm/day faster.

Level One

Finally we proceed to Level One, where the height of Plant #11 is modeled as a linear function of time $(1.568 + .189\text{time}_{11k})$ with a normally distributed residual ϵ_{11k} at each time point k. Four residuals (one for each time point) are sampled independently from a normal distribution with mean 0 and standard deviation .287 – the standard deviation again coming from parameter estimates from fitting Model C.1 to the actual data as reported in Section 10.5. Suppose we obtain residuals of $\epsilon_{111} = -.311$, $\epsilon_{112} = .119$, $\epsilon_{113} = .241$, and $\epsilon_{114} = -.066$. In that case, our parametrically generated data for Plant #11 from Pot #1 would look like:

$$\begin{aligned} Y_{111} &= 1.568 + .189(0) - .311 &= 1.257 \\ Y_{112} &= 1.568 + .189(5) + .119 &= 2.632 \\ Y_{113} &= 1.568 + .189(10) + .241 &= 3.699 \\ Y_{114} &= 1.568 + .189(15) - .066 &= 4.337 \end{aligned}$$

We would next turn to Plant #12 from Pot #1 ($i = 1$ and $j = 2$). Fixed effects would remain the same, as would coefficients for Pot #1, $a_1 = 1.232$ and $b_1 = .160$, at Level Three. We would, however, sample new residuals u_{12} and v_{12} at Level Two, producing a different intercept a_{12} and slope b_{12} than those observed for Plant #11. Four new independent residuals ϵ_{12k} would also be selected at Level One, from the same normal distribution as before with mean 0 and standard deviation .287.

Once an entire set of simulated heights for every pot, plant, and time point have been generated based on Model C.1, two models are fit to this data:

- Model C.1 – the correct (null) model that was actually used to generate the responses
- Model C – the incorrect (full) model that contains two extra variance components, $\sigma_{\tilde{v}}^2$ and $\sigma_{\tilde{u}\tilde{v}}$, that were not actually used when generating the responses

10.6 Parametric Bootstrap Testing

```
# Generate 1 set of bootstrapped data and run chi-square test
#   (will also work if use REML models, but may take longer)
set.seed(3333)
d <- drop(simulate(modelc10.ml))
m2 <- refit(modelc1.ml, newresp=d)
m1 <- refit(modelc10.ml, newresp=d)
drop_in_dev <- anova(m2, m1, test = "Chisq")
```

	npar	AIC	BIC	logLik	dev	Chisq	Df	pval
m1	13	588.5	640.8	-281.2	562.5	NA	NA	NA
m2	15	591.1	651.4	-280.5	561.1	1.405	2	0.4953

A likelihood ratio test statistic is calculated comparing Model C.1 to Model C. For example, after continuing as above to generate new Y_{ijk} values corresponding to all 413 leadplant height measurements, we fit both models to the "bootstrapped" data. Since the data was generated using Model C.1, we would expect the two extra terms in Model C ($\sigma_{\tilde{v}}^2$ and $\sigma_{\tilde{u}\tilde{v}}$) to contribute very little to the quality of the fit; Model C will have a slightly larger likelihood and loglikelihood since it contains every parameter from Model C.1 plus two more, but the difference in the likelihoods should be due to chance. In fact, that is what the output above shows. Model C does have a larger loglikelihood than Model C.1 (-280.54 vs. -281.24), but this small difference is not statistically significant based on a chi-square test with 2 degrees of freedom (p=.4953).

However, we are really only interested in saving the likelihood ratio test statistic from this bootstrapped sample ($2 * (-280.54 - (-281.24)) = 1.40$). By generating ("bootstrapping") many sets of responses based on estimated parameters from Model C.1 and calculating many likelihood ratio test statistics, we can observe how this test statistic behaves under the null hypothesis of $\sigma_{\tilde{v}}^2 = \sigma_{\tilde{u}\tilde{v}} = 0$, rather than making the (dubious) assumption that its behavior is described by a chi-square distribution with 2 degrees of freedom. Figure 10.15 illustrates the null distribution of the likelihood ratio test statistic derived by the parametric bootstrap procedure as compared to a chi-square distribution. A p-value for comparing our full and reduced models can be approximated by finding the proportion of likelihood ratio test statistics generated under the null model which exceed our observed likelihood ratio test (2.089). The parametric bootstrap provides a more reliable p-value in this case (.088 from table below); a chi-square distribution puts too much mass in the tail and not enough near 0, leading to overestimation of the p-value. Based on this test, we would still choose our simpler Model C.1, but we nearly had enough evidence to favor the more complex model.

```
bootstrapAnova(mA=modelcl.ml, m0=modelcl0.ml, B=1000)
```

```
   Df loglik    dev Chisq ChiDf pval_boot
m0 13 -277.8  555.6    NA    NA        NA
mA 15 -276.8  553.6 2.089     2     0.088
```

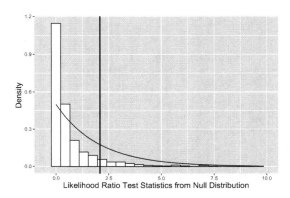

FIGURE 10.15: Null distribution of likelihood ratio test statistic derived using parametric bootstrap (histogram) compared to a chi-square distribution with 2 degrees of freedom (smooth curve). The vertical line represents the observed likelihood ratio test statistic.

Another way of testing whether or not we should stick with the reduced model or reject it in favor of the larger model is by generating parametric bootstrap samples, and then using those samples to produce 95% confidence intervals for both $\rho_{\tilde{u}\tilde{v}}$ and $\sigma_{\tilde{v}}$. From the output below, the 95% bootstrapped confidence interval for $\rho_{\tilde{u}\tilde{v}}$ (-1, 1) contains 0, and the interval for $\sigma_{\tilde{v}}$ (.00050, .0253) nearly contains 0, providing further evidence that the larger model is not needed.

```
confint(modelcl, method="boot", oldNames = F)
```

```
##                                    2.5 %     97.5 %
## sd_(Intercept)|plant            0.4546698  0.643747
## cor_time13.(Intercept)|plant   -0.0227948  0.624198
## sd_time13|plant                 0.0250352  0.042925
## sd_(Intercept)|pot              0.0000000  0.402128
## cor_time13.(Intercept)|pot     -1.0000000  1.000000
## sd_time13|pot                   0.0004998  0.025339
## sigma                           0.2550641  0.311050
## (Intercept)                     1.2413229  1.776788
## time13                          0.0827687  0.117109
```

```
## strl                       -0.3854266  0.216792
## cult                       -0.2519572  0.519487
## rem                        -0.2271840  0.474341
## time13:strl                 0.0370505  0.079057
## time13:cult                -0.0543212 -0.001854
## time13:rem                 -0.0590428 -0.009094
```

10.7 Exploding Variance Components

Our modeling task in Section 10.5 was simplified by the absence of covariates at Level Two. As multilevel models grow to include three or more levels, the addition of just a few covariates at lower levels can lead to a huge increase in the number of parameters (fixed effects and variance components) that must be estimated throughout the model. In this section, we will examine when and why the number of model parameters might explode, and we will consider strategies for dealing with these potentially complex models.

For instance, consider Model C, where we must estimate a total of 15 parameters: 8 fixed effects plus 7 variance components (1 at Level One, 3 at Level Two, and 3 at Level Three). By adding just a single covariate to the equations for a_{ij} and b_{ij} at Level Two in Model C (say, for instance, the size of each seed), we would now have a total of 30 parameters to estimate! The new multilevel model (Model C_plus) could be written as follows:

- Level One (timepoint within plant):

$$Y_{ijk} = a_{ij} + b_{ij}\text{time}_{ijk} + \epsilon_{ijk}$$

- Level Two (plant within pot):

$$a_{ij} = a_i + c_i\text{seedsize}_{ij} + u_{ij}$$
$$b_{ij} = b_i + d_i\text{seedsize}_{ij} + v_{ij}$$

- Level Three (pot):

$$a_i = \alpha_0 + \alpha_1 \text{strl}_i + \alpha_2 \text{cult}_i + \alpha_3 \text{rem}_i + \tilde{u}_i$$
$$b_i = \beta_0 + \beta_1 \text{strl}_i + \beta_2 \text{cult}_i + \beta_3 \text{rem}_i + \tilde{v}_i$$
$$c_i = \gamma_0 + \gamma_1 \text{strl}_i + \gamma_2 \text{cult}_i + \gamma_3 \text{rem}_i + \tilde{w}_i$$
$$d_i = \delta_0 + \delta_1 \text{strl}_i + \delta_2 \text{cult}_i + \delta_3 \text{rem}_i + \tilde{z}_i$$

or as a composite model:

$$\begin{aligned} Y_{ijk} = &[\alpha_0 + \alpha_1 \text{strl}_i + \alpha_2 \text{cult}_i + \alpha_3 \text{rem}_i + \gamma_0 \text{seedsize}_{ij} + \\ &\beta_0 \text{time}_{ijk} + \beta_1 \text{strl}_i \text{time}_{ijk} + \beta_2 \text{cult}_i \text{time}_{ijk} + \beta_3 \text{rem}_i \text{time}_{ijk} + \\ &\gamma_1 \text{strl}_i \text{seedsize}_{ij} + \gamma_2 \text{cult}_i \text{seedsize}_{ij} + \gamma_3 \text{rem}_i \text{seedsize}_{ij} + \\ &\delta_0 \text{seedsize}_{ij} \text{time}_{ijk} + \delta_1 \text{strl}_i \text{seedsize}_{ij} \text{time}_{ijk} + \\ &\delta_2 \text{cult}_i \text{seedsize}_{ij} \text{time}_{ijk} + \delta_3 \text{rem}_i \text{seedsize}_{ij} \text{time}_{ijk}] + \\ &[\tilde{u}_i + u_{ij} + \epsilon_{ijk} + \tilde{w}_i \text{seedsize}_{ij} + \tilde{v}_i \text{time}_{ijk} + v_{ij} \text{time}_{ijk} + \\ &\tilde{z}_i \text{seedsize}_{ij} \text{time}_{ijk}] \end{aligned}$$

where $\epsilon_{ijk} \sim N(0, \sigma^2)$,

$$\begin{bmatrix} u_{ij} \\ v_{ij} \end{bmatrix} \sim N \left(\begin{bmatrix} 0 \\ 0 \end{bmatrix}, \begin{bmatrix} \sigma_u^2 & \\ \sigma_{uv} & \sigma_v^2 \end{bmatrix} \right),$$

and

$$\begin{bmatrix} \tilde{u}_i \\ \tilde{v}_i \\ \tilde{w}_i \\ \tilde{z}_i \end{bmatrix} \sim N \left(\begin{bmatrix} 0 \\ 0 \\ 0 \\ 0 \end{bmatrix}, \begin{bmatrix} \sigma_{\tilde{u}}^2 & & & \\ \sigma_{\tilde{u}\tilde{v}} & \sigma_{\tilde{v}}^2 & & \\ \sigma_{\tilde{u}\tilde{w}} & \sigma_{\tilde{v}\tilde{w}} & \sigma_{\tilde{w}}^2 & \\ \sigma_{\tilde{u}\tilde{z}} & \sigma_{\tilde{v}\tilde{z}} & \sigma_{\tilde{w}\tilde{z}} & \sigma_{\tilde{z}}^2 \end{bmatrix} \right).$$

We would have 16 fixed effects from the four equations at Level Three, each with 4 fixed effects to estimate. And, with four equations at Level Three, the error covariance matrix at the pot level would be 4x4 with 10 variance components to estimate; each error term (4) has a variance associated with it, and each pair of error terms (6) has an associated correlation. The error structure at Levels One (1 variance term) and Two (2 variance terms and 1 correlation) would remain the same, for a total of 14 variance components.

Now consider adding an extra Level One covariate to the model in the previous paragraph. How many model parameters would now need to be estimated? (Try writing out the multilevel models and counting parameters.) The correct answer is 52 total parameters! There are 24 fixed effects (from 6 Level Three equations) and 28 variance components (1 at Level One, 6 at Level Two, and 21 at Level Three).

Estimating even 30 parameters as in Model C_plus from a single set of data is an ambitious task and computationally very challenging. Essentially we (or

10.7 Exploding Variance Components

the statistics package we are using) must determine which combination of values for the 30 parameters would maximize the likelihood associated with our model and the observed data. Even in the absolute simplest case, with only two options for each parameter, there would be over one billion possible combinations to consider! But if our primary interest is in fixed effects, we really only have 5 covariates (and their associated interactions) in our model. What can be done to make fitting a 3-level model with 1 covariate at Level One, 1 at Level Two, and 3 at Level Three more manageable? Reasonable options include:

- Reduce the number of variance components by assuming all error terms to be independent; that is, set all correlation terms to 0.
- Reduce the number of variance components by removing error terms from certain Levels Two and Three equations. Often, researchers will begin with a **random intercepts model**, in which only the first equation at Level Two and Three has an error term. With the leadplant data, we would account for variability in Day 13 height (intercept) among plants and pots, but assume that the effects of time and seed size are constant among pots and plants.
- Reduce the number of fixed effects by removing interaction terms that are not expected to be meaningful. Interaction terms between covariates at different levels can be eliminated simply by reducing the number of terms in certain equations at Levels Two and Three. There is no requirement that all equations at a certain level contain the same set of predictors. Often, researchers will not include covariates in equations beyond the intercept at a certain level unless there's a compelling reason.

By following the options above, our potential 30-parameter model (C_plus) can be simplified to this 9-parameter model:

- Level One:

$$Y_{ijk} = a_{ij} + b_{ij}\text{time}_{ijk} + \epsilon_{ijk}$$

- Level Two:

$$a_{ij} = a_i + c_i\text{seedsize}_{ij} + u_{ij}$$
$$b_{ij} = b_i$$

- Level Three:

$$a_i = \alpha_0 + \alpha_1 \text{strl}_i + \alpha_2 \text{cult}_i + \alpha_3 \text{rem}_i + \tilde{u}_i$$
$$b_i = \beta_0$$
$$c_i = \gamma_0$$

where $\epsilon_{ijk} \sim N(0, \sigma^2)$, $u_{ij} \sim N(0, \sigma_u^2)$, and $\tilde{u}_i \sim N(0, \sigma_{\tilde{u}}^2)$. Or, in terms of a composite model:

$$Y_{ijk} = [\alpha_0 + \alpha_1 \text{strl}_i + \alpha_2 \text{cult}_i + \alpha_3 \text{rem}_i + \gamma_0 \text{seedsize}_{ij} + \beta_0 \text{time}_{ijk}] +$$
$$[\tilde{u}_i + u_{ij} + \epsilon_{ijk}]$$

According to the second option, we have built a random intercepts model with error terms only at the first (intercept) equation at each level. Not only does this eliminate variance terms associated with the missing error terms, but it also eliminates correlation terms between errors (as suggested by Option 1) since there are no pairs of error terms that can be formed at any level. In addition, as suggested by Option 3, we have eliminated predictors (and their fixed effects coefficients) at every equation other than the intercept at each level.

The simplified 9-parameter model essentially includes a random effect for pot ($\sigma_{\tilde{u}}^2$) after controlling for sterilization and soil type, a random effect for plant within pot (σ_u^2) after controlling for seed size, and a random effect for error about the time trend for individual plants (σ^2). We must assume that the effect of time is the same for all plants and all pots, and it does not depend on seed size, sterilization, or soil type. Similarly, we must assume that the effect of seed size is the same for each pot and does not depend on sterilization or soil type. While somewhat proscriptive, a **random intercepts model** such as this can be a sensible starting point, since the simple act of accounting for variability of observational units at Levels Two and Three can produce better estimates of fixed effects of interest and their standard errors.

10.8 Building to a Final Model

In Model C we considered the main effects of soil type and sterilization on leadplant initial height and growth rate, but we did not consider interactions—even though biology researchers expect that sterilization will aid growth in certain soil types more than others. Thus, in Model D we will build Level Three interaction terms into Model C.1:

- Level One:

10.8 Building to a Final Model

$$Y_{ijk} = a_{ij} + b_{ij}\text{time}_{ijk} + \epsilon_{ijk}$$

- Level Two:

$$a_{ij} = a_i + u_{ij}$$
$$b_{ij} = b_i + v_{ij}$$

- Level Three:

$$a_i = \alpha_0 + \alpha_1\text{strl}_i + \alpha_2\text{cult}_i + \alpha_3\text{rem}_i + \alpha_4\text{strl}_i\text{rem}_i + \alpha_5\text{strl}_i\text{cult}_i + \tilde{u}_i$$
$$b_i = \beta_0 + \beta_1\text{strl}_i + \beta_2\text{cult}_i + \beta_3\text{rem}_i + \beta_4\text{strl}_i\text{rem}_i + \beta_5\text{strl}_i\text{cult}_i$$

where error terms are defined as in Model C.1.

From the output below, we see that the interaction terms were not especially helpful, except possibly for a differential effect of sterilization in remnant and reconstructed prairies on the growth rate of leadplants. But it's clear that Model D can be simplified through the removal of certain fixed effects with low t-ratios.

```
# Model D - add interactions to Model C.1
modeld10 <- lmer(hgt ~ time13 + strl + cult + rem +
  time13:strl + time13:cult + time13:rem + strl:cult +
  strl:rem + time13:strl:cult + time13:strl:rem +
  (time13|plant) + (1|pot), REML=T, data=leaddata)
```

```
##  Groups   Name        Variance Std.Dev. Corr
##  plant    (Intercept) 0.29271  0.5410
##           time13      0.00128  0.0357   0.20
##  pot      (Intercept) 0.07056  0.2656
##  Residual             0.08214  0.2866

##  Number of Level Two groups =   107
##  Number of Level Three groups =  32

##                  Estimate Std. Error t value
##  (Intercept)      1.53125  0.153905   9.9493
##  time13           0.09492  0.008119  11.6917
##  strl            -0.12651  0.222671  -0.5682
##  cult            -0.06185  0.329926  -0.1875
##  rem              0.12425  0.239595   0.5186
##  time13:strl      0.07279  0.012005   6.0633
##  time13:cult     -0.02262  0.020836  -1.0858
```

```
## time13:rem          -0.02047  0.013165 -1.5552
## strl:cult            0.27618  0.407111  0.6784
## strl:rem            -0.07139  0.378060 -0.1888
## time13:strl:cult    -0.01608  0.024812 -0.6481
## time13:strl:rem     -0.05199  0.023951 -2.1707
```

To form Model F, we begin by removing all covariates describing the intercept (Day 13 height), since neither sterilization nor soil type nor their interaction appears to be significantly related to initial height. However, sterilization, remnant prairie soil, and their interaction appear to have significant influences on growth rate, although the effect of cultivated soil on growth rate did not appear significantly different from that of restored prairie soil (the reference level). This means we may have a **three-way interaction** in our final composite model between sterilization, remnant prairies, and time. Three-way interactions show that the size of an interaction between two predictors differs depending on the level of a third predictor. Whew!

Our final model (Model F), with its constraints on Level Three error terms, can be expressed level-by-level as:

- Level One:

$$Y_{ijk} = a_{ij} + b_{ij}\text{time}_{ijk} + \epsilon_{ijk}$$

- Level Two:

$$a_{ij} = a_i + u_{ij}$$
$$b_{ij} = b_i + v_{ij}$$

- Level Three:

$$a_i = \alpha_0 + \tilde{u}_i$$
$$b_i = \beta_0 + \beta_1 \text{strl}_i + \beta_2 \text{rem}_i + \beta_3 \text{strl}_i \text{rem}_i$$

where $\epsilon_{ijk} \sim N(0, \sigma^2)$,

$$\begin{bmatrix} u_{ij} \\ v_{ij} \end{bmatrix} \sim N\left(\begin{bmatrix} 0 \\ 0 \end{bmatrix}, \begin{bmatrix} \sigma_u^2 & \\ \sigma_{uv} & \sigma_v^2 \end{bmatrix}\right),$$

and $\tilde{u}_i \sim N(0, \sigma_{\tilde{u}}^2)$.

10.8 Building to a Final Model

In composite form, we have:

$$Y_{ijk} = [\alpha_0 + \beta_0 \text{time}_{ijk} + \beta_1 \text{strl}_i \text{time}_{ijk} + \beta_2 \text{rem}_i \text{time}_{ijk}$$
$$+ \beta_3 \text{strl}_i \text{rem}_i \text{time}_{ijk}] + [\tilde{u}_i + u_{ij} + \epsilon_{ijk} + v_{ij} \text{time}_{ijk}]$$

```
# Model F - simplify and move toward "final model"
modelf10 = lmer(hgt ~ time13 + time13:strl +
  time13:rem + time13:strl:rem +
  (time13|plant) + (1|pot), REML=T, data=leaddata)
```

```
##  Groups   Name         Variance Std.Dev. Corr
##  plant    (Intercept)  0.29404  0.5423
##           time13       0.00143  0.0378   0.16
##  pot      (Intercept)  0.04871  0.2207
##  Residual              0.08205  0.2864

##  Number of Level Two groups = 107
##  Number of Level Three groups = 32

##                    Estimate Std. Error t value
## (Intercept)         1.52909   0.071336  21.435
## time13              0.09137   0.007745  11.798
## time13:strl         0.05967   0.010375   5.751
## time13:rem         -0.01649   0.013238  -1.246
## time13:strl:rem    -0.03939   0.023847  -1.652
```

A likelihood ratio test shows no significant difference between Models D and F (chi-square test statistic = 11.15 on 7 df, p=.1323), supporting the use of simplified Model F.

```
drop_in_dev <- anova(modelf10, modeld10, test = "Chisq")
```

```
          npar   AIC   BIC  logLik    dev  Chisq Df    pval
modelf10    10 581.2 621.4  -280.6  561.2     NA NA      NA
modeld10    17 584.0 652.4  -275.0  550.0  11.15  7  0.1323
```

We mentioned in Section 10.6 that we could also compare Model F and Model D, which differ only in their fixed effects terms, using the parametric bootstrap approach. In fact, in Section 10.6 we suggested that the p-value using the chi-square approximation (.1323) may be a slight under-estimate of the true p-value, but probably in the ballpark. In fact, when we generated 1000 bootstrapped samples of plant heights under Model F, and produced 1000 simulated likelihood ratios comparing Models D and F, we produced a p-value of .201. In Figure 10.16, we see that the chi-square distribution has too much area in the peak and too little area in the tails, although in general it approximates the parametric bootstrap distribution of the likelihood ratio pretty nicely.

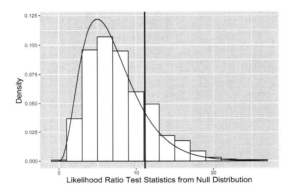

FIGURE 10.16: Null distribution of likelihood ratio test statistic derived using parametric bootstrap (histogram) compared to a chi-square distribution with 7 degrees of freedom (smooth curve). The vertical line represents the observed likelihood ratio test statistic.

```
bootstrapAnova(mA=modeld10, m0=modelf10, B=1000)
```

```
   Df logLik   dev Chisq ChiDf pval_boot
m0 10 -280.6 561.2    NA    NA        NA
mA 17 -275.0 550.0 11.15     7     0.201
```

The effects of remnant prairie soil and the interaction between remnant soil and sterilization appear to have marginal benefit in Model F, so we remove those two terms to create Model E. A likelihood ratio test comparing Models E and F, however, shows that Model F significantly outperforms Model E (chi-square test statistic = 9.40 on 2 df, p=.0090). Thus, we will use Model F as our "Final Model" for generating inference.

10.8 Building to a Final Model

```
drop_in_dev <- anova(modele10, modelf10, test = "Chisq")
```

```
         npar   AIC   BIC logLik   dev Chisq Df
modele10    8 586.6 618.8 -285.3 570.6    NA NA
modelf10   10 581.2 621.4 -280.6 561.2 9.401  2
         pval
modele10   NA
modelf10 0.009091
```

Estimates of model parameters can be interpreted in the following manner:

- $\hat{\sigma} = .287 =$ the standard deviation in within-plant residuals after accounting for time.
- $\hat{\sigma}_u = .543 =$ the standard deviation in Day 13 heights between plants from the same pot.
- $\hat{\sigma}_v = .037 =$ the standard deviation in rates of change in height between plants from the same pot.
- $\hat{\rho}_{uv} = .157 =$ the correlation in plants' Day 13 height and their rate of change in height.
- $\hat{\sigma}_{\tilde{u}} = .206 =$ the standard deviation in Day 13 heights between pots.
- $\hat{\alpha}_0 = 1.529 =$ the mean height for leadplants 13 days after planting.
- $\hat{\beta}_0 = 0.091 =$ the mean daily change in height from 13 to 28 days after planting for leadplants from reconstructed prairies or cultivated lands (rem=0) with no sterilization (strl=0) .
- $\hat{\beta}_1 = 0.060 =$ the increase in mean daily change in height for leadplants from using sterilized soil instead of unsterilized soil in reconstructed prairies or cultivated lands. Thus, leadplants grown in sterilized soil from reconstructed prairies or cultivated lands have an estimated daily increase in height of 0.151 mm.
- $\hat{\beta}_2 = -0.017 =$ the decrease in mean daily change in height for leadplants from using unsterilized soil from remnant prairies, rather than unsterilized soil from reconstructed prairies or cultivated lands. Thus, leadplants grown in unsterilized soil from remnant prairies have an estimated daily increase in height of 0.074 mm.
- $\hat{\beta}_3 = -0.039 =$ the decrease in mean daily change in height for leadplants from sterilized soil from remnant prairies, compared to the expected daily change based on $\hat{\beta}_1$ and $\hat{\beta}_2$. In this case, we might focus on how the interaction between remnant prairies and time differs for unsterilized and sterilized soil. Specifically, the negative effect of remnant prairies on growth rate (compared to reconstructed prairies or cultivated lands) is larger in sterilized soil than unsterilized; in sterilized soil, plants from remnant prairie soil grow .056 mm/day slower on average than plants from other soil types (.095 vs. .151 mm/day), while in unsterilized soil, plants from remnant prairie soil grow just .017 mm/day slower than plants from other soil types (.074 vs. .091

mm/day). Note that the difference between .056 and .017 is our three-way interaction coefficient. Through this three-way interaction term, we also see that leadplants grown in sterilized soil from remnant prairies have an estimated daily increase in height of 0.095 mm.

Based on t-values produced by Model F, sterilization has the most significant effect on leadplant growth, while there is some evidence that growth rate is somewhat slower in remnant prairies, and that the effect of sterilization is also somewhat muted in remnant prairies. Sterilization leads to an estimated 66% increase in growth rate of leadplants from Days 13 to 28 in soil from reconstructed prairies and cultivated lands, and an estimated 28% increase in soil from remnant prairies. In unsterilized soil, plants from remnant prairies grow an estimated 19% slower than plants from other soil types.

10.9 Covariance Structure (optional)

As in Chapter 9, it is important to be aware of the covariance structure implied by our chosen models (focusing initially on Model B). Our three-level model, through error terms at each level, defines a specific covariance structure at both the plant level (Level Two) and the pot level (Level Three). For example, our standard model implies a certain level of correlation among measurements from the same plant and among plants from the same pot. Although three-level models are noticeably more complex than two-level models, it is still possible to systematically determine the implication of our standard model; by doing this, we can evaluate whether our fitted model agrees with our exploratory analyses, and we can also decide if it's worth considering alternative covariance structures.

We will first consider Model B with $\tilde{v}_i$ at Level Three, and then we will evaluate the resulting covariance structure that results from removing $\tilde{v}_i$, thereby restricting $\sigma_{\tilde{v}}^2 = \sigma_{\tilde{u}\tilde{v}} = 0$. The composite version of Model B has been previously expressed as:

$$Y_{ijk} = [\alpha_0 + \beta_0 \text{time}_{ijk}] + [\tilde{u}_i + u_{ij} + \epsilon_{ijk} + (\tilde{v}_i + v_{ij})\text{time}_{ijk}]$$

where $\epsilon_{ijk} \sim N(0, \sigma^2)$,

$$\begin{bmatrix} u_{ij} \\ v_{ij} \end{bmatrix} \sim N\left(\begin{bmatrix} 0 \\ 0 \end{bmatrix}, \begin{bmatrix} \sigma_u^2 & \\ \sigma_{uv} & \sigma_v^2 \end{bmatrix}\right),$$

and

10.9 Covariance Structure (optional)

$$\begin{bmatrix} \tilde{u}_i \\ \tilde{v}_i \end{bmatrix} \sim N\left(\begin{bmatrix} 0 \\ 0 \end{bmatrix}, \begin{bmatrix} \sigma_{\tilde{u}}^2 & \\ \sigma_{\tilde{u}\tilde{v}} & \sigma_{\tilde{v}}^2 \end{bmatrix} \right).$$

In order to assess the implied covariance structure from our standard model, we must first derive variance and covariance terms for related observations (i.e., same timepoint and same plant, different timepoints but same plant, different plants but same pot). Each derivation will rely on the random effects portion of the composite model, since there is no variability associated with fixed effects. For ease of notation, we will let $t_k = \text{time}_{ijk}$, since all plants were planned to be observed on the same 4 days.

The variance for an individual observation can be expressed as:

$$Var(Y_{ijk}) = (\sigma^2 + \sigma_u^2 + \sigma_{\tilde{u}}^2) + 2(\sigma_{uv} + \sigma_{\tilde{u}\tilde{v}})t_k + (\sigma_v^2 + \sigma_{\tilde{v}}^2)t_k^2, \qquad (10.1)$$

and the covariance between observations taken at different timepoints (k and k') from the same plant (j) is:

$$Cov(Y_{ijk}, Y_{ijk'}) = (\sigma_u^2 + \sigma_{\tilde{u}}^2) + (\sigma_{uv} + \sigma_{\tilde{u}\tilde{v}})(t_k + t_{k'}) + (\sigma_v^2 + \sigma_{\tilde{v}}^2)t_k t_{k'}, \quad (10.2)$$

and the covariance between observations taken at potentially different times (k and k') from different plants (j and j') from the same pot (i) is:

$$Cov(Y_{ijk}, Y_{ij'k'}) = \sigma_{\tilde{u}}^2 + \sigma_{\tilde{u}\tilde{v}}(t_k + t_{k'}) + \sigma_{\tilde{v}}^2 t_k t_{k'}. \qquad (10.3)$$

Based on these variances and covariances, the covariance matrix for observations over time from the same plant (j) from pot i can be expressed as the following 4x4 matrix:

$$Cov(\mathbf{Y}_{ij}) = \begin{bmatrix} \tau_1^2 & & & \\ \tau_{12} & \tau_2^2 & & \\ \tau_{13} & \tau_{23} & \tau_3^2 & \\ \tau_{14} & \tau_{24} & \tau_{34} & \tau_4^2 \end{bmatrix},$$

where $\tau_k^2 = Var(Y_{ijk})$ and $\tau_{kk'} = Cov(Y_{ijk}, Y_{ijk'})$. Note that τ_k^2 and $\tau_{kk'}$ are both independent of i and j so that $Cov(\mathbf{Y}_{ij})$ will be constant for all plants from all pots. That is, every plant from every pot will have the same set of variances over the four timepoints and the same correlations between heights at different timepoints. But, the variances and correlations can change depending on the timepoint under consideration as suggested by the presence of t_k terms in Equations (10.1) through (10.3).

Similarly, the covariance matrix between observations from plants j and j' from pot i can be expressed as this 4x4 matrix:

$$Cov(\mathbf{Y}_{ij}, \mathbf{Y}_{ij'}) = \begin{bmatrix} \tilde{\tau}_{11} & & & \\ \tilde{\tau}_{12} & \tilde{\tau}_{22} & & \\ \tilde{\tau}_{13} & \tilde{\tau}_{23} & \tilde{\tau}_{33} & \\ \tilde{\tau}_{14} & \tilde{\tau}_{24} & \tilde{\tau}_{34} & \tilde{\tau}_{44} \end{bmatrix},$$

where $\tilde{\tau}_{kk} = Cov(Y_{ijk}, Y_{ij'k}) = \sigma_{\tilde{u}}^2 + 2\sigma_{\tilde{u}\tilde{v}}t_k + \sigma_{\tilde{v}}^2 t_k^2$ and $\tilde{\tau}_{kk'} = Cov(Y_{ijk}, Y_{ij'k'})$ as derived above. As we saw with $Cov(\mathbf{Y}_{ij})$, $\tilde{\tau}_{kk}$ and $\tilde{\tau}_{kk'}$ are both independent of i and j so that $Cov(\mathbf{Y}_{ij}, \mathbf{Y}_{ij'})$ will be constant for all pairs of plants from all pots. That is, any pair of plants from the same pot will have the same correlations between heights at any two timepoints. As with any covariance matrix, we can convert $Cov(\mathbf{Y}_{ij}, \mathbf{Y}_{ij'})$ into a correlation matrix if desired.

Now that we have the general covariance structure implied by the standard multilevel model in place, we can examine the specific structure suggested by the estimates of variance components in Model B. Restricted maximum likelihood (REML) in Section 10.4 produced the following estimates for variance components: $\hat{\sigma}^2 = .0822$, $\hat{\sigma}_u^2 = .299$, $\hat{\sigma}_v^2 = .00119$, $\hat{\sigma}_{uv} = \hat{\rho}_{uv}\sqrt{\hat{\sigma}_u^2 \hat{\sigma}_v^2} = .00528$, $\hat{\sigma}_{\tilde{u}}^2 = .0442$, $\hat{\sigma}_{\tilde{v}}^2 = .00126$, $\hat{\sigma}_{\tilde{u}\tilde{v}} = \hat{\rho}_{\tilde{u}\tilde{v}}\sqrt{\hat{\sigma}_{\tilde{u}}^2 \hat{\sigma}_{\tilde{v}}^2} = -.00455$. Based on these estimates and the derivations above, the within-plant correlation structure over time is estimated to be:

$$Corr(\mathbf{Y}_{ij}) = \begin{bmatrix} 1 & & & \\ .76 & 1 & & \\ .65 & .82 & 1 & \\ .54 & .77 & .88 & 1 \end{bmatrix}$$

for all plants j and all pots i, and the correlation structure between different plants from the same pot is estimated to be:

$$Corr(\mathbf{Y}_{ij}, \mathbf{Y}_{ij'}) = \begin{bmatrix} .104 & & & \\ .047 & .061 & & \\ -.002 & .067 & .116 & \\ -.037 & .068 & .144 & .191 \end{bmatrix}.$$

The within-plant correlation structure suggests that measurements taken closer in time tend to be more highly correlated than those with more separation in time, and later measurements tend to be more highly correlated than earlier measurements. Examination of standard deviation terms by timepoint suggests that variability increases over time within a plant (estimated SDs of .652, .703, .828, and .999 for Days 13, 18, 23, and 28, respectively). The correlation structure between plants from the same pot depicts a fairly low level of correlation; even for measurements taken at the same timepoint, the largest correlation between plants from the same pot occurs on Day 28 (r=.191) while the smallest correlation occurs on Day 18 (r=.061).

10.9 Covariance Structure (optional)

We can use these results to estimate within-plant and within-pot correlation structure after imposing the same constraints on Model B that we did on Model F (i.e., $\sigma_{\tilde{v}}^2 = \sigma_{\tilde{u}\tilde{v}} = 0$). Using the same REML variance components estimates as above except that $\hat{\sigma}_{\tilde{v}}^2 = 0$ rather than .00126 and $\hat{\sigma}_{\tilde{u}\tilde{v}} = \hat{\rho}_{\tilde{u}\tilde{v}}\sqrt{\hat{\sigma}_{\tilde{u}}^2 \hat{\sigma}_{\tilde{v}}^2} = 0$ rather than $-.00455$, the within-plant correlation structure is estimated to be:

$$Corr(\mathbf{Y}_{ij}) = \begin{bmatrix} 1 & & & \\ .80 & 1 & & \\ .75 & .84 & 1 & \\ .70 & .82 & .88 & 1 \end{bmatrix}$$

for all plants j and all pots i, and the correlation structure between different plants from the same pot is estimated to be:

$$Corr(\mathbf{Y}_{ij}, \mathbf{Y}_{ij'}) = \begin{bmatrix} .104 & & & \\ .095 & .087 & & \\ .084 & .077 & .068 & \\ .073 & .067 & .059 & .052 \end{bmatrix}.$$

Our model restrictions produced slightly higher estimated within-plant correlations, especially for observations separated by longer periods of time. Standard deviation terms by timepoint are very similar (estimated SDs of .652, .713, .806, and .923 for Days 13, 18, 23, and 28, respectively). In terms of the relationship between heights of different plants from the same pot, our model restrictions produced slightly higher correlation estimates with Day 13 height, but slightly lower correlation estimates associated with heights at Days 23 and 28. For measurements taken at the same timepoint, the largest correlation between plants from the same pot now occurs on Day 13 (r=.104) while the smallest correlation now occurs on Day 28 (r=.052). None of the differences in the covariance structure, however, should have a large impact on final conclusions, especially regarding fixed effects, so our strategy for dealing with boundary constraints appears very reasonable. In addition, the covariance structure implied by our standard 3-level model appears to model the correlation structure we observed in Figure 10.13 during our exploratory analyses very nicely. Even the variability over time implied by the standard model matches well with the raw observed variability in height by time period (respective standard deviations of .64, .64, .86, and .91). Thus, we feel well justified in fitting models based on the standard covariance structure.

10.9.1 Details of Covariance Structures

In this section, we present additional details regarding implications of our standard covariance structure for 3-level models. We will focus on Model B; derivations for Model F would proceed in a similar fashion.

The variance for an individual observation can be derived as:

$$Var(Y_{ijk}) = Var(\epsilon_{ijk} + u_{ij} + \tilde{u}_i + (v_{ij} + \tilde{v}_i)\text{time}_{ijk})$$
$$= (\sigma^2 + \sigma_u^2 + \sigma_{\tilde{u}}^2) + 2(\sigma_{uv} + \sigma_{\tilde{u}\tilde{v}})t_k + (\sigma_v^2 + \sigma_{\tilde{v}}^2)t_k^2$$

The covariance between observations taken at different timepoints from the same plant is:

$$Cov(Y_{ijk}, Y_{ijk'}) = Cov(\epsilon_{ijk} + u_{ij} + \tilde{u}_i + (v_{ij} + \tilde{v}_i)t_k,$$
$$\epsilon_{ijk'} + u_{ij} + \tilde{u}_i + (v_{ij} + \tilde{v}_i)t_{k'})$$
$$= (\sigma_u^2 + \sigma_{\tilde{u}}^2) + (\sigma_{uv} + \sigma_{\tilde{u}\tilde{v}})(t_k + t_{k'}) + (\sigma_v^2 + \sigma_{\tilde{v}}^2)t_k t_{k'}$$

The covariance between observations taken from different plants from the same pot is:

$$Cov(Y_{ijk}, Y_{ij'k'}) = Cov(\epsilon_{ijk} + u_{ij} + \tilde{u}_i + (v_{ij} + \tilde{v}_i)t_k,$$
$$\epsilon_{ij'k'} + u_{ij'} + \tilde{u}_i + (v_{ij'} + \tilde{v}_i)t_{k'})$$
$$= \sigma_{\tilde{u}}^2 + \sigma_{\tilde{u}\tilde{v}}(t_k + t_{k'}) + \sigma_{\tilde{v}}^2 t_k t_{k'}$$

Based on these variances and covariances and the expressions for $Cov(\mathbf{Y}_{ij})$ and $Cov(\mathbf{Y}_{ij}, \mathbf{Y}_{ij'})$ in Section 10.9, the complete covariance matrix for observations from pot i can be expressed as the following 24x24 matrix (assuming 4 observations over time for each of 6 plants):

$$Cov(\mathbf{Y}_i) = \begin{bmatrix} Cov(\mathbf{Y}_{i1}) & & & & & \\ Cov(\mathbf{Y}_{i1}, \mathbf{Y}_{i2}) & Cov(\mathbf{Y}_{i2}) & & & & \\ Cov(\mathbf{Y}_{i1}, \mathbf{Y}_{i3}) & Cov(\mathbf{Y}_{i2}, \mathbf{Y}_{i3}) & Cov(\mathbf{Y}_{i3}) & & & \\ Cov(\mathbf{Y}_{i1}, \mathbf{Y}_{i4}) & Cov(\mathbf{Y}_{i2}, \mathbf{Y}_{i4}) & Cov(\mathbf{Y}_{i3}, \mathbf{Y}_{i4}) & Cov(\mathbf{Y}_{i4}) & & \\ Cov(\mathbf{Y}_{i1}, \mathbf{Y}_{i5}) & Cov(\mathbf{Y}_{i2}, \mathbf{Y}_{i5}) & Cov(\mathbf{Y}_{i3}, \mathbf{Y}_{i5}) & Cov(\mathbf{Y}_{i4}, \mathbf{Y}_{i5}) & Cov(\mathbf{Y}_{i5}) & \\ Cov(\mathbf{Y}_{i1}, \mathbf{Y}_{i6}) & Cov(\mathbf{Y}_{i2}, \mathbf{Y}_{i6}) & Cov(\mathbf{Y}_{i3}, \mathbf{Y}_{i6}) & Cov(\mathbf{Y}_{i4}, \mathbf{Y}_{i6}) & Cov(\mathbf{Y}_{i5}, \mathbf{Y}_{i6}) & Cov(\mathbf{Y}_{i6}) \end{bmatrix}.$$

A covariance matrix for our entire data set, therefore, would be block diagonal, with $Cov(\mathbf{Y}_i)$ matrices along the diagonal reflecting within pot correlation and 0's off-diagonal reflecting the assumed independence of observations from plants from different pots. As with any covariance matrix, we can convert the $Cov(\mathbf{Y}_{ij}, \mathbf{Y}_{ij'})$ blocks for two different plants from the same pot into correlation matrices by dividing covariance terms by the product of corresponding standard deviations. Specifically, for $Cov(\mathbf{Y}_{ij}, \mathbf{Y}_{ij'})$, the diagonal terms in a correlation matrix are formed by $Corr(Y_{ijk}, Y_{ij'k}) = \frac{\tilde{\tau}_{kk}}{\sqrt{Var(Y_{ijk})Var(Y_{ij'k})}} = \frac{\tilde{\tau}_{kk}}{\tau_k^2}$ and the off-diagonal terms are formed by $Corr(Y_{ijk}, Y_{ij'k'}) = \frac{\tilde{\tau}_{kk'}}{\sqrt{Var(Y_{ijk})Var(Y_{ij'k'})}} = \frac{\tilde{\tau}_{kk'}}{\tau_k \tau_{k'}}$.

We calculated estimated covariance and correlation matrices within plant and between plants in Section 10.9 based on the standard covariance structure for three-level models. However, it can sometimes be helpful to consider alternative covariance structures and evaluate the robustness of results to changes in assumptions. A couple of natural covariance structures to fit in the Seed Germination case study, given the observed structure in our data, are the heterogeneous compound symmetry and heterogeneous AR(1) models. We fit both structures, along with the toeplitz structure, and compared the resulting models with our standard three-level model. In all cases, the AIC and BIC from the standard model (615.2 and 651.4, respectively) are considerably lower than the corresponding performance measures from the models with alternative covariance structures. Thus, we feel justified in fitting models based on the standard covariance structure.

10.10 Notes on Using R (optional)

When fitting three-level models in `lmer()`, note that an estimated variance at Level One (σ^2) comes for "free", but variance terms at Level Two (σ_u^2) and Level Three ($\sigma_{\tilde{u}}^2$) must be specified separately in Model A through (1 | plant) and (1 | pot). Specifying (time13 | plant) in Model B is equivalent to specifying (1+time13 | plant) and produces a total of 3 variance components to estimate: variances for error terms associated with the intercept (σ_u^2 comes for "free") and slope (σ_v^2 comes from the time13 term) at the plant level, along with a covariance or correlation (σ_{uv} or ρ_{uv}) between those two error terms. To restrict $\sigma_{uv} = \rho_{uv} = 0$, you could specify each error term separately: (1 | plant) + (0+time13 | plant).

Also note that to fit Model C.1 in R, the random error components are written as (time13 | plant) + (1 | pot), indicating you'd like error terms associated with the intercept and slope at the plant level, but only for the intercept term at the pot level. The fixed effects in Model C.1 just reflect all fixed effects in the composite model.

In Section 10.6 we sought to perform a significance test comparing Models C and C.1, where Model C.1 restricted two variance components from Model C to be 0. Our initial attempt used the `anova()` function in R, which created two problems: (a) the `anova()` function uses full maximum likelihood estimates rather than REML estimates of model parameters and performance, which is fine when two models differ in fixed effects but not, as in this case, when two models differ only in random effects; and, (b) the likelihood ratio test statistic is often not well approximated by a chi-square distribution. Therefore, we implemented the parametric bootstrap method to simulate the distribution of the likelihood ratio test statistic and obtain a more reliable p-value, also illustrating that the chi-square distribution would produce an artificially large p-value.

10.11 Exercises

10.11.1 Conceptual Exercises

1. **Seed germination.** In Sections 10.3.1 and 10.3.2, why must we be careful excluding plants with no height data? Why can't we simply leave those plants in our 3-level analysis with all missing heights set to 0 mm?

2. Give an example of a Level Two covariate that might have been recorded in this study.

3. In Figure 10.2, would using mean heights by pot be reasonable as well? How about for Figure 10.3?

4. Explain how "plant-to-plant variability can be estimated by averaging standard deviations from each pot ... while pot-to-pot variability can be estimated by finding the standard deviation of average intercept or slope within pot."

5. Shouldn't we subtract the mean number of days rather than 13 to calculate centered time? Why does the lower correlation between intercepts and slopes produced by centered time result in a more stable model?

6. Explain why an autoregressive error structure is suggested for leadplant data at the end of Section 10.3.2.

7. The experimental factors of interest in the seed germination study are Level Three covariates, yet the unconditional means model shows only 4.6% of total variability in plant heights to be at Level Three. Does that mean a multilevel model will not be helpful? Would it be

10.11 Exercises

better to find the mean height for each pot and just use those 72 values to examine the effects of experimental factors?

8. Explain why a likelihood ratio test is appropriate for comparing Models B and C.

9. Should we be concerned that $\hat{\sigma}_u^2$ increased from Model A to B? Why or why not?

10. Explain the idea of boundary constraints in your own words. Why can it be a problem in multilevel models?

11. In Model C, we initially addressed boundary constraints by removing the Level Three correlation between error terms from our multilevel model. What other model adjustments might we have considered?

12. How does Figure 10.15 show that a likelihood ratio test using a chi-square distribution would be biased?

13. In Section 10.7, a model with 52 parameters is described: (a) illustrate that the model does indeed contain 52 parameters; (b) explain how to minimize the total number of parameters using ideas from Section 10.7; (c) what assumptions have you made in your simplification in (b)?

14. In Section 10.8, Model F (the null model) is compared to Model D using a parametric bootstrap test. As in Section 10.6, show in detail how bootstrapped data would be generated under Model F for, say, Plant # 1 from Pot # 1. For the random parts, tell what distribution the random pieces are coming from and then select a random value from that distribution. Finally, explain how the parametric bootstrap test would be carried out.

15. Section 10.8 contains an interpretation for the coefficient of a three-way interaction term, $\hat{\beta}_3$. Provide an alternative interpretation for $\hat{\beta}_3$ by focusing on how the sterilization-by-soil type interaction differs over time.

16. **Collective efficacy and violent crime.** In a 1997 *Science* article, Sampson et al. [1997] studied the effects on violent crime of a neighborhood's collective efficacy, defined as "social cohesion among neighbors combined with their willingness to intervene on behalf of the common good." Multiple items related to collective efficacy were collected from 8782 Chicago residents from 343 neighborhood clusters. For this study, give the observational units at Levels One, Two, and Three.

17. Table 10.4 shows Table 3 from Sampson et al. [1997]. Provide inter-

TABLE 10.4: Correlates of collective efficacy from Table 3 of Sampson et al. (1997).

Variable	Coefficient	SE	t ratio
Intercept	3.523	0.013	263.20
Person-level predictors			
Female	-0.012	0.015	-0.76
Married	-0.005	0.021	-0.25
Separated or divorced	-0.045	0.026	-1.72
Single	-0.026	0.024	-1.05
Homeowner	**0.122**	0.020	6.04
Latino	0.042	0.028	1.52
Black	-0.029	0.030	-0.98
Mobility	-0.025	0.007	-3.71
Age	**0.0021**	0.0006	3.47
Years in neighborhood	6e-04	0.0008	0.78
SES	**0.035**	0.008	4.64
Neighborhood-level predictors			
Concentrated disadvantage	-0.172	0.016	-10.74
Immigrant concentration	**-0.037**	0.014	-2.66
Residential stability	**0.074**	0.013	5.61
Variance components			
Within neighborhoods	0.32		
Between neighborhoods	0.026		
Percent of variance explained			
Within neighborhoods	3.2		
Between neighborhoods	70.3		

10.11 Exercises

TABLE 10.5: Neighborhood correlates of perceived neighborhood violence from a portion of Table 4 from Sampson et al. (1997).

	Model 1			Model 2		
	Social composition			Social comp and collective efficacy		
Variable	Coefficient	SE	t	Coefficient	SE	t
Concentrated disadvantage	0.277	0.021	13.30	0.171	0.024	7.24
Immigrant concentration	0.041	0.017	2.44	0.018	0.016	1.12
Residential stability	-0.102	0.015	-6.95	-0.056	0.016	-3.49
Collective efficacy				-0.618	0.104	-5.95

pretations of the following coefficients in context: homeowner, age, SES, immigrant concentration, and residential stability.

18. Based on Table 10.4, let Y_{ijk} be the response of person j from neighborhood i to item k regarding collective efficacy; these are (difficulty-adjusted) responses to 10 items per person about collective efficacy. Then (a) write out the three-level model that likely produced this table, and (b) write out the corresponding composite model. Assume there was also an unreported variance component estimating item-to-item variability within person.

19. Suggest valuable exploratory data analysis plots to complement Table 10.4.

20. If the model suggested by Table 10.4 were expanded to include all potential variance components, how many total parameters would need to be estimated? If it were expanded further to include the 3 neighborhood-level covariates as predictors in all Level Three equations, how many total parameters would need to be estimated?

21. At the bottom of Table 10.4, the percent of variance explained is given within and between neighborhoods. Explain what these values likely represent and how they were calculated.

22. Table 10.5 shows a portion of Table 4 from Sampson et al. [1997]. Describe the multilevel model that likely produced this table. State the primary result from this table in context. [Note that collective efficacy is a Level Three covariate in this table, summarized over an entire neighborhood.] Estimates of neighborhood-level coefficients control for gender, marital status, homeownership, ethnicity, mobility, age, years in neighborhood, and SES of those interviewed. Model 1 accounts for 70.5% of the variation between neighborhoods in perceived violence, whereas Model 2 accounts for 77.8% of the variation.

10.11.2 Guided Exercises

1. **Tree tubes.** A student research team at St. Olaf College contributed to the efforts of biologist Dr. Kathy Shea to investigate a rich data set concerning forestation in the surrounding land [Eisinger et al., 2011]. Tubes were placed on trees in some locations or *transects* but not in others. Interest centers on whether tree growth is affected by the presence of tubes. The data is currently stored in long format in treetube.csv. Each row represents one tree in a given year. Key variables include:
   - TRANSECT: The id of the transect housing the tree
   - TUBEX: 1 if the tree had a tube, 0 if not
   - ID: The tree's unique id
   - SPECIES: The tree's species
   - YEAR: Year of the observation
   - HEIGHT: The tree's height in meters

   a. Perform basic exploratory data analysis. For example, which variables are correlated with the heights of the trees?
   b. Explore patterns of missing data. Consider looking for patterns between transects. If you found any patterns of missing data, how might this affect your modeling?
   c. We wish to fit a three-level model for a tree's height. What would be observational units at Level Three? At Level Two? At Level One? What are our Level Three variables? Level Two variables? Level One variables?
   d. Generate spaghetti plots showing how height varies over time for trees which had tubes and for trees which did not have tubes. What do you notice?
   e. Fit Model A, a three-level, unconditional means model for height. Write out the model at levels three, two, and one, as well as the composite model.
   f. Create a new variable, TIME, which represents the number of years since 1990. Use this to fit Model B, an unconditonal growth model for height. Write out this model at levels three, two, and one. What is the correlation of random effects at level two? What does this mean? (Section 10.5 may help.)
   g. In response to part (f), fit a new unconditional growth model (Model C) with $\rho_{uv} = 0$ and $\rho_{\hat{u}\hat{v}} = 0$ (uncorrelated random effects at levels two and three). Write out the model at levels two and three.
   h. If we wanted to compare Model B and Model C, it would not be appropriate to use a likelihood ratio test. Why is this?
   i. Use the parametric bootstrap to compare Model B and Model C. Which model does the test favor?

10.11 Exercises

j. Regardless of your answer to (i), we will build off of Model C for the rest of the analysis. With that in mind, test the hypothesis that trees' growth rates are affected by tubes by adding an interaction between TUBEX and TIME (Model D). Interpret the fitted estimate of this interaction.

k. Perform a likelihood ratio test comparing Model C and Model D. Interpret the results of this test in the context of this study.

2. **Kentucky math scores.** Data was collected from 48,058 eighth graders from Kentucky who took the California Basic Educational Skills Test [Bickel, 2007]. These students attended 235 different middle schools from 132 different districts, and the following variables were collected and can be found in kentucky.csv:
   - dis_id = District Identifier
   - sch_id = School Identifier
   - stud_nm = Student Identifier
   - female = Coded 1 if Female and 0 if Male
   - nonwhite = Coded 0 if White and 1 otherwise
   - readn = California Test of Basic Skills Reading Score
   - mathn = California Test of Basic Skills Math Score
   - sch_size = School-Level Size (centered natural log)
   - sch_ses = School-Level SES (socio-economic status, centered)
   - dis_size = District-Level Size (centered natural log)
   - dis_ses = District-Level SES (socio-economic status, centered)

   The primary research questions are whether or not math scores of eighth graders in Kentucky differ based on gender or ethnicity, and if any gender gap differs by ethnicity, or if any ethnicity gap differs by gender. Researchers wanted to be sure to adjust for important covariates at the school and district levels (e.g., school/district size and socio-economic status).

   a. Conduct a short exploratory data analysis. Which variables are most strongly related to mathn? Illustrate with summary statistics and plots. [Hints: your plots should accommodate the large number of observations in the data, and your summary statistics will have to handle occasional missing values.]

   b. (Model A) Run an unconditional means model with 3 levels and report the percentage of variability in math scores that can be explained at each level.

   c. (Model B) Add female, nonwhite, and their interaction at Level One.
      - Write out the complete three-level model. How many parameters must be estimated?
      - Run the model (be patient – it may take a few minutes!). Report and interpret a relevant pseudo R-squared value. Is

there evidence (based on the t-value) of a significant interaction? In layman's terms, what can you conclude based on the test for interaction?

d. (Model C) Subtract the `female`-by-`nonwhite` interaction from Model B and add `sch_ses`, where `sch_ses` is a predictor in all Level Two equations. How many parameters must be estimated? Break your count down into variance components at Level One + varcomps at Level Two + fixed effects + varcomps at Level Three. (No need to write out the model or run the model unless you want to.)

e. (Model D) Subtract `nonwhite` from Model C and add `dis_size`, where `dis_size` is a predictor in all Level Three equations.
   - Write out the complete three-level model. Also write out the composite model. How many parameters must be estimated?
   - Run the model (be patient!) and then re-run it with an error term only on the intercept equation at Level Three. What are the implications of using this error structure? Does it change any conclusions about fixed effects? Which of the two models would you choose and why?
   - (Optional) Explore the nature of the 3-way interaction between female, sch_ses, and dis_size by finding the predicted math score for 5 cases based on Model D with an error term only on the intercept equation at Level Three. Comment on trends you observe.
     - Female vs. male with average `sch_ses` and `dis_size`
     - Female vs. male with `sch_ses` at Q1 and `dis_size` at Q1
     - Female vs. male with `sch_ses` at Q1 and `dis_size` at Q3
     - Female vs. male with `sch_ses` at Q3 and `dis_size` at Q1
     - Female vs. male with `sch_ses` at Q3 and `dis_size` at Q3
   - (Optional) Create two scatterplots to illustrate the 3-way interaction between `female`, `sch_ses`, and `dis_size`.

10.11.3 Open-Ended Exercises

1. **Seed germination: coneflowers.** Repeat the exploratory data analyses and model fitting from this chapter using coneflowers rather than leadplants.

2. **Mudamalai leaf growth.** Plant growth is influenced by many environmental factors, among them sunlight and water availability.

10.11 Exercises

A study conducted in the Mudamalai Wildlife Sanctuary in India in October of 2008 had as its purpose to "broaden the specific knowledge of certain tree species and climate zones located in the Nilgiri Hills of South India, and to enhance the overall knowledge of the dynamic relationship between plant growth and its environment" [Pray, 2009].

Study researchers collected 1,960 leaves from 45 different trees (5 trees from each of 3 species within each of 3 climate zones). Within each tree, 3 branches were randomly chosen from each of 3 strata (high, medium, and low), and 5 leaves were randomly selected from each branch. Three different descriptive climatic zones were chosen for analysis—dry thorn, dry deciduous, and moist deciduous—and three different species of trees were analyzed—*Cassia fistula* (golden shower tree), *Anogeissus latifolia* (axlewood tree), and *Diospyros montana* (mountain ebony). Height and girth were measured for each tree, and length was assessed for each branch. Measurements taken on each leaf included length, width, surface area (determined carefully for 25 leaves per species and then by linear regression using length and width as predictors for the rest), pedial length (the length of the stem), pedial width, and percent herbivory (an eyeball estimate of the percent of each leaf that had been removed or eaten by herbivores). In addition, stomata density was measured using a compound scope to examine nail polish impressions of leaves for each strata of each tree (135 measurements).

Here is a description of available variables in `mudamalai.csv`:

- `Species` = tree species (*Cassia fistula*, *Anogeissus latifolia*, or *Diospyros montana*)
- `Zone` = climate zone (dry thorn, dry deciduous, or moist deciduous)
- `Tree` = tree number (1-5) within climate zone
- `Tree.height` = tree height (m)
- `Tree.girth` = tree girth (cm)
- `Strata` = height of branch from ground (high, medium, or low)
- `Branch` = branch number (1-3) within tree and strata
- `Branch.length` = length of branch (cm)
- `Length` = length of leaf (cm)
- `Width` = width of leaf (cm)
- `Area` = surface area of leaf (sq cm)
- `Pedial.length` = length of the leaf stem (mm)
- `Pedial.width` = width of the leaf stem (mm)
- `Herbivory` = percent of leaf removed or eaten by herbivores
- `Stomata` = density of specialized openings that allow for gas exchange on leaves

Biology researchers were interested in determining "optimal physical characteristics for growth and survival of these trees in the various areas" to further conservation efforts. Construct a multilevel model to address the researchers' questions, focusing on leaf area as the response of interest. Defend your final model, and interpret fixed effect and variance component estimates produced by your model.

11
Multilevel Generalized Linear Models

11.1 Learning Objectives

After finishing this chapter, students should be able to:
- Recognize when multilevel generalized linear models (multilevel GLMs) are appropriate.
- Understand how multilevel GLMs essentially combine ideas from earlier chapters.
- Apply exploratory data analysis techniques to multilevel data with non-normal responses.
- Recognize when crossed random effects are appropriate and how they differ from nested random effects.
- Write out a multilevel generalized linear statistical model, including assumptions about variance components.
- Interpret model parameters (including fixed effects and variance components) from a multilevel GLM.
- Generate and interpret random effect estimates.

```
# Packages required for Chapter 11
library(gridExtra)
library(lme4)
library(pander)
library(ggmosaic)
library(knitr)
library(kableExtra)
library(tidyverse)
```

11.2 Case Study: College Basketball Referees

An article by Anderson and Pierce [2009] describes empirical evidence that officials in NCAA men's college basketball tend to "even out" foul calls over the course of a game, based on data collected in 2004-2005. Using logistic regression to model the effect of foul differential on the probability that the next foul called would be on the home team (controlling for score differential, conference, and whether or not the home team had the lead), Anderson and Pierce found that "the probability of the next foul being called on the visiting team can reach as high as 0.70." More recently, Moskowitz and Wertheim, in their book *Scorecasting*, argue that the number one reason for the home field advantage in sports is referee bias. Specifically, in basketball, they demonstrate that calls over which referees have greater control—offensive fouls, loose ball fouls, ambiguous turnovers such as palming and traveling—were more likely to benefit the home team than more clearcut calls, especially in crucial situations [Moskowitz and Wertheim, 2011].

Data have been gathered from the 2009-2010 college basketball season for three major conferences to investigate the following questions [Noecker and Roback, 2012]:

- Does evidence that college basketball referees tend to "even out" calls still exist in 2010 as it did in 2005?
- How do results change if our analysis accounts for the correlation in calls from the same game and the same teams?
- Is the tendency to even out calls stronger for fouls over which the referee generally has greater control? Fouls are divided into offensive, personal, and shooting fouls, and one could argue that referees have the most control over offensive fouls (for example, where the player with the ball knocks over a stationary defender) and the least control over shooting fouls (where an offensive player is fouled in the act of shooting).
- Are the actions of referees associated with the score of the game?

11.3 Initial Exploratory Analyses

11.3.1 Data Organization

Examination of data for Case Study 11.2 reveals the following key variables in `basketball0910.csv`:

11.3 Initial Exploratory Analyses

TABLE 11.1: Key variables from the first 10 rows of data from the College Basketball Referees Case Study. Each row represents a different foul called; we see all 8 first-half fouls from Game 1 followed by the first 2 fouls called in Game 2.

game	visitor	hometeam	foul.num	foul.home	foul.diff	score.diff	lead.home	foul.type	time
1	IA	MN	1	0	0	7	1	Personal	14.167
1	IA	MN	2	1	-1	10	1	Personal	11.433
1	IA	MN	3	1	0	11	1	Personal	10.233
1	IA	MN	4	0	1	11	1	Personal	9.733
1	IA	MN	5	0	0	14	1	Shooting	7.767
1	IA	MN	6	0	-1	22	1	Shooting	5.567
1	IA	MN	7	1	-2	25	1	Shooting	2.433
1	IA	MN	8	1	-1	23	1	Offensive	1.000
2	MI	MIST	1	0	0	2	1	Shooting	18.983
2	MI	MIST	2	1	-1	2	1	Personal	17.200

- game = unique game identification number
- date = date game was played (YYYYMMDD)
- visitor = visiting team abbreviation
- hometeam = home team abbreviation
- foul.num = cumulative foul number within game
- foul.home = indicator if foul was called on the home team
- foul.vis = indicator if foul was called on the visiting team
- foul.diff = the difference in fouls before the current foul was called (home - visitor)
- score.diff = the score differential before the current foul was called (home - visitor)
- lead.vis = indicator if visiting team has the lead
- lead.home = indicator if home team has the lead
- previous.foul.home = indicator if previous foul was called on the home team
- previous.foul.vis = indicator if previous foul was called on the visiting team
- foul.type = categorical variable if current foul was offensive, personal, or shooting
- shooting = indicator if foul was a shooting foul
- personal = indicator if foul was a personal foul
- offensive = indicator if foul was an offensive foul
- time = number of minutes left in the first half when foul called

Data was collected for 4972 fouls over 340 games from the Big Ten, ACC, and Big East conference seasons during 2009-2010. We focus on fouls called during the first half to avoid the issue of intentional fouls by the trailing team at the end of games. Table 11.1 illustrates key variables from the first 10 rows of the data set.

For our initial analysis, our primary response variable is foul.home, and our primary hypothesis concerns evening out foul calls. We hypothesize that the probability a foul is called on the home team is inversely related to the foul differential; that is, if more fouls have been called on the home team than the visiting team, the next foul is less likely to be on the home team.

The structure of this data suggests a couple of familiar attributes combined in an unfamiliar way. With a binary response variable, a generalized linear model is typically applied, especially one with a logit link function (indicating logistic regression). But, with covariates at multiple levels—some at the individual foul level and others at the game level—a multilevel model would also be sensible. So what we need is a multilevel model with a non-normal response; in other words, a **multilevel generalized linear model (multilevel GLM)**. We will investigate what such a model might look like in the next section, but we will still begin by exploring the data with initial graphical and numerical summaries.

As with other multilevel situations, we will begin with broad summaries across all 4972 foul calls from all 340 games. Most of the variables we have collected can vary with each foul called; these Level One variables include:

- whether or not the foul was called on the home team (our response variable),
- the game situation at the time the foul was called (the time remaining in the first half, who is leading and by how many points, the foul differential between the home and visiting team, and who the previous foul was called on), and
- the type of foul called (offensive, personal, or shooting).

Level Two variables, those that remain unchanged for a particular game, then include only the home and visiting teams, although we might consider attributes such as attendance, team rankings, etc.

11.3.2 Exploratory Analyses

In Figure 11.1, we see histograms for the continuous Level One covariates (time remaining, foul differential, and score differential). These plots treat each foul within a game as independent even though we expect them to be correlated, but they provide a sense for the overall patterns. We see that time remaining is reasonably uniform. Score differential and foul differential are both bell-shaped, with a mean slightly favoring the home team in both cases – on average, the home team leads by 2.04 points (SD 7.24) and has 0.36 fewer previous fouls (SD 2.05) at the time a foul is called.

Summaries of the categorical response (whether the foul was called on the home team) and categorical Level One covariates (whether the home team has the lead and what type of foul was called) can be provided through tables of

11.3 Initial Exploratory Analyses

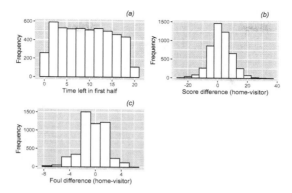

FIGURE 11.1: Histograms showing distributions of the 3 continuous Level One covariates: (a) time remaining, (b) score difference, and (c) foul difference.

proportions. More fouls are called on visiting teams (52.1%) than home teams, the home team is more likely to hold a lead (57.1%), and personal fouls are most likely to be called (51.6%), followed by shooting fouls (38.7%) and then offensive fouls (9.7%).

For an initial examination of Level Two covariates (the home and visiting teams), we can take the number of times, for instance, Minnesota (MN) appears in the long data set (with one row per foul called as illustrated in Table 11.1) as the home team and divide by the number of unique games in which Minnesota is the home team. This ratio (12.1), found in Table 11.2, shows that Minnesota is among the bottom three teams in the average total number of fouls in the first halves of games in which it is the home team. That is, games at Minnesota have few total fouls relative to games played elsewhere. Accounting for the effect of home and visiting team will likely be an important part of our model, since some teams tend to play in games with twice as many fouls called as others, and other teams see a noticeable disparity in the total number of fouls depending on if they are home or away.

Next, we inspect numerical and graphical summaries of relationships between Level One model covariates and our binary model response. As with other multilevel analyses, we will begin by observing broad trends involving all 4972 fouls called, even though fouls from the same game may be correlated. The conditional density plots in the first row of Figure 11.2 examine continuous Level One covariates. Figure 11.2a provides support for our primary hypothesis about evening out foul calls, indicating a very strong trend for fouls to be more often called on the home team at points in the game when more fouls had previously been called on the visiting team. Figures 11.2b and 11.2c then show that fouls were somewhat more likely to be called on the home team when the home team's lead was greater and (very slightly) later in the

TABLE 11.2: Average total number of fouls in the first half over all games in which a particular team is home or visitor. The left columns show the top 3 and bottom 3 teams according to total number of fouls (on both teams) in first halves of games in which they are the home team. The middle columns correspond to games in which the listed teams are the visitors, and the right columns show the largest differences (in both directions) between total fouls in games in which a team is home or visitor.

	Home		Visitor		Difference	
	Team	Fouls	Team	Fouls	Team	Fouls
Top 3	Duke	20.0	WVa	21.4	Duke	4.0
	VaTech	19.4	Nova	19.0	Wisc	2.6
	Nova	19.1	Wake	18.6	Pitt	2.3
Bottom 3	Mich	10.6	Wisc	10.4	WVa	-6.9
	Ill	11.6	Mich	11.1	Mia	-2.7
	MN	12.1	PSU	11.3	Clem	-2.6

half. Conclusions from the conditional density plots in Figures 11.2a-c are supported with associated empirical logit plots in Figures 11.2d-f. If a logistic link function is appropriate, these plots should be linear, and the stronger the linear association, the more promising the predictor. We see in Figure 11.2d further confirmation of our primary hypothesis, with lower log-odds of a foul called on the home team associated with a greater number of previous fouls the home team had accumulated compared to the visiting team. Figure 11.2e shows that game score may play a role in foul trends, as the log-odds of a foul on the home team grows as the home team accumulates a bigger lead on the scoreboard, and Figure 11.2f shows a very slight tendency for greater log-odds of a foul called on the home team as the half proceeds (since points on the right are closer to the beginning of the game).

Conclusions about continuous Level One covariates are further supported by summary statistics calculated separately for fouls called on the home team and those called on the visiting team. For instance, when a foul is called on the home team, there is an average of 0.64 additional fouls on the visitors at that point in the game, compared to an average of 0.10 additional fouls on the visitors when a foul is called on the visiting team. Similarly, when a foul is called on the home team, they are in the lead by an average of 2.7 points, compared to an average home lead of 1.4 points when a foul is called on the visiting team. As expected, the average time remaining in the first half at the time of the foul is very similar for home teams and visitors (9.2 vs. 9.5 minutes, respectively).

11.3 Initial Exploratory Analyses

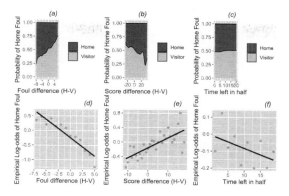

FIGURE 11.2: Conditional density and empirical logit plots of the binary model response (foul called on home or visitor) vs. the three continuous Level One covariates (foul differential, score differential, and time remaining). The dark shading in a conditional density plot shows the proportion of fouls called on the home team for a fixed value of (a) foul differential, (b) score differential, and (c) time remaining. In empirical logit plots, estimated log odds of a home team foul are calculated for each distinct foul (d) and score (e) differential, except for differentials at the high and low extremes with insufficient data; for time (f), estimated log odds are calculated for two-minute time intervals and plotted against the midpoints of those intervals.

The mosaic plots in Figure 11.3 examine categorical Level One covariates, indicating that fouls were more likely to be called on the home team when the home team was leading, when the previous foul was on the visiting team, and when the foul was a personal foul rather than a shooting foul or an offensive foul. A total of 51.8% of calls go against the home team when it is leading the game, compared to only 42.9% of calls when it is behind; 51.3% of calls go against the home team when the previous foul went against the visitors, compared to only 43.8% of calls when the previous foul went against the home team; and, 49.2% of personal fouls are called against the home team, compared to only 46.9% of shooting fouls and 45.7% of offensive fouls. Eventually we will want to examine the relationship between foul type (personal, shooting, or offensive) and foul differential, examining our hypothesis that the tendency to even out calls will be even stronger for calls over which the referees have greater control (personal fouls and especially offensive fouls).

The exploratory analyses presented above are an essential first step in understanding our data, seeing univariate trends, and noting bivariate relationships between variable pairs. However, our important research questions (a) involve the effect of foul differential after adjusting for other significant predictors of which team is called for a foul, (b) account for potential correlation between foul calls within a game (or within a particular home or visiting team), and (c)

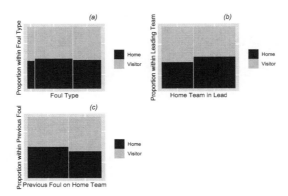

FIGURE 11.3: Mosaic plots of the binary model response (foul called on home or visitor) vs. the three categorical Level One covariates (foul type (a), team in the lead (b), and team called for the previous foul (c)). Each bar shows the percentage of fouls called on the home team vs. the percentage of fouls called on the visiting team for a particular category of the covariate. The bar width shows the proportion of fouls at each of the covariate levels.

determine if the effect of foul differential is constant across game conditions. In order to address research questions such as these, we need to consider multilevel, multivariate statistical models for a binary response variable.

11.4 Two-Level Modeling with a Generalized Response

11.4.1 A GLM Approach

One quick and dirty approach to analysis might be to run a multiple logistic regression model on the entire long data set of 4972 fouls. In fact, Anderson and Pierce ran such a model in their 2009 paper, using the results of their multiple logistic regression model to support their primary conclusions, while justifying their approach by confirming a low level of correlation within games and the minimal impact on fixed effect estimates that accounting for clustering would have. Output from one potential multiple logistic regression model is shown below; this initial modeling attempt shows significant evidence that referees tend to even out calls (i.e., that the probability of a foul called on the home team decreases as total home fouls increase compared to total visiting team fouls—that is, as `foul.diff` increases) after accounting for score differential and time remaining (Z=-3.078, p=.002). The extent of the effect of

11.4 Two-Level Modeling with a Generalized Response

foul differential also appears to grow (in a negative direction) as the first half goes on, based on an interaction between time remaining and foul differential (Z=-2.485, p=.013). We will compare this model with others that formally account for clustering and correlation patterns in our data.

```
# Logistic regression model (not multilevel)
mod0 = glm(foul.home ~ foul.diff + score.diff + lead.home +
    time + foul.diff:time + lead.home:time,
    family = binomial, data = refdata)
```

```
##                   Estimate Std. Error z value Pr(>|z|)
## (Intercept)      -0.103416   0.101056 -1.0234 0.306138
## foul.diff        -0.076599   0.024888 -3.0777 0.002086
## score.diff        0.020062   0.006660  3.0123 0.002593
## lead.home        -0.093227   0.160082 -0.5824 0.560318
## time             -0.013141   0.007948 -1.6533 0.098271
## foul.diff:time   -0.007492   0.003014 -2.4855 0.012938
## lead.home:time    0.021837   0.011161  1.9565 0.050404

##    Residual deviance =    6748   on   4965 df
```

11.4.2 A Two-Stage Modeling Approach

As we saw in Section 8.4.2, to avoid clustering we could consider fitting a separate regression model to each unit of observation at Level Two (each game in this case). Since our primary response variable is binary (Was the foul called on the home team or not?), we would fit a logistic regression model to the data from each game. For example, consider the 14 fouls called during the first half of the March 3, 2010, game featuring Virginia at Boston College (Game 110) in Table 11.3.

Is there evidence from this game that referees tend to "even out" foul calls when one team starts to accumulate more fouls? Is the score differential associated with the probability of a foul on the home team? Is the effect of foul differential constant across all foul types during this game? We can address these questions through multiple logistic regression applied to only data from Game 110.

First, notation must be defined. Let Y_{ij} be an indicator variable recording if the jth foul from Game i was called on the home team (1) or the visiting team (0). We can consider Y_{ij} to be a Bernoulli random variable with parameter p_{ij}, where p_{ij} is the true probability that the jth foul from Game i was called on the home team. As in Chapter 6, we will begin by modeling the logit function of p_{ij} for Game i as a linear function of Level One covariates. For instance, if we were to consider a simple model with foul differential as the sole predictor,

TABLE 11.3: Key variables from the March 3, 2010, game featuring Virginia at Boston College (Game 110).

visitor	hometeam	foul.num	foul.home	foul.diff	score.diff	foul.type	time
VA	BC	1	0	0	0	Shooting	19.783
VA	BC	2	0	-1	4	Personal	18.950
VA	BC	3	0	-2	7	Shooting	16.917
VA	BC	4	0	-3	10	Personal	14.883
VA	BC	5	1	-4	10	Personal	14.600
VA	BC	6	1	-3	6	Offensive	9.750
VA	BC	7	1	-2	4	Shooting	9.367
VA	BC	8	0	-1	4	Personal	9.200
VA	BC	9	0	-2	6	Shooting	7.667
VA	BC	10	0	-3	12	Shooting	2.500
VA	BC	11	1	-4	13	Personal	2.083
VA	BC	12	1	-3	13	Personal	1.967
VA	BC	13	0	-2	12	Shooting	1.733
VA	BC	14	1	-3	14	Personal	0.700

we could model the probability of a foul on the home team in Game 110 with the model:

$$\log\left(\frac{p_{110j}}{1 - p_{110j}}\right) = a_{110} + b_{110}\text{foul.diff}_{110j} \quad (11.1)$$

where i is fixed at 110. Note that there is no separate error term or variance parameter, since the variance is a function of p_{ij} with a Bernoulli random variable.

Maximum likelihood estimators for the parameters in this model (a_{110} and b_{110}) can be obtained through statistical software. $e^{a_{110}}$ represents the odds that a foul is called on the home team when the foul totals in Game 110 are even, and $e^{b_{110}}$ represents the multiplicative change in the odds that a foul is called on the home team for each additional foul for the home team relative to the visiting team during the first half of Game 110.

For Game 110, we estimate $\hat{a}_{110} = -5.67$ and $\hat{b}_{110} = -2.11$ (see output below). Thus, according to our simple logistic regression model, the odds that a foul is called on the home team when both teams have an equal number of fouls in Game 110 is $e^{-5.67} = 0.0035$; that is, the probability that a foul is called on the visiting team (0.9966) is $1/0.0035 = 289$ times higher than the probability a foul is called on the home team (0.0034) in that situation. While these parameter estimates seem quite extreme, reliable estimates are difficult to obtain with 14 observations and a binary response variable, especially in a case like this where the fouls were only even at the start of the game. Also, as the gap between home and visiting fouls increases by 1, the odds that a foul is

11.4 Two-Level Modeling with a Generalized Response

called on the visiting team increases by a multiplicative factor of more than 8 (since $1/e^{-2.11} = 8.25$). In Game 110, this trend toward referees evening out foul calls is statistically significant at the 0.10 level (Z=-1.851, p=.0642).

```
lreg.game110 <- glm(foul.home ~ foul.diff,
                    family = binomial, data = game110)
```

```
##                 Estimate Std. Error z value Pr(>|z|)
## (Intercept)     -5.668       3.131   -1.810   0.07030
## foul.diff       -2.110       1.140   -1.851   0.06415

##  Residual deviance =  11.75   on    12  df
```

As in Section 8.4.2, we can proceed to fit similar logistic regression models for each of the 339 other games in our data set. Each model will yield a new estimate of the intercept and the slope from Equation (11.1). These estimates are summarized graphically in Figure 11.4. There is noticeable variability among the 340 games in the fitted intercepts and slopes, with an IQR for intercepts ranging from -1.45 to 0.70, and an IQR for slopes ranging from -1.81 to -0.51. The majority of intercepts are below 0 (median = -0.41), so that in most games a foul is less likely to be called on the home team when foul totals are even. In addition, almost all of the estimated slopes are below 0 (median = -0.89), indicating that as the home team's foul total grows in relation to the visiting team's foul total, the odds of a foul on the home team continues to decrease.

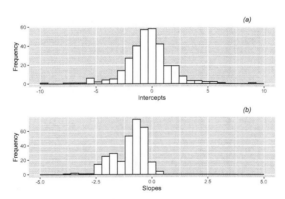

FIGURE 11.4: Histograms of (a) intercepts and (b) slopes from fitting simple logistic regression models by game. Several extreme outliers have been cut off in these plots for illustration purposes.

At this point, you might imagine expanding model building efforts in a couple of directions: (a) continue to improve the Level One model in Equation (11.1) by controlling for covariates and adding potential interaction terms, or (b) build Level Two models to explain the game-to-game variability in intercepts or slopes using covariates which remain constant from foul to foul within a game (like the teams playing). While we could pursue these two directions independently, we can accomplish our modeling goals in a much cleaner and more powerful way by proceeding as in Chapter 8 and building a unified multilevel framework under which all parameters are estimated simultaneously and we remain faithful to the correlation structure inherent in our data.

11.4.3 A Unified Multilevel Approach

As in Chapters 8 and 9, we will write out a composite model after first expressing Level One and Level Two models. That is, we will create Level One and Level Two models as in Section 11.4.2, but we will then combine those models into a composite model and estimate all model parameters simultaneously. Once again Y_{ij} is an indicator variable recording if the jth foul from Game i was called on the home team (1) or the visiting team (0), and p_{ij} is the true probability that the jth foul from Game i was called on the home team. Our Level One model with foul differential as the sole predictor is given by Equation (11.1) generalized to Game i:

$$\log\left(\frac{p_{ij}}{1-p_{ij}}\right) = a_i + b_i \text{foul.diff}_{ij}$$

Then we include no fixed covariates at Level Two, but we include error terms to allow the intercept and slope from Level One to vary by game, and we allow these errors to be correlated:

$$a_i = \alpha_0 + u_i$$
$$b_i = \beta_0 + v_i,$$

where the error terms at Level Two can be assumed to follow a multivariate normal distribution:

$$\begin{bmatrix} a_i \\ b_i \end{bmatrix} \sim N\left(\begin{bmatrix} 0 \\ 0 \end{bmatrix}, \begin{bmatrix} \sigma_u^2 & \\ \sigma_{uv} & \sigma_v^2 \end{bmatrix}\right)$$

Our composite model then looks like:

$$\log\left(\frac{p_{ij}}{1-p_{ij}}\right) = a_i + b_i \text{foul.diff}_{ij}$$
$$= (\alpha_0 + u_i) + (\beta_0 + v_i)\text{foul.diff}_{ij}$$
$$= [\alpha_0 + \beta_0 \text{foul.diff}_{ij}] + [u_i + v_i \text{foul.diff}_{ij}]$$

11.4 Two-Level Modeling with a Generalized Response

The major changes when moving from a normally distributed response to a binary response are the form of the response variable (a logit function) and the absence of an error term at Level One.

Again, we can use statistical software to obtain parameter estimates for this unified multilevel model using all 4972 fouls recorded from the 340 games. For example, the glmer() function from the lme4 package in R extends the lmer() function to handle generalized responses and to account for the fact that fouls are not independent within games. Results are given below for the two-level model with foul differential as the sole covariate and Game as the Level Two observational unit.

```
# Multilevel model with only foul.diff and errors on slope
#    and int and 1 RE
model.b1 <- glmer(foul.home ~ foul.diff + (foul.diff|game),
            family = binomial, data = refdata)
```

```
##  Groups Name        Variance Std.Dev. Corr
##  game   (Intercept) 0.29414  0.5424
##         foul.diff   0.00124  0.0351   -1.00

##  Number of Level Two groups =  340

##              Estimate Std. Error z value  Pr(>|z|)
## (Intercept)  -0.1568    0.04637   -3.382  7.190e-04
## foul.diff    -0.2853    0.03835   -7.440  1.006e-13

##   AIC =   6791 ;  BIC =   6824
```

When parameter estimates from the multilevel model above are compared with those from the naive logistic regression model assuming independence of all observations (below), there are noticeable differences. For instance, each additional foul for the visiting team is associated with a 33% increase ($1/e^{-.285}$) in the odds of a foul called on the home team under the multilevel model, but the single level model estimates the same increase as only 14% ($1/e^{-.130}$). Also, estimated standard errors for fixed effects are greater under multilevel generalized linear modeling, which is not unusual after accounting for correlated observations, which effectively reduces the sample size.

```
# Logistic regression model (not multilevel) with
#   only foul.diff
mod0a <- glm(foul.home ~ foul.diff, family = binomial,
            data = refdata)
```

```
##              Estimate Std. Error z value  Pr(>|z|)
## (Intercept)   -0.1300    0.02912  -4.466 7.983e-06
## foul.diff     -0.1305    0.01426  -9.148 5.804e-20

## Residual deviance =  6798  on  4970 df
```

11.5 Crossed Random Effects

In the College Basketball Referees case study, our two primary Level Two covariates are home team and visiting team. In Section 11.3.2 we showed evidence that the probability a foul is called on the home team changes if we know precisely who the home and visiting teams are. However, if we were to include an indicator variable for each distinct team, we would need 38 indicator variables for home teams and 38 more for visiting teams. That's a lot of degrees of freedom to spend! And adding those indicator variables would complicate our model considerably, despite the fact that we're not very interested in specific coefficient estimates for each team—we just want to control for teams playing to draw stronger conclusions about referee bias (focusing on the foul.diff variable). One way, then, of accounting for teams is by treating them as **random effects** rather than fixed effects. For instance, we can assume the effect that Minnesota being the home team has on the log odds of foul call on the home team is drawn from a normal distribution with mean 0 and variance σ_v^2. Maybe the estimated random effect for Minnesota as the home team is -0.5, so that Minnesota is slightly less likely than the typical home team to have fouls called on it, and the odds of a foul on the home team are somewhat correlated whenever Minnesota is the home team. In this way, we account for the complete range of team effects, while only having to estimate a single parameter (σ_v^2) rather than coefficients for 38 indicator variables. The effect of visiting teams can be similarly modeled.

How will treating home and visiting teams as random effects change our multilevel model? Another way we might view this situation is by considering that Game is not the only Level Two observational unit we might have selected. What if we instead decided to focus on Home Team as the Level Two observational unit? That is, what if we assumed that fouls called on the same home team across all games must be correlated? In this case, we could redefine our Level One model from Equation (11.1). Let Y_{hj} be an indicator variable recording if the jth foul from Home Team h was called on the home team (1) or the visiting team (0), and p_{hj} be the true probability that the jth foul from Home Team h was called on the home team. Now, if we were to consider a simple model with foul differential as the sole predictor, we could model the probability of a foul on the home team for Home Team h with the model:

11.5 Crossed Random Effects

$$\log\left(\frac{p_{hj}}{1-p_{hj}}\right) = a_h + b_h \text{foul.diff}_{hj}$$

In this case, e^{a_h} represents the odds that a foul is called on the home team when total fouls are equal between both teams in a game involving Home Team h, and e^{b_h} represents the multiplicative change in the odds that a foul is called on the home team for every extra foul on the home team compared to the visitors in a game involving Home Team h. After fitting logistic regression models for each of the 39 teams in our data set, we see in Figure 11.5 variability in fitted intercepts (mean=-0.15, sd=0.33) and slopes (mean=-0.22, sd=0.12) among the 39 teams, although much less variability than we observed from game-to-game. Of course, each logistic regression model for a home team was based on about 10 times more foul calls than each model for a game, so observing less variability from team-to-team was not unexpected.

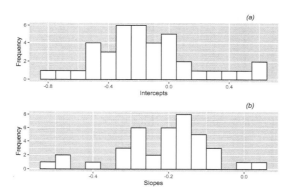

FIGURE 11.5: Histograms of (a) intercepts and (b) slopes from fitting simple logistic regression models by home team.

From a modeling perspective, accounting for clustering by game and by home team (not to mention by visiting team) brings up an interesting issue we have not yet considered—can we handle random effects that are not nested? Since each foul called is associated with only one game (or only one home team and one visiting team), foul is considered nested in game (or home or visiting team). However, a specific home team is not associated with a single game; that home team appears in several games. Therefore, any effects of game, home team, and visiting team are considered **crossed random effects**.

A two-level model which accounts for variability among games, home teams, and visiting teams would take on a slightly new look. First, the full subscripting would change a bit. Our primary response variable would now be written as $Y_{i[gh]j}$, an indicator variable recording if the jth foul from Game i was called on the home team (1) or the visiting team (0), where Game i pitted Visiting

Team g against Home Team h. Square brackets are introduced since g and h are essentially at the same level as i, whereas we have assumed (without stating so) throughout this book that subscripting without square brackets implies a movement to lower levels as the subscripts move left to right (e.g., ij indicates i units are at Level Two, while j units are at Level One, nested inside Level Two units). We can then consider $Y_{i[gh]j}$ to be a Bernoulli random variable with parameter $p_{i[gh]j}$, where $p_{i[gh]j}$ is the true probability that the jth foul from Game i was called on Home Team h rather than Visiting Team g. We will include the crossed subscripting only where necessary.

Typically, with the addition of crossed effects, the Level One model will remain familiar and changes will be seen at Level Two, especially in the equation for the intercept term. In the model formulation below we allow, as before, the slope and intercept to potentially vary by game:

- Level One:

$$\log\left(\frac{p_{i[gh]j}}{1 - p_{i[gh]j}}\right) = a_i + b_i \text{foul.diff}_{ij} \quad (11.2)$$

- Level Two:

$$a_i = \alpha_0 + u_i + v_h + w_g$$
$$b_i = \beta_0,$$

Therefore, at Level Two, we assume that a_i, the log odds of a foul on the home team when the home and visiting teams in Game i have an equal number of fouls, depends on four components:

- α_0 is the population average log odds across all games and fouls (fixed)
- u_i is the effect of Game i (random)
- v_h is the effect of Home Team h (random)
- w_g is the effect of Visiting Team g (random)

where error terms (random effects) at Level Two can be assumed to follow independent normal distributions:

$$u_i \sim N\left(0, \sigma_u^2\right)$$
$$v_h \sim N\left(0, \sigma_v^2\right)$$
$$w_g \sim N\left(0, \sigma_w^2\right).$$

We could include terms that vary by home or visiting team in other Level

11.5 Crossed Random Effects

Two equations, but often adjusting for these random effects on the intercept is sufficient. The advantages to including additional random effects are three-fold. First, by accounting for additional sources of variability, we should obtain more precise estimates of other model parameters, including key fixed effects. Second, we obtain estimates of variance components, allowing us to compare the relative sizes of game-to-game and team-to-team variability. Third, as outlined in Section 11.8, we can obtain estimated random effects which allow us to compare the effects on the log-odds of a home foul of individual home and visiting teams.

Our composite model then looks like:

$$\log\left(\frac{p_{i[gh]j}}{1-p_{i[gh]j}}\right) = [\alpha_0 + \beta_0 \text{foul.diff}_{ij}] + [u_i + v_h + w_g].$$

We will refer to this as Model A3, where we look at the effect of foul differential on the odds a foul is called on the home team, while accounting for three crossed random effects at Level Two (game, home team, and visiting team). Parameter estimates for Model A3 are given below:

```
# Model A3
model.a3 <- glmer(foul.home ~ foul.diff + (1|game) +
    (1|hometeam) + (1|visitor),
    family = binomial, data = refdata)
```

```
##  Groups    Name         Variance  Std.Dev.
##  game      (Intercept)  0.1716    0.414
##  hometeam  (Intercept)  0.0681    0.261
##  visitor   (Intercept)  0.0232    0.152

##  Number of games = 340
##  Number of hometeams = 39
##  Number of visitors = 39

##                Estimate  Std. Error  z value  Pr(>|z|)
## (Intercept)    -0.1878   0.06331     -2.967   3.011e-03
## foul.diff      -0.2638   0.03883     -6.795   1.085e-11

##  AIC = 6780 ; BIC = 6813
```

From this output, we obtain estimates of our five model parameters:

- $\hat{\alpha}_0 = -0.188 =$ the mean log odds of a home foul at the point where total fouls are equal between teams. In other words, when fouls are balanced between teams, the probability that a foul is called on the visiting team (.547) is 20.7% ($1/e^{-.188} = 1.207$) higher than the probability a foul is called on the home team (.453).
- $\hat{\beta}_0 = -0.264 =$ the decrease in mean log odds of a home foul for each 1 foul increase in the foul differential. More specifically, the odds the next foul is called on the visiting team rather than the home team increases by 30.2% with each additional foul called on the home team ($1/e^{-.264} = 1.302$).
- $\hat{\sigma}_u^2 = 0.172 =$ the variance in intercepts from game-to-game.
- $\hat{\sigma}_v^2 = 0.068 =$ the variance in intercepts among different home teams.
- $\hat{\sigma}_w^2 = 0.023 =$ the variance in intercepts among different visiting teams.

Based on the t-value (-6.80) and p-value ($p < .001$) associated with foul differential in this model, we have significant evidence of a negative association between foul differential and the odds of a home team foul. That is, we have significant evidence that the odds that a foul is called on the home team shrinks as the home team has more total fouls compared with the visiting team. Thus, there seems to be preliminary evidence in the 2009-2010 data that college basketball referees tend to even out foul calls over the course of the first half. Of course, we have yet to adjust for other significant covariates.

11.6 Parametric Bootstrap for Model Comparisons

Our estimates of variance components provide evidence of the relative variability in the different Level Two random effects. For example, an estimated 65.4% of variability in the intercepts is due to differences from game-to-game, while 25.9% is due to differences among home teams, and 8.7% is due to differences among visiting teams. At this point, we could reasonably ask: if we use a random effect to account for differences among games, does it really pay off to also account for which team is home and which is the visitor?

To answer this, we could compare models with and without the random effects for home and visiting teams (i.e., Model A3, the full model, vs. Model A1, the reduced model) using a likelihood ratio test. In Model A1, Level Two now looks like:

$$a_i = \alpha_0 + u_i$$
$$b_i = \beta_0,$$

The likelihood ratio test (see below) provides significant evidence (LRT=16.074,

11.6 Parametric Bootstrap for Model Comparisons

df=2, p=.0003) that accounting for variability among home teams and among visiting teams improves our model.

```
drop_in_dev <- anova(model.a3, model.a1, test = "Chisq")
```

```
         npar  AIC  BIC logLik  dev Chisq Df      pval
model.a1    3 6793 6812  -3393 6787    NA NA        NA
model.a3    5 6780 6813  -3385 6770 16.07  2 0.0003233
```

In Section 8.5.4 we noted that REML estimation is typically used when comparing models that differ only in random effects for multilevel models with normal responses, but with multilevel generalized linear models, full maximum likelihood (ML) estimation procedures are typically used regardless of the situation (including in the function `glmer()` in R). So the likelihood ratio test above is based on ML methods—can we trust its accuracy? In Section 9.6.4 we introduced the parametric bootstrap as a more reliable method in many cases for conducting hypothesis tests compared to the likelihood ratio test, especially when comparing full and reduced models that differ in their variance components. What would a parametric bootstrap test say about testing $H_0 : \sigma_v^2 = \sigma_w^2 = 0$ vs. H_A : at least one of σ_v^2 or σ_w^2 is not equal to 0? Under the null hypothesis, since the two variance terms are being set to a value (0) on the boundary constraint, we would not expect the chi-square distribution to adequately approximate the behavior of the likelihood ratio test statistic.

Figure 11.6 illustrates the null distribution of the likelihood ratio test statistic derived by the parametric bootstrap procedure with 100 samples as compared to a chi-square distribution. As we observed in Section 9.6.4, the parametric bootstrap provides a more reliable p-value in this case ($p < .001$ from output below) because a chi-square distribution puts too much mass in the tail and not enough near 0. However, the parametric bootstrap is computationally intensive, and it can take a long time to run even with moderately complex models. With this data, we would select our full Model A3 based on a parametric bootstrap test.

```
bootstrapAnova(mA=model.a3, m0=model.a1, B=1000)
```

```
   Df logLik  dev Chisq ChiDf pval_boot
m0  3  -3393 6787    NA    NA        NA
mA  5  -3385 6770 16.07     2         0
```

We might also reasonably ask: is it helpful to allow slopes (coefficients for foul

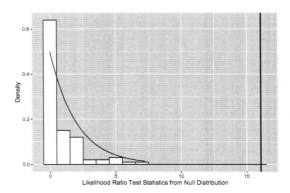

FIGURE 11.6: Null distribution of likelihood ratio test statistic comparing Models A3 and A1 derived using parametric bootstrap with 100 samples (histogram) compared to a chi-square distribution with 2 degrees of freedom (smooth curve). The vertical line represents the observed likelihood ratio test statistic.

differential) to vary by game, home team, and visiting team as well? Again, since we are comparing models that differ in random effects, and since the null hypothesis involves setting random effects at their boundaries, we use the parametric bootstrap. Formally, we are comparing Model A3 to Model B3, which has the same Level One equation as Model A3:

$$\log\left(\frac{p_{i[gh]j}}{1 - p_{i[gh]j}}\right) = a_i + b_i \text{foul.diff}_{ij}$$

but 6 variance components to estimate at Level Two:

$$a_i = \alpha_0 + u_i + v_h + w_g$$
$$b_i = \beta_0 + z_i + r_h + s_g,$$

where error terms (random effects) at Level Two can be assumed to follow independent normal distributions:

$$u_i \sim N\left(0, \sigma_u^2\right)$$
$$z_i \sim N\left(0, \sigma_z^2\right)$$
$$v_h \sim N\left(0, \sigma_v^2\right)$$
$$r_h \sim N\left(0, \sigma_r^2\right)$$
$$w_g \sim N\left(0, \sigma_w^2\right)$$
$$s_g \sim N\left(0, \sigma_s^2\right).$$

11.6 Parametric Bootstrap for Model Comparisons

Thus our null hypothesis for comparing Model A3 vs. Model B3 is $H_0 : \sigma_z^2 = \sigma_r^2 = \sigma_s^2 = 0$. We do not have significant evidence (LRT=0.349, df=3, p=.46 by parametric bootstrap) of variability among slopes, so we will only include random effects for game, home team, and visiting team for the intercept going forward. Figure 11.7 illustrates the null distribution of the likelihood ratio test statistic derived by the parametric bootstrap procedure as compared to a chi-square distribution, again showing that the tails are too heavy in the chi-square distribution.

```
bootstrapAnova(mA=model.b3, m0=model.a3, B=1000)
```

```
    Df logLik  dev  Chisq ChiDf pval_boot
m0   5  -3385 6770    NA    NA        NA
mA   8  -3385 6770 0.349     3      0.46
```

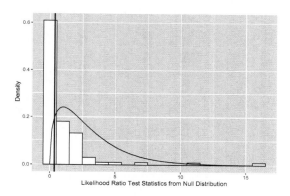

FIGURE 11.7: Null distribution of likelihood ratio test statistic comparing Models A3 and B3 derived using parametric bootstrap with 100 samples (histogram) compared to a chi-square distribution with 3 degrees of freedom (smooth curve). The vertical line represents the observed likelihood ratio test statistic.

Note that we could have also allowed for a correlation between the error terms for the intercept and slope by game, home team, or visiting team – i.e., assume, for example:

$$\begin{bmatrix} u_i \\ z_i \end{bmatrix} \sim N\left(\begin{bmatrix} 0 \\ 0 \end{bmatrix}, \begin{bmatrix} \sigma_u^2 & \\ \sigma_{uz} & \sigma_z^2 \end{bmatrix}\right)$$

while error terms by game, home team, or visiting team are still independent. Here, the new model would have 6 additional parameters when compared with Model A3 (3 variance terms and 3 covariance terms). By the parametric bootstrap, there is no significant evidence that the model with 6 additional parameters is necessary (LRT=6.49, df=6, p=.06 by parametric bootstrap). The associated p-value based on a likelihood ratio test with approximate chi-square distribution (and restricted maximum likelihood estimation) is .370, reflecting once again the overly heavy tails of the chi-square distribution.

11.7 A Final Model for Examining Referee Bias

In constructing a final model for this college basketball case study, we are guided by several considerations. First, we want to estimate the effect of foul differential on the odds of a home foul, after adjusting for important covariates. Second, we wish to find important interactions with foul differential, recognizing that the effect of foul differential might depend on the game situation and other covariates. Third, we want to account for random effects associated with game, home team, and visiting team. What follows is one potential final model that follows these guidelines:

- Level One:

$$\log\left(\frac{p_{i[gh]j}}{1-p_{i[gh]j}}\right) = a_i + b_i\text{foul.diff}_{ij} + c_i\text{score.diff}_{ij} + d_i\text{lead.home}_{ij}$$
$$+ f_i\text{time}_{ij} + k_i\text{offensive}_{ij} + l_i\text{personal}_{ij}$$
$$+ m_i\text{foul.diff}_{ij}\text{ offensive}_{ij} + n_i\text{foul.diff}_{ij}\text{ personal}_{ij}$$
$$+ o_i\text{foul.diff}_{ij}\text{ time}_{ij} + q_i\text{lead.home}_{ij}\text{ time}_{ij}$$

- Level Two:

11.7 A Final Model for Examining Referee Bias

$$a_i = \alpha_0 + u_i + v_h + w_g$$
$$b_i = \beta_0$$
$$c_i = \gamma_0$$
$$d_i = \delta_0$$
$$f_i = \phi_0$$
$$k_i = \kappa_0$$
$$l_i = \lambda_0$$
$$m_i = \mu_0$$
$$n_i = \nu_0$$
$$o_i = \omega_0$$
$$q_i = \xi_0,$$

where error terms at Level Two can be assumed to follow independent normal distributions:

$$u_i \sim N\left(0, \sigma_u^2\right)$$
$$v_h \sim N\left(0, \sigma_v^2\right)$$
$$w_g \sim N\left(0, \sigma_w^2\right).$$

Our composite model then looks like:

$$\log\left(\frac{p_{i[gh]j}}{1 - p_{i[gh]j}}\right) = [\alpha_0 + \beta_0 \text{foul.diff}_{ij} + \gamma_0 \text{score.diff}_{ij} + \delta_0 \text{lead.home}_{ij}$$
$$+ \phi_0 \text{time}_{ij} + \kappa_0 \text{offensive}_{ij} + \lambda_0 \text{personal}_{ij}$$
$$+ \mu_0 \text{foul.diff}_{ij} \text{offensive}_{ij} + \nu_0 \text{foul.diff}_{ij} \text{personal}_{ij}$$
$$+ \omega_0 \text{foul.diff}_{ij} \text{time}_{ij} + \xi_0 \text{lead.home}_{ij} \text{time}_{ij}]$$
$$+ [u_i + v_h + w_g].$$

Using the composite form of this multilevel generalized linear model, the parameter estimates for our 11 fixed effects and 3 variance components are given in the output below:

```
# Model F (potential final model)
model.f <- glmer(foul.home ~ foul.diff + score.diff +
  lead.home + time + offensive + personal +
  foul.diff:offensive + foul.diff:personal +
  foul.diff:time + lead.home:time + (1|game) +
```

```
  (1|hometeam) + (1|visitor),
  family = binomial, data = refdata)
```

```
##  Groups    Name         Variance Std.Dev.
##  game      (Intercept)  0.1846   0.430
##  hometeam  (Intercept)  0.0783   0.280
##  visitor   (Intercept)  0.0431   0.208

##  Number of games =       340
##  Number of hometeams =    39
##  Number of visitors =     39

##                        Estimate  Std. Error  z value
##  (Intercept)          -0.246475   0.133956   -1.8400
##  foul.diff            -0.171469   0.045363   -3.7800
##  score.diff            0.033522   0.008236    4.0702
##  lead.home            -0.150619   0.177204   -0.8500
##  time                 -0.008747   0.008560   -1.0218
##  offensive            -0.080795   0.111232   -0.7264
##  personal              0.067200   0.065397    1.0276
##  foul.diff:offensive  -0.103574   0.053869   -1.9227
##  foul.diff:personal   -0.055634   0.031948   -1.7414
##  foul.diff:time       -0.008689   0.003274   -2.6540
##  lead.home:time        0.026007   0.012171    2.1368
##                        Pr(>|z|)
##  (Intercept)          6.577e-02
##  foul.diff            1.569e-04
##  score.diff           4.697e-05
##  lead.home            3.953e-01
##  time                 3.069e-01
##  offensive            4.676e-01
##  personal             3.041e-01
##  foul.diff:offensive  5.452e-02
##  foul.diff:personal   8.161e-02
##  foul.diff:time       7.955e-03
##  lead.home:time       3.262e-02

##  AIC =   6731  ;  BIC =    6822
```

In general, we see a highly significant negative effect of foul differential—a strong tendency for referees to even out foul calls when one team starts amassing more fouls than the other. Important covariates to control for (because of their effects on the odds of a home foul) include score differential, whether the home team held the lead, time left in the first half, and the type of foul called.

Furthermore, we see that the effect of foul differential depends on type of foul called and time left in the half—the tendency for evening out foul calls is

11.7 A Final Model for Examining Referee Bias

stronger earlier in the half, and when offensive and personal fouls are called instead of shooting fouls. The effect of foul type supports the hypothesis that if referees are consciously or subconsciously evening out foul calls, the behavior will be more noticeable for calls over which they have more control, especially offensive fouls (which are notorious judgment calls) and then personal fouls (which don't affect a player's shot, and thus a referee can choose to let them go uncalled). Evidence like this can be considered **dose response**, since higher "doses" of referee control are associated with a greater effect of foul differential on their calls. A dose response effect provides even stronger indication of referee bias.

Analyses of data from 2004-2005 [Noecker and Roback, 2012] showed that the tendency to even out foul calls was stronger when one team had a large lead, but we found no evidence of a foul differential by score differential interaction in the 2009-2010 data, although home team fouls are more likely when the home team has a large lead, regardless of the foul differential.

Here are specific interpretations of key model parameters:

- $\exp(\hat{\alpha}_0) = \exp(-0.247) = 0.781$. The odds of a foul on the home team is 0.781 at the end of the first half when the score is tied, the fouls are even, and the referee has just called a shooting foul. In other words, only 43.9% of shooting fouls in those situations will be called on the home team.
- $\exp(\hat{\beta}_0) = \exp(-0.172) = 0.842$. Also, $0.842^{-1} = 1.188$. As the foul differential decreases by 1 (the visiting team accumulates another foul relative to the home team), the odds of a home foul increase by 18.8%. This interpretation applies to shooting fouls at the end of the half, after controlling for the effects of score differential and whether the home team has the lead.
- $\exp(\hat{\gamma}_0) = \exp(0.034) = 1.034$. As the score differential increases by 1 (the home team accumulates another point relative to the visiting team), the odds of a home foul increase by 3.4%, after controlling for foul differential, type of foul, whether or not the home team has the lead, and time remaining in the half. Referees are more likely to call fouls on the home team when the home team is leading, and vice versa. Note that a change in the score differential could result in the home team gaining the lead, so that the effect of score differential experiences a non-linear "bump" at 0, where the size of the bump depends on the time remaining (this would involve the interpretation for $\hat{\xi}_0$).
- $\exp(\hat{\mu}_0) = \exp(-0.103) = 0.902$. Also, $0.902^{-1} = 1.109$. The effect of foul differential increases by 10.9% if a foul is an offensive foul rather than a shooting foul, after controlling for score differential, whether the home team has the lead, and time remaining. As hypothesized, the effect of foul differential is greater for offensive fouls, over which referees have more control when compared with shooting fouls. For example, midway through the half (time=10), the odds that a shooting foul is on the home team increase by 29.6% for each extra foul on the visiting team, while the odds that an offensive foul is on the home team increase by 43.6%.

- $\exp(\hat{\nu}_0) = \exp(-0.056) = 0.946$. Also, $0.946^{-1} = 1.057$. The effect of foul differential increases by 5.7% if a foul is a personal foul rather than a shooting foul, after controlling for score differential, whether the home team has the lead, and time remaining. As hypothesized, the effect of foul differential is greater for personal fouls, over which referees have more control when compared with shooting fouls. For example, midway through the half (time=10), the odds that a shooting foul is on the home team increase by 29.6% for each extra foul on the visiting team, while the odds that an personal foul is on the home team increase by 36.9%.
- $\exp(\hat{\omega}_0) = \exp(-0.0087) = 0.991$. Also, $0.991^{-1} = 1.009$. The effect of foul differential increases by 0.9% for each extra minute that is remaining in the half, after controlling for foul differential, score differential, whether the home team has the lead, and type of foul. Thus, the tendency to even out foul calls is strongest earlier in the game. For example, midway through the half (time=10), the odds that a shooting foul is on the home team increase by 29.6% for each extra foul on the visiting team, while at the end of the half (time=0) the odds increase by 18.8% for each extra visiting foul.
- $\hat{\sigma}_u^2 = 0.185 =$ the variance in log-odds intercepts from game-to-game after controlling for all other covariates in the model.

11.8 Estimated Random Effects

Our final model includes random effects for game, home team, and visiting team, and thus our model accounts for variability due to these three factors without estimating fixed effects to represent specific teams or games. However, we may be interested in examining the relative level of the random effects used for different teams or games. The presence of random effects allows different teams or games to begin with different baseline odds of a home foul. In other words, the intercept—the log odds of a home foul at the end of the first half when the score is tied, the fouls are even, and a shooting foul has been called—is adjusted by a random effect specific to the two teams playing and to the game itself. Together, the estimated random effects for, say, home team, should follow a normal distribution as we assumed: centered at 0 and with variance given by σ_v^2. It is sometimes interesting to know, then, which teams are typical (with estimated random effects near 0) and which fall in the upper and lower tails of the normal distribution.

Figure 11.8 shows the 39 estimated random effects associated with each home team (see the discussion on empirical Bayes estimates in Chapter 8). As expected, the distribution is normal with standard deviation near $\hat{\sigma}_v = 0.28$. To compare individual home teams, Figure 11.9 shows the estimated random effect and associated prediction interval for each of the 39 home teams. Although there is a great deal of uncertainty surrounding each estimate, we see that, for instance, DePaul and Seton Hall have higher baseline odds of home fouls than Purdue or Syracuse. Similar histograms and prediction intervals plots can be generated for random effects due to visiting teams and specific games.

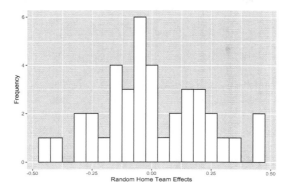

FIGURE 11.8: Histogram of estimated random effects for 39 home teams in Model F.

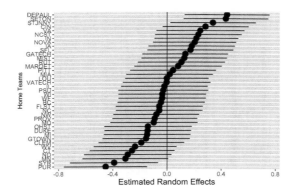

FIGURE 11.9: Estimated random effects and associated prediction intervals for 39 home teams in Model F.

11.9 Notes on Using R (optional)

Here we discuss in more detail how to fit Model B3 from Section 11.6. Note that, in `glmer()` or `lmer()`, if you have two equations at Level Two and want to fit a model with an error term for each equation, but you also want to assume that the two error terms are independent, the error terms must be requested separately. For example, (1 | hometeam) allows the Level One intercept to vary by home team, while (0+foul.dff | hometeam) allows the Level One effect of foul.diff to vary by home team. Under this formulation, the correlation between those two error terms is assumed to be 0; a non-zero correlation could have been specified with (1+foul.diff | hometeam).

The R code below shows how fixef() can be used to extract the estimated fixed effects from a multilevel model. Even more, it shows how ranef() can be used to illustrate estimated random effects by Game, Home Team, and Visiting Team, along with prediction intervals for those random effects. These estimated random effects are sometimes called Empirical Bayes estimators. In this case, random effects are placed only on the [["(Intercept)"]] term; the phrase "Intercept" could be replaced with other Level One covariates whose values are allowed to vary by game, home team, or visiting team in our model.

```
# Get estimated random effects based on Model F
re.int <- ranef(model.f)$`game`[["(Intercept)"]]
hist(re.int, xlab = "Random Effect",
     main = "Random Effects for Game")
Home.re <- ranef(model.f)$`hometeam`[["(Intercept)"]]
hist(Home.re, xlab = "Random Effect",
     main = "Random Effects for Home Team")
Visiting.re <- ranef(model.f)$`visitor`[["(Intercept)"]]
hist(Visiting.re, xlab = "Random Effect",
  main = "Random Effects for the Visiting Team",
  xlim = c(-0.5,0.5))
cbind(Home.re, Visiting.re)   # 39x2 matrix of REs by team

# Prediction intervals for random effects based on Model F
ranef1 <- dotplot(ranef(model.f, postVar = TRUE),
                  strip = FALSE)
print(ranef1[[3]], more = TRUE) ##HOME
print(ranef1[[2]], more = TRUE) ##VIS
print(ranef1[[1]], more = TRUE)
```

11.10 Exercises

11.10.1 Conceptual Exercises

1. Give an example of a data set and an associated research question that would best be addressed with a multilevel model for a Poisson response.

2. **College basketball referees.** Explain to someone unfamiliar with the plots in Figure 11.2 how to read both a conditional density plot and an empirical logit plot. For example, explain what the dark region in a conditional density plot represents, what each point in an empirical logit plot represents, etc.

3. With the strength of the evidence found in Figure 11.2, plots (a) and (d), is there any need to run statistical models to convince someone that referee tendencies are related to the foul differential?

4. In Section 11.4.2, why don't we simply run a logistic regression model for each of the 340 games, collect the 340 intercepts and slopes, and fit a model to those intercepts and slopes?

5. Explain in your own words the difference between crossed and nested random effects (Section 11.5).

6. In the context of Case Study 9.2, describe a situation in which crossed random effects would be needed.

7. Assume that we added z_i, r_h, and s_g to the Level Two equation for b_i in Equations (11.2). (a) Give interpretations for those 3 new random effects. (b) How many additional parameters would need to be estimated (and name the new model parameters)?

8. In Section 11.6, could we use a likelihood ratio test to determine if it would be better to add either a random effect for home team or a random effect for visiting team to the model with a random effect for game (assuming we're going to add one or the other)? Why or why not?

9. Describe how we would obtain the 1000 values representing the parametric bootstrap distribution in the histogram in Figure 11.6. In particular, describe how to simulate responses for the first two rows of the data set (the first two fouls in the IA at MN game from 03.07.2010).

10. In Figure 11.6, why is it a problem that "a chi-square distribution puts too much mass in the tail" when using a likelihood ratio test to compare models?

11. What would be the implications of using the R expression (foul.diff | game) + (foul.diff | hometeam) + (foul.diff | visitor) in Model B6 (Section 11.6) to obtain our variance components?

12. Explain the implications of having no error terms in any Level Two equations in Section 11.7 except the one for a_i.

13. In the interpretation of $\hat{\beta}_0$, explain why this applies to "shooting fouls at the end of the half". Couldn't we just say that we controlled for type of foul and time elapsed?

14. In the interpretation of $\hat{\gamma}_0$, explain the idea that "the effect of score differential experiences a non-linear 'bump' at 0, where the size of the bump depends on time remaining." Consider, for example, the effect of a point scored by the home team with 10 minutes left in the first half, depending on whether the score is tied or the home team is ahead by 2.

15. In the interpretation of $\hat{\mu}_0$, verify the odds increases of 29.6% for shooting fouls and 43.6% for offensive fouls. Where does the stated 10.9% increase factor in?

16. We could also interpret the interaction between two quantitative variables described by $\hat{\omega}_0$ as "the effect of time remaining increases by 0.9% for each extra foul on the visiting team, after controlling for ..." Numerically illustrate this interpretation by considering foul differentials of 2 (the home team has 2 more fouls than the visitors) and -2 (the visitors have 2 more fouls than the home team).

17. Provide interpretations in context for $\hat{\phi}_0$, $\hat{\kappa}_0$, and $\hat{\xi}_0$ in Model F.

18. In Section 11.8, why isn't the baseline odds of a home foul for DePaul considered a model parameter?

19. **Heart attacks in Aboriginal Australians.** Randall et al. published a 2014 article in *Health and Place* entitled, "Exploring disparities in acute myocardial infarction events between Aboriginal and non-Aboriginal Australians: Roles of age, gender, geography and area-level disadvantage." They used multilevel Poisson models to compare rates of acute myocardial infarction (AMI) in the 199 Statistical Local Areas (SLAs) in New South Wales. Within SLA, AMI rates (number of events over population count) were summarized by subgroups determined by year (2002-2007), age (in 10-year groupings), sex, and Aboriginal status; these are then our Level One variables. Analyses also incorporated remoteness (classified by quartiles) and socio-economic status (classified by quintiles) assessed at the SLA level (Level Two) [Randall et al., 2014]. For this study, give the observational units at Levels One and Two.

11.10 Exercises

TABLE 11.4: Adjusted rate ratios for individual-level variables from the multilevel Poisson regression model with random intercept for area from Table 2 in Randall et al. (2014).

	RR	95% CI	p-Value
Aboriginal			
No(ref)	1.00		<0.01
Yes	2.10	1.98-2.23	
Age Group			
25-34 (ref)	1.00		<0.01
35-44	6.01	5.44-6.64	
45-54	19.36	17.58-21.31	
55-64	40.29	36.67-44.26	
65-74	79.92	72.74-87.80	
75-84	178.75	162.70-196.39	
Sex			
Male (ref)	1.00		<0.01
Female	0.45	0.44-0.45	
Year			
2002 (ref)	1.00		<0.01
2003	1.00	0.98-1.03	
2004	0.97	0.95-0.99	
2005	0.91	0.89-0.94	
2006	0.88	0.86-0.91	
2007	0.88	0.86-0.91	

20. Table 11.4 shows Table 2 from Randall et al. [2014]. Let Y_{ij} be the number of acute myocardial infarctions in subgroup j from SLA i; write out the multilevel model that likely produced Table 11.4. How many fixed effects and variance components must be estimated?

21. Provide interpretations in context for the following rate ratios, confidence intervals, and p-values in Table 11.4: RR of 2.1 and CI of 1.98 - 2.23 for Aboriginal = Yes; p-value of <.01 for Age Group; CI of 5.44 - 6.64 for Age Group = 35-44; RR of 0.45 for Sex = Female; CI of 0.86 - 0.91 for Year = 2007.

22. Given the rate ratio and 95% confidence interval reported for Aboriginal Australians in Table 11.4, find the estimated model fixed effect for Aboriginal Australians from the multilevel model along with its standard error.

23. How might the p-value for Age Group have been produced?

TABLE 11.5: Adjusted rate ratios for area-level variables from the multilevel Poisson regression model with random intercept for area from Table 3 in Randall et al. (2014). Area-level factors added one at a time to the fully adjusted individual-level model (adjusted for Aboriginal status, age, sex and year) due to being highly associated.

	RR	95% CI	p-Value
Remoteness of Residence			
Major City	1.00		<0.01
Inner Regional	1.16	1.04-1.28	
Outer Regional	1.11	1.01-1.23	
Remote/very remote	1.22	1.02-1.45	
SES quintile			
1 least disadvantaged	1.00		<0.01
2	1.26	1.11-1.43	
3	1.40	1.24-1.58	
4	1.46	1.30-1.64	
5 most disadvantaged	1.70	1.52-1.91	

24. Randall et al. [2014] report that, "we identified a significant interaction between Aboriginal status and age group ($p < 0.01$) and Aboriginal status and sex ($p < 0.01$), but there was no significant interaction between Aboriginal status and year (p=0.94)." How would the multilevel model associated with Table 11.4 need to have been adjusted to allow these interactions to be tested?

25. Table 11.5 shows Table 3 from Randall et al. [2014]. Describe the changes to the multilevel model for Table 11.4 that likely produced this new table.

26. Provide interpretations in context for the following rate ratios, confidence intervals, and p-values in Table 11.5: p-value of <.01 for Remoteness of Residence; RR of 1.22 for Remote/very remote; CI of 1.52 - 1.91 for SES quintile $= 5$.

27. Randall et al. [2014] also report the results of a single-level Poisson regression model: "After adjusting for age, sex and year of event, the rate of AMI events in Aboriginal people was 2.3 times higher than in non-Aboriginal people (95% CI: 2.17-2.44)." Compare this to the results of the multilevel Poisson model; what might explain any observed differences?

28. Randall et al. [2014] claim that, "our application of multilevel modelling techniques allowed us to account for clustering by area of

11.10 Exercises

residence and produce 'shrunken' small-area estimates, which are not as prone to random fluctuations as crude or standardised rates." Explain what they mean by this statement.

11.10.2 Open-Ended Exercises

1. **Airbnb in Chicago.** Trinh and Ameri [2018] collected data on 1561 Airbnb listings in Chicago from August 2016, and then they merged in information from the neighborhood (out of 43 in Chicago) where the listing was located. We can examine traits that are associated with listings that achieve an overall satisfaction of 5 out of 5 vs. those that do not (i.e., treat satisfaction as a binary variable). Conduct an EDA, build a multilevel generalized model, and interpret model coefficients to answer questions such as: What are characteristics of a higher rated listing? Are the most influential traits associated with individual listings or entire neighborhoods? Are there intriguing interactions where the effect of one variable depends on levels of another?

   The following variables can be found in airbnb.csv or derived from the variables found there:

   - overall_satisfaction = rating on a 0-5 scale.
   - satisfaction = 1 if overall_satisfaction is 5, 0 otherwise
   - price = price for one night (in dollars)
   - reviews = number of reviews posted
   - room_type = Entire home/apt., Private room, or Shared room
   - accommodates = number of people the unit can hold
   - bedrooms = number of bedrooms
   - minstay = minimum length of stay (in days)
   - neighborhood = neighborhood where unit is located (1 of 43)
   - district = district where unit is located (1 of 9)
   - WalkScore = quality of the neighborhood for walking (0-100)
   - TransitScore = quality of the neighborhood for public transit (0-100)
   - BikeScore = quality of the neighborhood for biking (0-100)
   - PctBlack = proportion of black residents in a neighborhood
   - HighBlack = 1 if PctBlack above .60, 0 otherwise

2. **Seed germination.** We will return to the data from Angell [2010] used in Case Study 10.2, but whether or not a seed germinated will be considered the response variable rather than the heights of germinated plants. We will use the wide data set (one row per plant) seeds2.csv for this analysis as described in Section 10.3.1, although we will ignore plant heights over time and focus solely

on if the plant germinated at any time. Use multilevel generalized linear models to determine the effects of soil type and sterilization on germination rates; perform separate analyses for coneflowers and leadplants, and describe differences between the two species. Support your conclusions with well-interpreted model coefficients and insightful graphical summaries.

3. **Book banning.** Provocative literature constantly risks being challenged and subsequently banned. Reasons for censorship have changed over time, with modern day challenges more likely to be for reasons relating to obscenity and child protection rather than overt political and religious objections [Jenkins, 2008]. Many past studies have addressed chiefly the reasons listed by those who challenge a book, but few examine the overall context in which books are banned—for example, the characteristics of the states in which they occur.

A team of students assembled a data set by starting with information on book challenges from the American Library Society [Fast and Hegland, 2011]. These book challenges—an initiation of the formal procedure for book censorship—occurred in U.S. States between January 2000 and November 2010. In addition, state-level demographic information was obtained from the U.S. Census Bureau and the Political Value Index (PVI) was obtained from the Cook Political Report.

We will consider a data set with 931 challenges and 18 variables. All book challenges over the nearly 11-year period are included except those from the State of Texas; Texas featured 683 challenges over this timeframe, nearly 5 times the number in the next largest state. Thus, the challenges from Texas have been removed and could be analyzed separately. Here, then, is a description of available variables in `bookbanningNoTex.csv`:

- `book` = unique book ID number
- `booktitle` = name of book
- `author` = name of author
- `state` = state where challenge made
- `removed` = 1 if book was removed (challenge was successful); 0 otherwise
- `pvi2` = state score on the Political Value Index, where positive indicates a Democratic leaning, negative indicates a Republican leaning, and 0 is neutral
- `cperhs` = percentage of high school graduates in a state (grand mean centered)
- `cmedin` = median state income (grand median centered)

11.10 Exercises

- `cperba` = percentage of college graduates in a state (grand mean centered)
- `days2000` = date challenge was made, measured by number of days after January 1, 2000
- `obama` = 1 if challenge was made while Barack Obama was president; 0 otherwise
- `freqchal` = 1 if book was written by a frequently challenged author (10 or more challenges across the country); 0 otherwise
- `sexexp` = 1 if reason for challenge was sexually explicit material; 0 otherwise
- `antifamily` = 1 if reason for challenge was antifamily material; 0 otherwise
- `occult` = 1 if reason for challenge was material about the occult; 0 otherwise
- `language` = 1 if reason for challenge was inappropriate language; 0 otherwise
- `homosexuality` = 1 if reason for challenge was material about homosexuality; 0 otherwise
- `violence` = 1 if reason for challenge was violent material; 0 otherwise

The primary response variable is `removed`. Certain potential predictors are measured at the state level (e.g., `pvi2` and `cperhs`), at least one is at the book level (`freqchal`), and several are specific to an individual challenge (e.g., `obama` and `sexexp`). In addition, note that the same book can be challenged for more than one reason, in different states, or even at different times in the same state.

Perform exploratory analyses and then run multilevel models to examine significant determinants of successful challenges. Write a short report comparing specific reasons for the challenge to the greater context in which a challenge was made.

4. **Yelp restaurant reviews.** Janusz and Mohr [2018] assembled a data set of Yelp restaurant reviews in Madison, WI, from 2005 through 2017 based on the Yelp Dataset Challenge on Kaggle[1]. The data in `yelp.csv` contains almost 60,000 reviews on 888 restaurants from over 20,000 reviewers, and it contains a selection of variables on the reviewer (e.g., total reviews, average stars), the restaurant (e.g., neighborhood, average stars, category), and the review itself (e.g., stars, year, useful ratings, actual text).

There are various questions that could be pursued with this data. Here are just a few ideas:

---

[1] https://www.kaggle.com/yelp-dataset/yelp-dataset

- how can we model number of stars in the rating, or whether or not the rating was 5 stars?
- how can we model whether or not at least one person thought the review was useful?
- how can we model whether or not at least one person thought the review was cool?

A few things to keep in mind while building models to answer your questions:

- user and restaurant can be considered crossed random effects
- convergence may be an issue. You may have to take a random sample of reviews, or a targeted sample of more frequently appearing users and/or restaurants.

Bibliography

E. H. Allison, W. N. Adger, M. C. Badjeck, K. Brown, D. Conway, N. K. Dulvy, A. Halls, A. Perry, and J. D. Reynolds. Effects of climate change on the sustainability of capture and enhancement fisheries important to the poor. *Fisheries Management Science Programme*, 2005.

Kyle J. Anderson and David A. Pierce. Officiating bias: the effect of foul differential on foul calls in NCAA basketball. *Journal of Sports Sciences*, 27 (7):687–94, 2009. doi: 10.1080/02640410902729733.

Diane Angell. Effects of soil type and sterilization on the growth of coneflowers and leadplants. St. Olaf College. Class data for Biology 261, 2010.

Annika Awad, Evan Lebo, and Anna Linden. Intercontinental comparative analysis of Airbnb booking factors. St. Olaf College. Statistics 316 Project, 2017.

S. G. Baer, D. J. Kitchen, J. M. Blair, and C. W. Rice. Changes in ecosystem structure and function along a chronosequence of restored grasslands. *Ecological Applications*, 12(6):1688–1701, 2002. doi: 10.1890/1051-0761(2002)012[1688:CIESAF]2.0.CO;2.

Douglas Bates, Martin Mächler, Ben Bolker, and Steve Walker. Fitting linear mixed-effects models using lme4. *Journal of Statistical Software, Articles*, 67 (1):1–48, 2015. ISSN 1548-7660. URL https://www.jstatsoft.org/v067/i01.

Ben Bayer and Michael Fitzgerald. Reconstructing Alabama: Reconstruction Era demographic and statistical research. St. Olaf College. Senior IR Project, 2011.

BBC News. Sisters 'make people happy', Apr 2 2009. URL http://news.bbc.co.uk/2/hi/health/7977454.stm.

Robert Bickel. *Multilevel Analysis for Applied Research: It's Just Regression!* Guilford Publications, New York, 2007.

J. A. Bishop. An experimental study of the cline of industrial melanism in biston betularia (l.) (lepidoptera) between urban liverpool and rural north wales. *Journal of Animal Ecology*, 41(1):209–243, 1972. doi: 10.2307/3513.

Margaret Blakeman, Tim Renier, and Rami Shandaq. Modeling Donald

Trump's voters in the 2016 Election. St. Olaf College. Statistics 316 Project, 2018.

H. Jane Brockmann. Satellite male groups in horseshoe crabs, limulus polyphemus. *Ethology*, 102(1):1–21, 1996. URL http://dx.doi.org/doi: 10.1111/j.1439-0310.1996.tb01099.x.

Kenneth Brown and Bulent Uyar. A hierarchical linear model approach for assessing the effects of house and neighborhood characteristics on housing prices. *Journal of Real Estate Practice and Education*, 7(1):15–24, 2004. URL http://aresjournals.org/doi/abs/10.5555/repe.7.1.f687057161743261.

Richard Buddin and Ron Zimmer. Student achievement in charter schools: A complex picture. *Journal of Policy Analysis and Management*, 24(2): 351–371, 2005. URL http://dx.doi.org/10.1002/pam.20093.

Bureau of Labor Statistics. National Longitudinal Surveys, 1997. URL https://www.bls.gov/nls/nlsy97.htm.

A.C. Cameron and P.K. Trivedi. Econometric models based on count data: Comparisons and applications of some estimators and tests. *Journal of Applied Econometrics*, 1:29–53, 1986.

Philip Camill, Mark J. McKone, Sean T. Sturges, William J. Severud, Erin Ellis, Jacob Limmer, Christopher B. Martin, Ryan T. Navratil, Amy J. Purdie, Brody S. Sandel, Shano Talukder, and Andrew Trout. Community- and ecosystem-level changes in a specie-rich tallgrass prairie restoration. *Ecological Applications*, 14(6):1680–1694, 2004. doi: 10.1890/03-5273.

Ann Cannon, George Cobb, Brad Hartlaub, Julie Legler, Robin Lock, Tom Moore, Allan Rossman, and Jeff Witmer. *Stat2: Modeling with Regression and ANOVA*. Macmillan, 2019.

Centers for Disease Control and Prevention. Youth Risk Behavior Survey data, 2009. URL http://www.cdc.gov/HealthyYouth/yrbs/index.htm.

Central Intelligence Agency. The World Factbook 2013, 2013. URL https://www.cia.gov/library/publications/download/download-2013/index.html.

Christopher Chapp, Paul Roback, Kendra Jo Johnson-Tesch, Adrian Rossing, and Jack Werner. Going vague: Ambiguity and avoidance in online political messaging. *Social Science Computer Review*, Aug 2018. URL https://doi.org/10.1177/0894439318791168.

Patrick J. Curran, Eric Stice, and Laurie Chassin. The relation between adolescent alcohol use and peer alcohol use: A longitudinal random coefficients model. *Journal of Consulting and Clinical Psychology*, 65(1):130–140, 1997. URL http://dx.doi.org/10.1037/0022-006X.65.1.130.

Samantha Dahlquist and Jin Dong. The effects of credit cards on tipping. St. Olaf College. Statistics 272 Project, 2011.

A. C. Davison and D. V. Hinkley. *Bootstrap Methods and Their Application.* Cambridge University Press, 1997.

P. J. Diggle, P. Heagarty, K.-Y. Liang, and S. L. Zeger. *Analysis of Longitudinal Data.* Oxford University Press, 2002.

Bradley Efron. Bayesian inference and the parametric bootstrap. *Ann. Appl. Stat.*, 6(4):1971–1997, Dec 2012. URL https://doi.org/10.1214/12-AOAS571.

Bradley Efron and R.J. Tibshirani. *An Introduction to the Bootstrap.* Chapman & Hall/CRC, Boca Raton, FL, 1993.

Robert Eisinger, Amanda Elling, and J.R. Stamp. Tree growth rates and mortality. In *Proceedings of the National Conference on Undergraduate Research (NCUR)*, Ithaca College, New York, 2011.

Brian S. Everitt and Torsten Hothorn. *A Handbook of Statistical Analyses using R.* Chapman & Hall/ CRC, Boca Raton, FL, 2006.

Julian Faraway. *Extending the Linear Model With R: Generalized Linear, Mixed Effects and Nonparametric Regression Models.* Chapman & Hall/ CRC, Boca Raton, FL, 2005.

Anthony Farrar and Thomas H. Bruggink. A new test of the moneyball hypothesis. *The Sport Journal*, May 2011. URL http://thesportjournal.org/article/a-new-test-of-the-moneyball-hypothesis/.

Shannon Fast and Thomas Hegland. Book challenges: A statistical examination. St. Olaf College. Statistics 316 Project, 2011.

Kara Finnigan, Nancy Adelman, Lee Anderson, Lynyonne Cotton, Mary Beth Donnelly, and Tiffany Price. *Evaluation of Public Charter Schools Program: Final Evaluation Report.* U.S. Department of Education, Washington, D.C., 2004.

Lisa Fisher, Katie Murney, and Tyler Radtke. Emergency department overcrowding and factors that contribute to ambulance diversion. St. Olaf College. Statistics 316 Project, 2019.

Richard A. Friedman. Standing up at your desk could make you smarter. *The New York Times*, Apr 20 2018.

Andrew Gelman, Jeffrey Fagan, and Alex Kiss. An analysis of the NYPD's stop-and-frisk policy in the context of claims of racial bias. *Journal of The American Statistical Association*, 102:813–823, Sept 2007.

Thomas Gilovich, Robert Vallone, and Amos Tversky. The hot hand in basketball: On the misperception of random sequences. *Cognitive Psychology*, 17(3): 295–314, 1985. URL https://doi.org/10.1016/0010-0285(85)90010-6.

I. B. Goldstein and D. Shapiro. Ambulatory blood pressure in women: Family history of hypertension and personality. *Psychology, Health & Medicine*, 5 (3):227–240, 2000. URL https://doi.org/10.1080/713690197.

Shelly Grabe, Janet Shibley Hyde, and L. Moniquee Ward. The role of the media in body image concerns among women: A meta-analysis of experimental and correlational studies. *Pyschological Bulletin*, 134(3):460–476, 2008. URL https://doi.org/10.1037/0033-2909.134.3.460.

Preston C. Green III, Bruce D. Baker, and Joseph O. Oluwole. Having it both ways: How charter schools try to obtain funding of public schools and the autonomy of private schools. *Emory Law Journal*, 63(2):303–337, 2003.

Tim C. Hesterberg. What teachers should know about the bootstrap: Resampling in the undergraduate statistics curriculum. *The American Statistician*, 69(4):371–386, 2015. URL https://doi.org/10.1080/00031305.2015.1089789. PMID: 27019512.

Walt Hickey. The dollar-and-cents case against Hollywood's exclusion of women. FiveThirtyEight, Apr 2014. URL https://fivethirtyeight.com/features/the-dollar-and-cents-case-against-hollywoods-exclusion-of-women/.

P. A. Holst, D. Kromhout, and R. Brand. For debate: pet birds as an independent risk factor for lung cancer. *British Medical Journal*, 297(6659): 1319–1321, Nov 1988. doi: 10.1136/bmj.297.6659.1319.

Brooke Janusz and Michael Mohr. Predicting user Yelp star ratings based on restaurant attributes. St. Olaf College. Statistics 316 Project, 2018.

Christine A. Jenkins. Book challenges, challenging books, and young readers: The research picture. *Language Arts*, 85(3):228–36, Jan 2008.

Kaggle. House sales in King County, USA, 2018a. URL https://www.kaggle.com/harlfoxem/housesalesprediction/home.

Kaggle. NBA enhanced box scores and standings, 2018b. URL https://www.kaggle.com/pablote/nba-enhanced-stats.

KIPP. KIPP North Star Academy, 2018. URL http://www.kipp.org/school/kipp-north-star-academy/.

Johannes M. H. Knops and David Tilman. Dynamics of soil nitrogen and carbon accumulation for 61 years after agricultural abandonment. *Ecology*, 81 (1):88–98, 2000. doi: 10.1890/0012-9658(2000)081[0088:DOSNAC]2.0.CO;2.

Gina Kolata. Picture emerging on genetic risks of ivf. *The New York Times*, Feb 16 2009.

Jan Komdeur, Serge Daan, Joost Tinbergen, and Christa Mateman. Extreme adaptive modification in sex ratio of the Seychelles warbler's eggs. *Nature*, 385:522–525, Feb 1997. URL http://dx.doi.org/10.1038/385522a0.

Nan M. Laird. Missing data in longitudinal studies. *Statistics in Medicine*, 7(1-2):305–315, 1988. URL http://dx.doi.org/10.1002/sim.4780070131.

Michael M. Lewis. *Moneyball: The Art of Winning an Unfair Game*. W. W. Norton & Company, 2003.

M Martinsen, S Bratland-Sanda, A K Eriksson, and J Sundgot-Borgen. Dieting to win or to be thin? A study of dieting and disordered eating among adolescent elite athletes and non-athlete controls. *British Journal of Sports Medicine*, 44(1):70–76, 2009. URL http://bjsm.bmj.com/content/44/1/70.

T.J. Mathews and Brady E. Hamilton. Trend analysis of the sex ratio at birth in the United States. *National Vital Statistics Reports*, 53(20):1–20, 06 2005. URL https://www.cdc.gov/nchs/data/nvsr/nvsr53/nvsr53_20.pdf.

Peter McCullagh and John Ashworth Nelder. *Generalized Linear Models*. Chapman & Hall/ CRC, Boca Raton, Florida, 2nd edition, 1989.

Minnesota Department of Education. Minnesota Department of Education data center, 2018. URL https://education.mn.gov/MDE/Data/.

Tobias J. Moskowitz and L. Jon Wertheim. *Scorecasting: The Hidden Influences Behind How Sports Are Played and Games Are Won*. Crown Archetype, New York, 2011.

Per Nafstad, Jorgen A. Hagen, Leif Oie, Per Magnus, and Jouni J. K. Jaakkola. Day care centers and respiratory health. *Pediatrics*, 103(4):753–758, 1999. URL http://pediatrics.aappublications.org/content/103/4/753.

National Center for Education Statistics. The Integrated Postsecondary Education Data System, 2018. URL https://nces.ed.gov/ipeds/.

John Ashworth Nelder and Robert William Maclagan Wedderburn. Generalized linear models. *Journal of the Royal Statistical Society. Series A (General)*, 135(3):370–384, 1972. URL http://www.jstor.org/stable/2344614.

Cecilia A. Noecker and Paul Roback. New insights on the tendency of NCAA basketball officials to even out foul calls. *Journal of Quantitative Analysis in Sports*, 8(3):1–23, Oct 2012.

Philippine Statistics Authority. Family income and expenditure survey, 2015. URL https://www.kaggle.com/grosvenpaul/family-income-and-expenditure.

Joyce H. Poole. Mate guarding, reproductive success and female choice in African elephants. *Animal Behaviour*, 37:842–849, 1989. URL http://www.sciencedirect.com/science/article/pii/0003347289900687.

Ian Pray. Effects of rainfall and sun exposure on leaf characteristics. St. Olaf College. Bio in South India Project, 2009.

J Proudfoot, D Goldberg, A Mann, B Everitt, I Marks, and J A Gray. Computerized, interactive, multimedia cognitive-behavioural program for anxiety and depression in general practice. *Psychological Medicine*, 33(2):217–27, Feb 2003. doi: 10.1017/s0033291702007225.

R Core Team. *R: A Language and Environment for Statistical Computing*. R Foundation for Statistical Computing, Vienna, Austria, 2020. URL https://www.R-project.org.

Fred Ramsey and Daniel Schafer. *The Statistical Sleuth: A course in methods of data analysis*. Brooks/Cole Cengage, Boston, Massachusetts, 2nd edition, 2002.

D.A. Randall, L.R. Jorm, S. Lujic, S.J. Eades, T.R. Churches, A.J. O'Loughlin, and A.H. Leyland. Exploring disparities in acute myocardial infarction events between Aboriginal and non-Aboriginal Australians: Roles of age, gender, geography and area-level disadvantage. *Health & Place*, 28:58–66, 2014. ISSN 1353-8292. doi: https://doi.org/10.1016/j.healthplace.2014.03.009.

Stephen W. Raudenbush and Anthony S. Bryk. *Hierarchical Linear Models: Applications and Data Analysis Methods*. SAGE Publications, Inc., Thousand Oaks, CA, 2nd edition, 2002.

Joseph Lee Rodgers and Debby Doughty. Does having boys or girls run in the family? *CHANCE*, 14(4):8–13, 2001. URL http://dx.doi.org/10.1080/09332480.2001.10542293.

Marieke Roskes, Daniel Sligte, Shaul Shalvi, and Carsten K. W. De Dreu. The right side? Under time pressure, approach motivation leads to right-oriented bias. *Psychology Science*, 22(11):1403–1407, 2011. URL https://doi.org/10.1177/0956797611418677.

Michael E. Sadler and Christopher J. Miller. Performance anxiety: A longitudinal study of the roles of personality and experience in musicians. *Social Psychological and Personality Science*, 1(3):280–287, 2010. URL http://dx.doi.org/10.1177/1948550610370492.

Robert J. Sampson, Stephen W. Raudenbush, and Felton Earls. Neighborhoods and violent crime: A multilevel study of collective efficacy. *Science*, 277 (5328):918–924, 1997. ISSN 0036-8075. doi: 10.1126/science.277.5328.918.

Joseph Scotto, Alfred W. Kopf, and Fredrick Urbach. Non-melanoma skin cancer among Caucasians in four areas of the United States. *Cancer*, 34(4):1333–1338, Oct 1974. URL https://doi.org/10.1002/1097-0142(197410)34:4<1333::AID-CNCR2820340447>3.0.CO;2-A.

Prabha Siddarth, Alison C. Burggren, Harris A. Eyre, Gary W. Small, and

Bibliography

David A. Merrill. Sedentary behavior associated with reduced medial temporal lobe thickness in middle-aged and older adults. *PLOS ONE*, 13(4): 1–13, Apr 2018. doi: 10.1371/journal.pone.0195549.

Judith D. Singer and John B. Willett. *Applied Longitudinal Data Analysis: Modeling Change and Event Occurrence.* Oxford University Press, Inc., New York, 1st edition, 2003.

Aaron Smith and Maeve Duggan. Online dating and relationships. Pew Research Center, Oct 2013. URL https://www.issuelab.org/resources/15934/15934.pdf.

Stephen M. Stigler. *Statistics on the Table: The History of Statistical Concepts and Methods.* Harvard University Press, 2002.

Adam Sturtz, Alex Lampert, and Trent Friedrich. Project 5183. St. Olaf College. Statistics 316 Project, 2013.

Ly Trinh and Pony Ameri. Airbnb price determinants: A multilevel modeling approach. St. Olaf College. Statistics 316 Project, 2018.

UCLA Statistical Consulting Group. Zero-inflated negative binomial regression: R data analysis examples, 2018. URL https://stats.idre.ucla.edu/r/dae/zinb/.

U.S. Department of Education. National charter school resource center, 2018. URL https://charterschoolcenter.ed.gov/faqs.

Q. H. Vuong. Likelihood ratio tests for model selection and non-nested hypotheses. *Econometrica*, 57:307–333, 1989. doi: 10.2307/1912557.

Chanequa J. Walker-Barnes and Craig A. Mason. Ethnic differences in the effect of parenting on gang involvement and gang delinquency: A longitudinal, hierarchical linear modeling perspective. *Child Development*, 72(6):1814–1831, 2001. URL http://dx.doi.org/10.1111/1467-8624.00380.

Robert E. Weiss. *Modeling Longitudinal Data.* Springer-Verlag, New York, 2005.

Wikipedia contributors. Kentucky Derby. In *Wikipedia*, 2018. URL https://en.wikipedia.org/wiki/Kentucky_Derby.

John Witte, David Weimer, Arnold Shober, and Paul Schlomer. The performance of charter schools in Wisconsin. *Journal of Policy Analysis and Management*, 26(3):557–573, 2007. URL http://dx.doi.org/10.1002/pam.20265.

Donald R. Zak, William E. Holmes, David C. White, Aaron D. Peacock, and David Tilman. Plant diversity, soil microbial communities, and ecosystem function. *Ecology*, 84(8):2042–2050, 2003. doi: 10.1890/02-0433.

Index

adjusted R-squared, 24
AIC, 24, 63

Bernoulli distribution, 72
Bernoulli process, 72, 152
beta distribution, 84
BIC, 24, 63
binary logistic regression, 171
binomial distribution, 73
binomial logistic regression, 155, 170
bootstrap distribution, 21
bootstrapping, 20
boundary constraint, 339

canonical link, 146
case resampling, 20
centered variable, 8, 13
centering, 246
chi-square distribution, 86
coded scatterplot, 11
composite model, 226
conditional probability, 50
covariance structure, 301, 358
cross-level interaction, 226
crossed random effects, 387

deviance, 104
deviance residuals, 111, 166
dose response, 196, 397
drop-in-deviance test, 105, 162, 245

effective sample size, 236
empirical Bayes estimates, 239, 398
empirical logit plot, 160
error, 12
exponential distribution, 80
extra-binomial variation, 167

F-distribution, 87
fixed effects, 196, 227

gamma distribution, 81
gamma function, 76
generalized linear models (GLMs), 2, 145, 158
geometric distribution, 74

hurdle model, 141
hypergeometric distribution, 77

independent, 43, 106
indicator variable, 8, 17
inference, 19
interaction, 22
intraclass correlation, 195
intraclass correlation coefficient, 235

lack-of-fit, 113, 165
lattice plot, 217, 269
level-one model, 223
level-two model, 223
levels, 195, 213
likelihood, 44, 106
likelihood ratio test (LRT), 60, 245
linear least squares regression (LLSR), 2, 3, 113, 170, 251
loess smoother, 270
logistic regression, 152, 155
logit, 155
longitudinal, 268

maximum likelihood estimate (MLE), 44, 106
missing data, 266
mixture model, 90, 129

multilevel data, 205
multilevel GLM, 376
multilevel model, 2, 248, 251, 254
multivariate normal distribution, 228

negative binomial distribution, 75
negative binomial regression, 124
nested models, 25, 58, 105
normal distribution, 83
null (reduced) model, 59, 105

odds ratio, 155
offset, 117
one-parameter exponential family, 145
overdispersion, 113, 121, 167, 198

parametric bootstrap, 295, 343, 391
Pearson residuals, 111, 166
percentile method, 21
Poisson distribution, 79
Poisson process, 79
Poisson regression, 94, 113
probability density function (pdf), 80
probability mass function (pmf), 72
profile likelihood, 164
pseudo R-squared, 239

quasi-Poisson, 121
quasibinomial, 168, 198
quasilikelihood, 121, 168

R-squared, 16, 24
random effects, 196, 227
random intercepts model, 234, 351
random slopes and intercepts model, 236
relative risk (rate ratio), 103
residual, 13
residual deviance, 112, 166
restricted maximum likelihood (REML), 229

simple linear regression, 3
spaghetti plot, 270

t-distribution, 87

three-level structure, 325
three-way interaction, 354
trellis graph, 272
Tukey's honestly significant differences, 119

unconditional growth model, 281, 335
unconditional means model, 234, 280, 333

variance components, 229, 349
Vuong test, 131

Wald-type confidence interval, 103, 163
Wald-type test, 104, 106, 162

zero-inflated Poisson, 127

For Product Safety Concerns and Information please contact our EU representative GPSR@taylorandfrancis.com Taylor & Francis Verlag GmbH, Kaufingerstraße 24, 80331 München, Germany

Printed and bound by CPI Group (UK) Ltd, Croydon, CR0 4YY
10/06/2025
01898565-0001